Graduate Texts in Physics

Graduate Texts in Physics publishes core learning/teaching material for graduate- and advanced-level undergraduate courses on topics of current and emerging fields within physics, both pure and applied. These textbooks serve students at the MS- or PhD-level and their instructors as comprehensive sources of principles, definitions, derivations, experiments and applications (as relevant) for their mastery and teaching, respectively. International in scope and relevance, the textbooks correspond to course syllabi sufficiently to serve as required reading. Their didactic style, comprehensiveness and coverage of fundamental material also make them suitable as introductions or references for scientists entering, or requiring timely knowledge of, a research field.

More information about this series at http://www.springer.com/series/8431

Ernst Meyer • Roland Bennewitz •
Hans J. Hug

Scanning Probe Microscopy

The Lab on a Tip

Second Edition

 Springer

Ernst Meyer
Department of Physics
University of Basel
Basel, Switzerland

Roland Bennewitz
Leibniz Institute for New Materials gGmbH
INM
Saarbrücken, Germany

Hans J. Hug
Empa
Dübendorf, Zürich, Switzerland

ISSN 1868-4513   ISSN 1868-4521   (electronic)
Graduate Texts in Physics
ISBN 978-3-030-37091-6   ISBN 978-3-030-37089-3   (eBook)
https://doi.org/10.1007/978-3-030-37089-3

This Springer imprint is published by the registered company Springer Nature Switzerland AG
The registered company address is: Gewerbestrasse 11, 6330 Cham, Switzerland

This book is dedicated to Hans-Joachim Güntherodt who has been a great tutor and has supported many young scientists in this field.

Preface

Three decades after its invention, scanning probe microscopy has become a widely used method in laboratories as diverse as industrial magnetic storage development or structural biology. Consequently, the community of users ranges from biologists and medical researchers to physicists and engineers, all of them exploiting the unrivalled resolution and profiting from the relative simplicity of the experimental implementation.

In recent years, the authors have taught numerous courses on scanning probe microscopy, normally in combination with hands-on student experiments. The audiences ranged from physics freshmen to biology post-docs and even high-school teachers. We found it of particular importance to cover not only the physical principles behind scanning probe microscopy but also questions of instrumental designs, basic features of the different imaging modes, and recurring artifacts. With this book, our intention is to provide a general textbook for all types of classes that address scanning probe microscopy. Third year undergraduates and beyond should be able to use it for self-study or as a textbook to accompany a course on probe microscopy. Furthermore, it will be valuable as a reference book in any scanning probe microscopy laboratory.

The book starts with a thorough introduction, which comprises aspects common to all scanned probe microscopes. These aspects range from the underlying concept of near-field interactions to the construction of mechanical damping systems for the experimental setup. The next three chapters describe in great detail scanning tunneling microscopy, scanning force microscopy, and magnetic force microscopy. In each chapter, a discussion of basic physical concepts is followed by a introduction of experimental procedures, complemented by application examples. For tunneling microscopy, the oldest of these methods, the applications are the focus of its description. The different operation modes are emphasized in the chapter on force microscopy, while the complex interpretation of the results is stressed in the chapter on magnetic force microscopy. The fifth chapter gives brief descriptions of other members of the family of scanning probe microscopes, be the scanning near-field optical microscopy or electrochemical scanning tunneling microscopy. Recurring artifacts in all modes of operation are addressed in Chap. 6. The book closes with a look at the future prospects of scanning probe microscopy, also discussing related techniques in nanoscience.

This book is the outcome of research and teaching at the Department of Physics of the University of Basel, the INM Leibniz-Institute for New Materials (INM) and the Swiss Federal Laboratories for Materials Science and Technology (EMPA). Financial support from the Swiss National Science Foundation (SNF), the Deutsche Forschungsgemeinschaft (DFG), and the Swiss Nanoscience Institute (SNI) is gratefully acknowledged. We also thank the European Research Council (ERC) under the European Unions Horizon 2020 research and innovation programme (ULTRADISS grant agreement No 834402). We would like to acknowledge the contributions of all collaborators, in particular of the mechanical and electronic workshops. We dedicate this book to Hans-Joachim Güntherodt who has been a great tutor and has supported many young scientists in this field. We also acknowledge Alexis Baratoff for his invaluable advice.

Basel, Switzerland   Ernst Meyer
                                                                                                   Roland Bennewitz
                                                                                                       Hans J. Hug

Contents

Abbreviations

AFM	Atomic force microscopy
AM	Amplitude modulation
ARPES	Angle-resolved photoemission spectroscopy
DFM	Dynamic force microscopy
DFT	Density functional theory
DMI	Dzyaloshinkii-Moriya interaction
ECSTM	Electrochemical scanning tunneling microscopy
ESR	Electron spin resonance
FFM	Friction force microscopy
FFT	Fast Fourier transformation
FM	Frequency modulation
GMR	Giant magnetoresistance
KPFM	Kelvin probe force microscopy
LDA	Local density approximation
LFM	Lateral force microscopy
LSTM	Laser scanning tuneling microscopy
MExFM	Magnetic exchange force microscopy
MFM	Magnetic force microscopy
MRFM	Magnetic resonance force microscopy
nc-AFM	Non-contact atomic force microscopy
NMR	Nuclear magnetic resonance
NVCM	Nitrogen vacancy center microscopy
PEMSA	Photoemission microscopy with scanning aperture
PIFM	Photon induced force microscopy
PLL	Phase locked loop
SCM	Scanning capacitance microscopy
SEM	Scanning electron microscope
SFM	Scanning force microscopy
SHPM	Scanning Hall probe microscopy
SICM	Scanning ion conductance microscopy
SNAM	Scanning near-field acoustic microscopy
SNM	Scanning noise microscopy
SNOM	Scanning near-field optical microscopy
SPM	Scanning probe microscopy

SPotM	Scanning potentiometry microscopy
SP-STM	Spin-polarized scanning tunneling microscopy
SQUID	Superconductive quantum interference device
SSRM	Scanning spreading resistance microscopy
STAP	Scanning tunneling atom probe
SThM	Scanning thermal microscopy
STM	Scanning tuneling microscopy
STMiP	STM with inverse photoemission
TERS	Tip enhanced Raman scattering
TMR	Tunneling magnetoresistance

Introduction to Scanning Probe Microscopy

1

Abstract

An introduction into the field of scanning probe microscopy is given. These aspects range from the underlying concept of near-field interactions to the construction of mechanical damping systems for the experimental setup.

Richard Feynman foresaw the enormous potential of studying the physics of structures at the nanometer scale in his talk 'There's Plenty of Room at the Bottom' at Caltech in 1959 [178]:

> But I am not afraid to consider the final question as to whether, ultimately – in the great future – we can arrange the atoms the way we want; the very atoms, all the way down! What would happen if we could arrange the atoms one by one the way we want them (within reason, of course; you can't put them so that they are chemically unstable, for example). [...]

> What could we do with layered structures with just the right layers? What would the properties of materials be if we could really arrange the atoms the way we want them? They would be very interesting to investigate theoretically. I can't see exactly what would happen, but I can hardly doubt that when we have some control of the arrangement of things on a small scale we will get an enormously greater range of possible properties that substances can have, and of different things that we can do.

His visionary talk covered a wide range of concepts and opportunities, on which we work today in the field of nanoscience that has since been established. For example, he pointed out the close relationship between physics and biology when it comes to nanostructures and the importance of quantum effects in structures built from a few atoms. His ideas on how to produce nanometer-sized structures included thin-film evaporation through masks, somehow predicting today's experiments with two-dimensional electron gases. For the construction of few-atom devices, Feynman envisioned a series of machines of decreasing length scale, each generation constructing the next smaller one. This extensive approach has been outrun by a simpler solution: with the invention of scanning probe microscopy, the large gap between the macroscopic world and single-atom manipulation has been bridged in one step.

© Springer Nature Switzerland AG 2021

E. Meyer et al., *Scanning Probe Microscopy*, Graduate Texts in Physics,

https://doi.org/10.1007/978-3-030-37089-3_1

Near-field interactions between micro-fabricated probes and the sample allow imaging, analysis, and manipulation on the atomic scale, fulfilling Feynman's ideas at least in the laboratory.

This book aims to give an introduction to the field of scanning probe microscopy. Basic concepts are described as well as experimental procedures, with emphasis on scanning tunneling and scanning force microscopy, while other members of this family of methods like scanning near-field optical microscopy are presented in a more general way.

1.1   Overview

Scanning probe microscopy (SPM) covers a lateral range of imaging from several $100\,\mu m$ to $10\,pm$. Surfaces of solids can be mapped with atomic resolution, revealing not only the structure of perfect crystalline surfaces but also the distribution of point defects, adsorbates, and structural defects like steps. Scanning probe microscopy has become an essential tool in the emerging field of nanoscience, as local experiments with single atoms or molecules can be performed. Force measurements of single chemical bonds or optical spectra of single molecules may serve as examples. Furthermore, the local probe can be used to manipulate single atoms or molecules and hence to form artificial structures on the atomic scale.

The starting point of SPM was the invention in 1982 of the scanning tunneling microscope (STM) by Binnig and Rohrer [80, 85], who were awarded the Nobel prize for physics in 1986. In the STM, a sharp metallic needle is scanned over the surface at a distance of less than 1 nm. This distance is controlled by the tunneling current between the tip and the conducting surface. The tunneling current is a quantum mechanical effect, with two properties important for STM: it flows between two electrodes even through a thin insulator or a vacuum gap, and it decays on the length scale of one atomic radius. In the STM the tunneling current flows from the very last atom of the tip apex to single atoms at the surface, inherently providing atomic resolution.

In a standard experiment (see Fig. 1.1), the tip is moved in three dimensions by piezoelectric actuators. An electronic controller guides the tip at a tip–sample distance corresponding to a constant preset tunneling current. This distance is recorded by a computer as a function of the lateral position and displayed as a microscope image. High mechanical stability of the experimental setup turns out to be a prerequisite for successful measurements on the atomic scale.

With this example of the STM, all elements of a scanning probe microscope have been introduced. A short-range interaction, yielding the desired resolution, is sensed by a local probe. The probe is scanned over the surface under study, and the measured quantities are recorded and processed in a computer system. The experiment needs a rigid construction and an effective vibrational isolation in order to allow reproducible positioning on the atomic scale.

The family of scanning probe microscopes has several members, based on a variety of tip–sample interactions. The first and most important extension of the STM

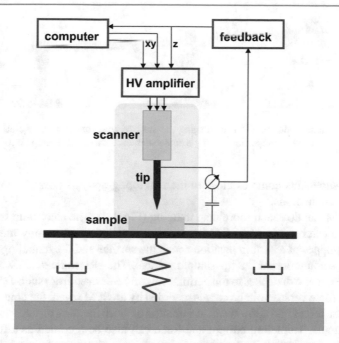

Fig. 1.1 Basic setup of an STM. The tunneling current is used to control the tip–sample distance z via a feedback circuit. The distance z is recorded by a computer as a function of the scanned coordinates x and y. A high-voltage amplifier is required to drive the piezoelectric scanner. Good vibrational isolation of the experiment is a prerequisite for high-resolution imaging

was the scanning force microscope (SFM), invented in 1986 by Binnig, Quate, and Gerber [86]. In this instrument, the tip height is controlled in such a way that the force between tip and sample is constant. While the use of the STM is restricted to conducting surfaces, the SFM is in principle capable of determining the topography of any surface, conducting or not. Based on the assumption that forces between the atoms at the tip apex and the atoms of the surface determine the resolution of this instrument, it is commonly called the atomic force microscope (AFM). We will discuss in Chap. 3 the extent to which this assumption is justified. The third distinguished member of the family of SPMs is the scanning near-field optical microscope (SNOM), which uses short-range components of the electromagnetic field as tip–sample interaction.

So far, we have mentioned the imaging capabilities of scanning probe microscopes. Generally, it is said that the movement of the tip at constant tunneling current or constant force reveals the topography of the sample surface. Some caution is required when using the term topography. In STM, it is actually a map of constant density of states that is recorded, and this may differ from the geometric topography. For example, a molecule adsorbed on top of a metal surface may reduce the local density of states and may actually be imaged as a depression. In force microscopy, the situation is even more complicated as different parts of the tip interact differently with features on the surface. The measured height of steps, for example, may deviate significantly from their geometric height. However, on homogenous surfaces

Fig. 1.2 Different modes of SPM. **a** Imaging the surface with true atomic resolution. **b** Local spectroscopy of surface properties of single molecules on surfaces. **c** Manipulation of surface structure

SPM measurements come as close to the real topography as possible with current experimental methods.

But SPM can do much more than imaging (Fig. 1.2). The electronic structure of the surface can be studied in STM using the so-called spectroscopy modes. Here the tip is stopped at a certain position above the surface and the tunneling current is recorded as a function of the tip–sample voltage. The electronic density of states at different energetic distances to the Fermi level can be derived from such $I(V)$ curves. Similarly, force versus voltage curves recorded by an SFM reveal the contribution of electrostatic forces and allow the determination of work function differences between tip and sample. Tunneling current and force can also be recorded as a function of the tip–sample distance, and additional information about the tip–sample interactions is obtained. The stability and sensitivity of these spectroscopic methods can be enhanced by employing lock-in techniques. A small oscillating voltage is added to the tip–sample voltage, and its effect on the tunneling current or force is analyzed with a lock-in amplifier. In the same manner, the tip–sample distance can be modulated in order to determine distance dependencies of current or force.

The third important strength of SPM beyond topographic imaging and local measurement of surface properties is the manipulation of surfaces. Single atoms of the surface or adsorbates on it have been systematically moved in STM in order to build nanometer-sized structures. This can be accomplished by pushing or pulling the atoms with the tip, or even by transfer of atoms to and from the tip. Such experiments establish a lithography on the molecular scale. STM is not the only tool for surface manipulation. The tip of an SFM can be used to deposit charges on insulating samples, or to study microscopic effects in wear by scratching the surface. Single molecules can be optically bleached by a SNOM. All these examples share the fact that the results and effects of manipulation are studied with the same tip that was used as a tool to perform it.

Scanning probe microscopy has found wide applications in Surface Science, where problems like surface structure, adsorption of molecules, or local electronic properties could be studied. The first nanostructures have been built in an atom-by-atom way and characterized. More industrial applications include surface control in Materials Science. Roughness and hardness are being measured on the nanometer scale. Magnetic structures on data storage devices can be analyzed as well as the optical quality of coatings. The microscopic origins of friction have been investigated by SFM. Force microscopy allows nanometer-scale imaging of biological materials which are not accessible to electron microscopy for preparative reasons. Beyond imaging, force measurements between functional molecular groups have stimulated great interest in biophysics. A series of monographs edited by R. Wiesendanger and H.-J. Güntherodt cover the development of SPM [243,691,692]. An encyclopedic introduction to procedures used in SPM has been edited by Colton et al. [130].

1.2 Basic Concepts

The basic physical principle of any scanning probe method is the interaction between the scanning probe and the sample. If this interaction has a near-field character, one can overcome the resolution limits of far-field techniques like optical microscopy or scanning electron microscopy (SEM), which are generally of the order of half a wavelength of the photons or electrons. However, the resolution in SPM is limited by the geometrical shape of the probe. The lateral resolution thus depends on the vertical amplitude of the structures on the surface. The typical probing tip can be envisaged as a cone with an opening angle and a finite radius at the apex. Images of steps on the surface with walls steeper than the opening angle will be smeared out due to the convolution of the tip shape with the surface structure. Holes in the surface with a diameter smaller than the tip radius will not be imaged at all or with reduced corrugation. These situations are depicted schematically in Fig. 1.3. On atomically flat surfaces or molecular layers, the resolution is determined by the atomic structure of the tip apex. The range of the interaction and its strength compared to other interactions will eventually determine the resolution of the relevant method. Artifacts arising from the shape of the probe will be discussed in detail in Chap. 6.

The most powerful near-field interaction is the tunneling current across a vacuum gap. There are two reasons for this:

- the decay length is as small as an atomic diameter,
- there are no other electrical currents flowing through the vacuum which could obscure the measurement.

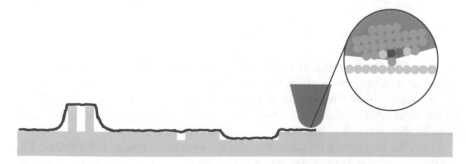

Fig. 1.3 Resolution limits of SPM. Images of steep walls and small holes are distorted when the size of features is comparable to that of the tip apex. On atomically flat surfaces, the resolution is determined by the atomic structure of the tip apex

Unfortunately, the situation is different in force microscopy. There are certainly chemical binding forces with an interaction range comparable to the decay length of the tunneling current. But additionally, there are always long-range forces like van der Waals forces between the mesoscopic tip and the sample. The short-range forces providing high resolution have to be measured on the background of the long-range forces, which do not contribute to the resolution at all. To overcome this problem, there are some tricks to reduce the effect of long-range forces and to enable high resolution imaging with the short-range forces in a very impressive way. The easiest way is to measure force variations at constant height, where submolecular resolution was achieved on planar molecules [231]. It turns out that in this mode, AFM can surpass the lateral resolution of STM. Other important experimental modifications are the attachment of CO molecules on the tip, the sharpening of the probing tip by focussed ion beam and to use small amplitudes to reduce the effects of long range forces and to be able to enter the short-range force regime without irreversible changes of tip or sample.

Not all SPM is based on near-field interactions in a strict interpretation of the concept. Electrostatic forces, for example, play an important role in SFM. The electrostatic force of an isolated charge with its inverse square distance dependence would not usually be called a near-field interaction. Nevertheless the imaging of localized charges or of the local contact potential by SFM is of great interest. In some cases the electrostatic potential can even become the source of a real near-field interaction for geometrical reasons. Consider the perfect surfaces of alkali halide crystals. The variation of the electrostatic potential above the alternating charges decays exponentially with a decay length of the order of the lattice constant. A similar phenomenon can be found in magnetic force microscopy (MFM). The decay length of the interaction between periodically arranged magnetic domains and the magnetic tip is typically of the order of the domain size.

Some experimental techniques use the setup and the distance control of SPM to record quantities locally which have no near-field character at all. For example, a scanned micropipette can collect local ion currents and scanned thermocouples map the surface temperature.

1.2.1   Local Probes

Tips for STM can be prepared as easily as cutting a PtIr wire with household scissors. However, reproducible sharp tips with a high aspect ratio for imaging of pits or steps are mostly electrochemically etched tungsten wires. Typical tunneling currents are of the order of 1 nA. Therefore, a low-noise current-to-voltage converter with high amplification is required in order to obtain suitable signals.

For SFM, the probing tip has to be mounted onto some type of force sensor. The standard solution is a microfabricated cantilever spring with integrated tip. Commercial products are made of silicon, SiO_2 or Si_3N_4. These tips have pyramidal shapes, with side walls defined by crystallographic planes of low etching rate. Silicon nitride cantilevers are produced on top of a silicon master and their tips are therefore hollow. Alternative force sensors include electrochemically etched metal tips. The wire itself can be used as a spring for sensing the force, or the metal tips can be attached to quartz tuning forks as dynamic force sensors. An additional sharpening of the probing tips by focussed ion beams is advantageous to improve the lateral resolution and to reduced the influence of long range forces.

The SFM tips can be covered with different materials for special experimental purposes. Diamond coatings have been realized to harden the tips for wear experiments. Metal coatings allow one to control the electrostatic potential, and magnetic coatings are used in MFM. The SFM tips can even be functionalized in a chemical sense to study specific molecular interactions. Recently, the unique mechanical and chemical properties of carbon nanotubes have attracted much interest as SFM tips. An important step for SFM operated at low temperatures is to pick up single CO molecules or noble gases, such as Xe or Ar. These molecules are not very reactive and give the opportunity to enter the short-range force regime.

Local optical near-field effects can be probed either by using very small light sources or by recording the light scattered from the gap between an SFM tip and structures at the sample surface. A widely used small light source is a sub-wavelength sized aperture in the metal coating at the apex of a tapered optical fiber. In combination with SFM, small optical apertures have been produced at the apex of hollow tips.

Even more complex structures on tips can be realized by lithographic techniques. In this way, scanning thermocouples for temperature mapping and even single electron transistors on the tip have been fabricated and operated.

1.2.2   Scanning and Control

1.2.2.1 Actuators

The scanning probe needs to be positioned with an accuracy of 1 pm if atomic resolution is required. This is usually achieved by means of piezoelectric actuators. Piezoelectric materials change their shape in an electric field due to their anisotropic crystal structure. Using the transverse piezoelectric effect, the length of a bar of material can be adjusted by applying a voltage to electrodes attached to its side walls. The standard material for piezoelectric actuators is PZT (lead zirconium titanate). The relative length variation $\Delta l / l$ of a bar is given by

$$\frac{\Delta l}{l} = d_{31} E ,$$

where E is the electric field across the bar and d_{31} the transverse piezoelectric coefficient. A typical value (PZT-5H) is $d_{31} = 0.262$ nm/V. In order to elongate a bar of length 1 cm by 1 μm, a voltage of 380 V has to be applied across a thickness of 1 mm. These numbers indicate that a high-voltage amplifier is necessary to drive the piezoelectric actuators. The material for actuators needs to be polarized with very high voltages to produce the desired effect. This polarization can be lost either gradually after long use, or suddenly when the material is heated above its Curie temperature. Therefore, the calibration of actuators has to be repeated regularly. The

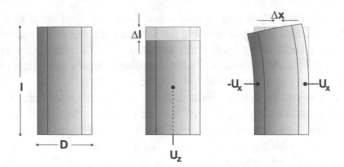

Fig. 1.4 Schematic side view of tube scanning. The outside electrode is divided into four parts. When a voltage U_z is applied to the inner electrode, the length of the tube is changed. When a pair of voltages U_x and $-U_x$ is applied to opposite electrodes outside the tube, the tube is bent and a lateral movement is realized. Note that the bending of the tube is largely exaggerated in order to visualize the effect. However, one should keep in mind that lateral displacements by tube scanners are always accompanied by height changes

material may have to be re-polarized by exposing it to a high voltage above the Curie temperature and cooling the sample with the voltage applied.

The three-dimensional positioning of the scanning probe can be obtained by three piezoelectric bars in an orthogonal assembly. However, the standard scanner of most SPMs is the so-called tube scanner (see Fig. 1.4). The elongation of a tube of piezoelectric material is used for the vertical movement, while the bending of the tube accomplishes the horizontal scanning movement. For this purpose, the outside of the tube is contacted by four symmetric electrodes separated along the tube. By applying equal but opposite voltages to opposing electrodes the tube will bend due to contraction and expansion. The inner wall is contacted by a single electrode for application of the actuating voltage for vertical movement. The vertical displacement Δl of the tube scanner can be estimated from

$$\Delta l = \frac{d_{31} l U_z}{h} \, ,$$

where l is the length of the tube, U_z the voltage applied to the inner electrode, and h the wall thickness of the tube. The lateral displacement Δx of the tube end, which is introduced by applying the voltages $-U_x$ and U_x to opposite side electrodes, is given by

$$\Delta x = \frac{2\sqrt{2} d_{31} l^2 U_x}{\pi D h} \, ,$$

where D is the average diameter of the tube. These formulas have been derived and tested in [109, 713].

The positioning performance of a piezoelectric tube scanner is limited by several problems:

- non-linearity of the tube scanner,
- hysteresis of the scanning movement,

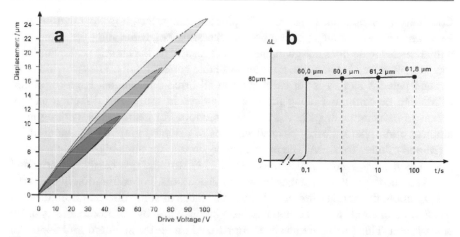

Fig. 1.5 a Hysteresis curves of a piezoactuator for various peak voltages. The hysteresis is related to the distance moved. **b** Creep of piezomotion after a 60 μm change in length as a function of time. Creep is of the order of 1% of the last commanded motion per time decade   (from https://www.physikinstrumente.com/en/technology/piezo-technology/properties-piezo-actuators/displacement-behavior/)

- creep of the material,
- noise and drift of the high voltage supply,
- thermal drift of the whole mechanical setup.

The non-linearity of the the bending scheme will add an offset to the vertical position when scanning horizontally. For very small scan ranges compared to the tube length, such as a few nanometers to one centimeter, these non-linear effects do not disturb the measurements. However, topography images recorded over 100 μm often show a significant curvature of the surface.

Like magnetic materials, piezoelectric materials also exhibit hysteretic behavior (Fig. 1.5). When applying a saw-tooth voltage, the position of the actuator at a certain voltage can deviate by as much as 15% between the forward and backward movements. Furthermore, piezoelectric material will creep after a sudden jump in the applied voltage. This ongoing creep of the position in the direction of the jump is logarithmic as a function of time. In SPM, the creep has two major effects: firstly, it aggravates the hysteresis at the turning points of the scanning movements, and secondly, the vertical position of the tip will creep after approach to the sample. Since the tip–sample distance is kept constant by the controller, the recorded surface topography will exhibit a logarithmic curvature in the slow scan direction. One of the resolution limits of the positioning in SPM is the noise of the high-voltage amplifiers. The other limit is the thermal drift, both of the voltage supply and the mechanical setup. The latter is suppressed in low-temperature experiments. A linear drift of the experiment can be compensated by continuously adding a voltage ramp to the actuating voltages.

Although the sensitivity of a tube scanner can be calculated in principle, a calibration of the scanner in both the lateral and vertical directions is indispensable.

Considering the non-linear nature of the piezoelectric effect, it is clear that for each scan range a corresponding calibration is necessary, i.e., calibrating the scanner on a micrometer scale does not guarantee correct results on the atomic scale. For large scan ranges, special calibration grids are available commercially. They provide well-defined patterns and steps for calibration in all three dimensions. High-resolution SPMs can be calibrated along the atomic lattices in the lateral direction and by means of monatomic steps in the vertical direction. For instruments working in an ambient atmosphere, freshly cleaved surfaces of oriented graphite can be used for STM and of mica for SFM. The most suitable surface for ultra-high vacuum experiments is the 7×7 reconstruction of the Si(111) surface. It is easily prepared and resolved due to the large distance between the adatoms. For calibration in the vertical direction, the height difference between terraces at some distance from the step has to be evaluated, since the step itself can be obscured by electronic effects or tip convolution. This procedure has been formalized under the standard ISO 5436. An absolute calibration of scanners is also possible using interferometric or capacitive position sensors. A knowledge of the actual position can then be used to construct closed-loop controlled scanners without creep or hysteresis.

Piezoelectric actuators need high voltages for operation. To overcome this requirement, new positioning schemes have been developed for low-cost SPM devices. Magnetic actuation has been realized by coils and permanent magnets, similarly to a loudspeaker.

1.2.2.2 Controller

The (x, y)-position of the scanning probe is usually calculated line-by-line by a computer, resulting in rectangular frames. The numbers are converted into voltages, which are amplified to meet the high-voltage requirements of the actuators. The z-position, i.e., the tip–sample separation, is controlled by a closed-loop feedback system. For the example of STM, the difference between the actual tunneling current and a preset setpoint current is used to correct the tip–sample distance. This difference is referred to as the error signal. The reaction is defined by multiplying the error by a proportional gain on the one hand, and integrating the error over a certain time weighted with an integral gain on the other hand. A suitable choice of setpoint, proportional gain, and integral gain is crucial for a successful measurement. When values of the proportional gain are too high, feedback oscillations may occur with the mechanical eigenfrequency of the scanner. When values are too low, details can be lost due to slow reaction of the feedback. For the integral gain, oscillations again occur when values are too high, while low values result in difficulties when steps on the surface have to be compensated. As a starting recipe, the proportional gain can be set to 2/3 of the value at which oscillations start to occur. The integral gain can then be increased up to the very first signs of reappearance of the oscillations. Although ideal values for the gains can be calculated in principle, the ever-changing distance characteristic of the current leaves the final choice to the experimentalist. Since the distance dependencies of the current in STM and the force in SFM are highly non-linear, a change of the setpoint always requires an adequate readjustment of the gains. In STM, this problem can be partly overcome by means of a logarithmic current amplifier, which compensates the exponential distance dependence.

1.2.3   Vibrational Isolation

Mechanical stability and vibrational isolation are prerequisites for the achievement of high-resolution images in SPM. The construction of the tip holder, sample holder, and actuators should be as rigid as possible, in order to increase the mechanical eigenfrequency of the instrument. The feedback control speed will always be limited by this eigenfrequency, which for typical tube scanner setups lies in the range 0.5–10 kHz. SPM instruments are very sensitive to acoustic noise and should always be operated under a solid box. Isolation from building vibrations can be achieved by suspending the instrument with springs, with rubber feet in various forms, or with inflated air-damped feet. It is very useful to construct a two-step damping system, for example a table with air-damped feet and an instrument base made of alternating stacks of viton rings and heavy metal plates. The advantage of such two-step damping is depicted in Fig. 1.6. Each damping step and the instrument itself have a transfer function describing the frequency-dependent response of the system to external noise. The two damping steps have resonance frequencies well below building and acoustic frequencies and are strongly damped (low Q-factor). If the

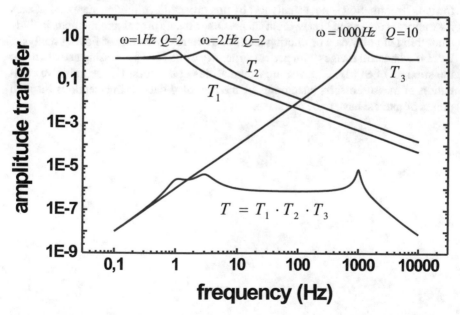

Fig. 1.6 Transfer functions of a two-step damping system with low resonance frequencies ω_0 (T_1, T_2). The final response of the instrument to external noise is given by the product $T_1T_2T_3$, where T_3 is the transfer function of the SPM with a typical resonance of 1 kHz

SPM has a resonance frequency of 1 kHz, all noise between 10 and 400 Hz can be effectively damped with a flat response function.

1.2.4   Computer Control and Image Processing

SPM is a computer-based method. The computer provides the signals for scanning and collects, stores, and displays the data. For some systems, the computer even takes over the feedback control. The repetitive tasks are often carried out by a digital signal processor running in parallel to the computer, which provides the user interface. As the speed of calculation in most computers is subject to unpredictable fluctuations, the performance should be checked by independent methods whenever real-time experiments are required, as in studies of the velocity dependence.

Two-dimensional data are usually displayed on the screen in a so-called false color representation, where for each (x, y)-value the measured signal is coded by a colored pixel. By convention, elevations in topography are displayed in a brighter color than depressions.

The collected data are usually processed for optimal contrast in display and publication. The instrumental drift and the creep of the piezoelectric actuators cause a tilt or even a curvature of the topographic data along the slow scan direction. This tilt can be compensated by subtracting an average plane or a second-order polynomial from the data. Information is not usually lost by this procedure, although stepped surfaces may appear corrugated in saw-tooth shape. Other image processing methods include smoothing of noisy data or filtering atomic structures by means of Fourier analysis. Even if wonderful images with perfect structure result, such processing usually hides the strength of SPM, i.e., revealing irregularities at surfaces. For quantitative evaluation of measurements, histograms of the recorded data or line-sections through points of interest have to be extracted.

Introduction to Scanning Tunneling Microscopy

2

Abstract

An introduction is given into scanning tunneling microscopy (STM), where a small tunneling current is measured between probing tip and sample. Various operation modes, such as constant tunneling and constant height modes as well as tunneling spectroscopy, are described and application examples are given.

Scanning tunneling microscopy (STM) was invented by Binnig and Rohrer (see Fig. 2.1) [80, 109]. Using the combination of a coarse approach and piezoelectric transducers, a sharp, metallic probing tip is brought into close proximity with the sample. The distance between tip and sample is only a few angstrom units, which means that the electron wave functions of tip and sample start to overlap. A bias voltage between tip and sample causes electrons to tunnel through the barrier. The tunneling current is in the range of pA to nA and is measured with a preamplifier.[1] This signal is the input signal of the feedback loop, which is designed to keep the tunneling current constant during (x, y)-scanning. The output signal is amplified (high voltage amplifier) and connected to the z-piezo. According to the feedback output voltage and the sensitivity of the piezo (typically nm/V) the tunneling tip is moved backwards or forwards and the tunneling current is kept constant during acquisition of the image. This operation mode is called constant current mode.

There exist other modes, such as the constant height mode, where the tip is moved at constant height and variations in the current are measured. The (x, y)-movement of the tip is controlled by a computer. The z-position (output of feedback loop) is measured at discrete (x, y)-positions. The data $z(x_i, y_j)$ can be displayed in several

---

[1] I–V converters are used to convert the tunneling current into a voltage. A possible realisation consists of an operational amplifier (e.g., Burr Brown OPA 111) and a resistor R, where the output voltage is given by $V_{out} = R I_t$. With a resistor of the order of $R = 100\,M\Omega - 1\,G\Omega$, currents of the order of nanoamperes, $I_t \approx 1\,nA$, are measurable. Smaller currents of the order of $0.1\,pA$ are more difficult to measure. Field-effect transistors are needed very close to the tip.

© Springer Nature Switzerland AG 2021
E. Meyer et al., *Scanning Probe Microscopy*, Graduate Texts in Physics,
https://doi.org/10.1007/978-3-030-37089-3_2

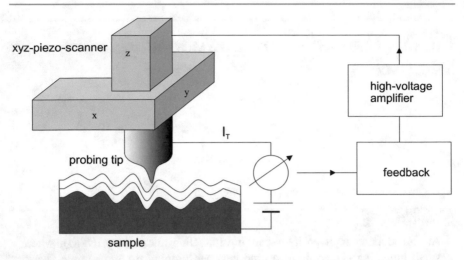

Fig. 2.1 Schematic diagram of the scanning tunneling microscope. An xyz-piezoelectric scanner moves the tip over the surface. A feedback loop can be used to keep the tunneling current constant

ways: line-scan image, grey-scale image or colour encoded image.[2] The line-scan image is the most natural way to represent the data, because each line represents the scan of the tip in the fast direction. However, grey-scale images or colour encoded images are more frequently used, because they are better adapted to human pattern recognition. Bright spots represent hillocks or protrusions and dark spots represent valleys or depressions.

Scan areas are limited by the choice of the piezoelectric scanner and the maximum output voltage V_{max} of the high voltage amplifier. With a typical sensitivity of 3 nm/V and $V_{max} = 250$ V, a maximum scan range of 750 nm can be achieved. Some commercial scanners are designed for ranges of up to 100 μm. However, these large scanners are often inadequate for atomic-scale imaging. Thus, the maximum scan range of high resolution STMs is in the range 1000 nm = 1 μm. Whether the microscope can achieve atomic resolution depends on the stability of the instrument and the vibrational isolation. In Chap. 1 the design of these components is explained in more detail. Generally, the mechanical construction of the STM should be rigid with a high resonance frequency ($\approx$1 kHz) and the vibrational isolation should have a low resonance frequency ($\approx$1 Hz) with low Q-factor ($Q \approx 1$).

2.1   Tunneling: A Quantum-Mechanical Effect

According to quantum mechanics, a particle with energy E can penetrate a barrier $\phi > E$ (see Fig. 2.2). In the classically forbidden region, the wave function ψ decays

---

[2]Selection of the number of data points is user-defined, e.g., $400 \times 400, 500 \times 500$ or 1000×1000. If fast fourier algorithms are to be used, powers of two, 2^N, are favourable, e.g., $256 \times 256, 512 \times 512$ or 1024×1024.

Fig. 2.2 One-dimensional tunneling junction. The wave functions of tip and sample overlap. When a bias voltage is applied, electrons can flow from the tip to the sample (tunneling into empty states of the sample) or from the sample to the tip (tunneling from filled states of the sample). The barrier height is approximately determined by the work function. However, the barrier height also depends on other effects, such as image forces. At very small distances the barrier may collapse

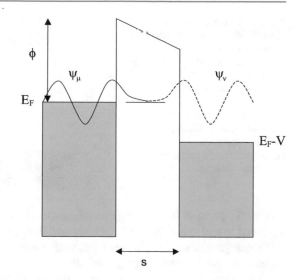

exponentially, viz.,

$$\psi(z) = \psi(0) \exp -\frac{\sqrt{2m(\phi - E)}z}{\hbar} \, , \tag{2.1}$$

where m is the mass of the particle and $\hbar = 1.05 \times 10^{-34}$ J s. In STM the barrier is given by the vacuum gap between sample and tip. Then the tunneling current I_t can be calculated by taking into account the density of states of the sample, $\rho_s(E_F)$, at the Fermi edge:

$$I_t \propto V \rho_s(E_F) \exp\left[-2\frac{\sqrt{2m(\Phi - E)}z}{\hbar}\right] \propto V \rho_s(E_F) e^{-1.025\sqrt{\Phi}z} \, , \tag{2.2}$$

where the barrier height Φ is in eV and z in angstrom units. With a typical barrier height of $\Phi = 5$ eV, which corresponds to the work function of gold, the tunneling current decays by an order of magnitude when the vacuum gap is changed by 1 Å.

Even though the theoretical prediction of vacuum tunneling was found in the 1930s [189], it took a long time to observe it experimentally. Young et al. [718, 719] presented the first evidence for vacuum tunneling. The current versus voltage curves acquired with their topographiner, a predecessor of the STM, showed a transition from the field emission regime to the tunneling regime. In 1982, Binnig et al. presented current versus distance curves with barrier heights of the order of 0.6–3 eV [84]. In 1983, Binnig and Rohrer presented the first successful imaging of Au(110) and Si(111)7 × 7. The latter surface yielded images of the arrangement of the Si adatoms [85]. It contradicted previous models of the surface and constituted an important step towards the complete structure analysis of this complex surface, which was finally determined by Takayanagi et al. [632] with a UHV-transmission electron microscopy analysis. More importantly, the work of Binnig and Rohrer started the field of real-space microscopy, opening the door to the nanometer world for science and technology. In the beginning of STM, many scientists were very sceptical about

Fig. 2.3 Schematics of the
Tersoff–Hamann model. The
tip wave function is
approximated by an s-wave.
The tunneling current is
proportional to the local
density of states of the
sample at the Fermi level,
LDOS(E_F), at the distance
$r_0 = R + s$, which
corresponds to the center of
curvature of the tip

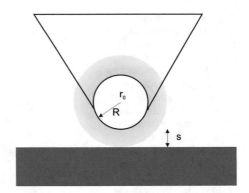

this work, because it was not generally expected that atomic resolution could be achieved in real space. However, Binnig and Rohrer succeeded in convincing more and more groups to work in this field, until finally, STM became one of the most important tools in modern surface science.

The model described above is valid for one dimension. However, STM corresponds to a three-dimensional geometry. It was Tersoff and Hamann [640,641] who first showed that (2.2) is valid in three dimensions under the assumption that the tunneling tip wave function can be approximated by a spherical s-wave function and $\rho_s(E_F)$ is the local density of states of the sample (LDOS) at the Fermi edge at a distance $s + R$, where s is the distance between the apex of the tip and the sample surface and R is the radius of curvature of the tip. Thus, it is the LDOS of the sample at the center of the tip (see Fig. 2.3). This model also explained the high resolution on metals, such as Au(110). However, the large corrugation heights on surfaces such as Al(111), Au(111) or Cu(111), could not be explained quantitatively.[3] This topic will be discussed further in the following sections. The Tersoff–Hamann model is the standard model for STM, constituting an elegant first-order model of STM.

2.1.1   Tersoff–Hamann Model

The Tersoff–Hamann model [640,641] is based upon the transfer Hamiltonian approach, which was introduced by Bardeen [49] to explain current versus voltage curves of oxide barriers between superconductors. Bardeen showed that the tunneling current between two electrodes separated by an insulator is given by

$$I_t = \frac{4\pi e}{\hbar} \int_{-\infty}^{\infty} \left[f(E_F - eV + \varepsilon) - f(E_F + \varepsilon) \right] \rho_s(E_F - eV + \varepsilon)\rho_t(E_F + \varepsilon)M^2 d\varepsilon ,$$

$$(2.3)$$

---

[3]In this context, the corrugation height corresponds to the height difference between the hollow site and the top site in constant current operation.

where

$$f(E) = \frac{1}{1 + e^{(E-E_F)/k_BT}} \tag{2.4}$$

is the Fermi function, ρ_s, ρ_t are the density of states of the sample and tip, e is the electron charge, $\hbar = h/(2\pi)$, h is Planck's constant and ε is an integration variable. Bardeen showed that the tunneling matrix element M is given by

$$M = \frac{\hbar}{2m} \int_{\text{surface}} \left(\psi_s^* \frac{\partial \psi_t}{\partial z} - \psi_s \frac{\partial \psi_t^*}{\partial z} \right) dS, \tag{2.5}$$

where ψ_s, ψ_t are the wave functions of the sample and the tip. Thus, the problem is solved for the sample and the tip separately and then the matrix element is determined according to (2.5). It is remarkable that the integral is to be calculated between the electrodes (surface integral). For low voltages, (2.5) simplifies to

$$I_t = \frac{4\pi e}{\hbar} \int_0^{eV} \rho_s(E_F - eV + \varepsilon)\rho_t(E_F + \varepsilon)M^2 d\varepsilon. \tag{2.6}$$

In the case of a flat LDOS tip, which means that ρ_t is constant in the studied energy range, the energy dependent part of the tunneling current is determined by the sample alone:

$$I_t \propto V\rho_s(E_F - eV), \tag{2.7}$$

which leads us back to the statement that the tunneling current is essentially determined by the LDOS of the sample at the Fermi energy. Extensions of the Tersoff–Hamann model can be found in the literature. Baratoff presented wave-vector-dependent tunneling current calculations [47] and Chen extended the Tersoff–Hamann model to d-wave functions [108]. Furthermore, the Green's function formalism has been introduced by several authors (see [109]).

2.2 Instrumental Aspects

2.2.1 Tunneling Tips

The preparation of tunneling tips is not a trivial subject and there are still many unknown parameters. The first recipe is to produce smooth tips with small curvature radius of the order of 100–1000 Å by electrochemical etching. Either AC or DC voltages can be used. A simple and reliable technique is AC-etching with voltages of 3–12 V. Preferably, only a short piece of the wire is exposed to the electrolyte, where the rest is isolated by teflon tubes. A DC method was already developed in the 1950s for field ion microscopy [653]. A tungsten wire (about 0.25 mm in diameter) is immersed in 1 M aqueous solution of NaOH. The counterelectrode is a piece of stainless steel or platinum, typically with a ring geometry. A positive voltage of 2–12 V is applied to the wire. Etching occurs at the liquid–air interface. As soon as the

Fig. 2.4 Examples of SEM pictures of tunneling tips. The tips were etched for 60 s with 12 V AC and then with 2 V AC until fracture occurred. Courtesy of A. Wetzel and V. Thommen

neck of the wire becomes thin enough, the weight of the lower part of the wire pulls the neck and finally fractures it. Both tips (falling part and fixed part) can be used. Some authors have controlled the cutoff time [303], which is the time for the etching current to cut off after the tip has fractured. An electronic circuit that measures the current and switches off when the wire fractures can reduce the cutoff time and the radius of curvature. An elegant way was proposed to use two lamellae (double lamellae drop off etching procedure), where the current is interrupted immediately as soon as the wire fractures [374]. After electrochemical etching, the tip has to be rinsed thoroughly with water in order to remove NaOH residuals. It is even advisable to use boiling water. Although these tips look very smooth in a scanning electron microscope (SEM) (see Fig. 2.4), they do not always yield atomic resolution. One problem is the formation of a surface oxide during etching, which has to be removed before tunneling. One way is to remove the oxide by resistive heating or electron bombardment [109].

Other mechanical preparation methods produce rougher tips. For example, PtIr tips can be mechanically cut or W tips can be mechanically grinded to provide immediate atomic resolution. However, the aspect ratio is rather poor and imaging of step sites or hillocks is not possible without tip convolution.

Several methods have been developed in field ion microscopy to produce tips with well-defined atomic structure. Fink developed a field evaporation procedure that yields single-atom tips [246]. Binh and Garcia produced nanotips in high fields [79].

Finally, there are some in situ tip preparation methods. Controlled collision has been introduced by Binnig and Rohrer on $Si(111)7 \times 7$ [80]. High-field treatments have been mentioned in the literature [56, 701]. The process is probably related to field enhanced diffusion of tungsten, which leads to a restructuring of the tip.

2.2.2 Implementation in Different Environments

The STM can be operated in different environments such as ultrahigh vacuum (UHV), liquids, or ambient pressure. Most STM work has been dedicated to UHV, because the preparation of conductive surfaces is often limited to this environment. Exceptions are gold, highly oriented pyrolitic graphite (HOPG) and other layered materials, such as transition metal dichalcogenides (e.g., TaS_2), which can be measured in ambient pressure. Active fields of application are electrolytes and low current STM on organic molecules. Another important parameter is the temperature. At low temperature, the diffusion of atoms is reduced, and this gives the opportunity to manipulate single physisorbed atoms [161]. Alternatively, high temperature gives the opportunity to observe dynamical processes [283,394]. Room temperature experiments are sometimes disturbed by temperature variations, which lead to thermal drift.

2.2.3 Operation Modes

The STM can be operated in the constant current mode or in the constant height mode (see Introduction of Sect. 2.2). Furthermore, there exist a number of spectroscopic modes which are described in this section.

Constant Current Mode The experimental setup is shown in Fig. 2.5. The first stage of the electronics is the preamplifier. As described in Sect. 2.2, gains of 10^8–10^9 are selected by the appropriate choice of resistor (10^8–10^9 Ω) in the I–V converter. Since the current depends exponentially on the distance between tip and sample, it is advisable to linearize the signal with a logarithmic amplifier. The feedback loop of the STM is set to a certain value, the set point, which is compared with the instantaneous value of the current. A PI-feedback [647] is generally used, which means that the

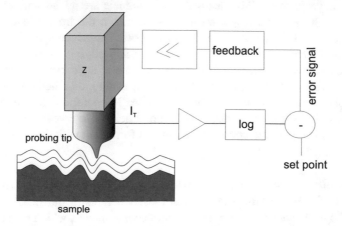

Fig. 2.5 Feedback loop of an STM consisting of a tunneling gap, preamplifier, logarithmic amplifier, PI-feedback, high voltage amplifier and z-piezo

error signal is multiplied by a proportional gain (P-part) and integrated (I-part) with a characteristic time constant. The P- and I-parts are added together. The output of the feedback is then amplified (high voltage amplifier) and fed directly to the z-piezo.

The STM user has to optimize the feedback parameters in each experiment, because the gain of the whole loop also depends on various factors, such as preamplifier gain, high voltage amplifier gain, probing tip characteristics (e.g., barrier height) and mechanical properties of the microscope. A simple way to optimize the feedback is the Ziegler–Nichols procedure [658,732]. The P-part is increased until small oscillations of the current are observed. Then, the P-part is reduced to $0.45 P_{crit}$. If the oscillation period (T_{crit}) is determined with an oscillograph, the I-part can be set to $0.85 T_{crit}$. The higher the selected gain, the better the feedback loop follows the contours of the constant tunneling current. The output of the feedback loop is digitized and gives the 'topography' image, $z(x, y)$. Actually, the term 'topography' has to be used with some caution, because the contours of constant tunneling current are not identical with the contours of constant total charge density, which would be the ideal topography of a surface. Rather, they are related to contours of constant local density of states at the Fermi level (LDOS), as explained in Sect. 2.1.1. On an atomic scale, images often depend on the applied voltage because of variations in the LDOS. Nevertheless, the constant current images are often very close to topography on a scale of a few nanometers where LDOS variations are small. Thus, step heights are relatively easy to interpret. Exceptions are inhomogenous surfaces, where surface states or variations in barrier height lead to strong variations in the current between the terraces [322].

Constant Height Mode In the constant height mode, the feedback loop is turned off and the tip scans at constant height over the surface. Variations in the tunneling current are digitized directly. The sample is often tilted and this has to be corrected electronically. The instrument is first operated in constant current mode and the tilt correction is performed. This means that x, y-signals are added with appropriate prefactors to the z-signal $z_{out} = \alpha x + \beta y + z_{in}$. Then the tip is moved to a certain z-position and the feedback loop is turned off. The advantage of the constant height mode is that high scanning rates can be achieved, e.g., to observe dynamical processes. However, thermal drift limits the time of the experiment and there is an increased risk of crashing the tip. At low temperature, this mode is widely used to image planar molecules in a combined STM/AFM operation. An extension is to acquire constant height mode images at different heights to reveal structural details at the optimum height or to compare the contrast formation with theoretical models.

Spectroscopic Modes Scanning tunneling spectroscopy (STS) experiments are important to determine properties such as barrier heights or the LDOS as a function of voltage. There exist a variety of spectroscopic modes, which means that one externally adjustable parameter, such as the voltage or the z-distance, is varied and an experimentally available property is measured, e.g., the current $I(V)$, $I(z)$, or

Fig. 2.6 Example of I versus z curves. Reprinted from Binnig, Rohrer, Gerber and Weibel [84], https://doi.org/10.1063/1.92999, with permission from AIP Publishing. *Curves* show different slopes, representing different barrier heights. This was achieved by successive cleaning procedures. These measurements represent the first vacuum tunneling experiment through a controllable gap

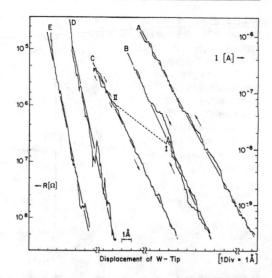

the derivatives of the current dI/dV, dI/dz. These curves can be measured either at a fixed lateral position or at various lateral positions (x_i, y_i). A simple way to get spectroscopic information is to acquire several images at different voltages or distances. Because of thermal drift, this approach is not ideal, although it remains feasible. Alternatively, the tip can be stopped at different positions (x_i, y_i), the voltage or distance is ramped, and then the tip moves to the next position. In an ideal STS experiment the tip DOS should be negligible, although this seems not always to be the case. Tip cleaning procedures for STS are described in the literature [109, 176].

An example is given in Fig. 2.6, which shows I versus z curves from Binnig et al. [84]. The slope of the curve $\log(I)$ versus z yields the barrier height:

$$\log(I) = -A\sqrt{\Phi}z + \text{Const.} , \tag{2.8}$$

where $A = 1.025\,\text{Å}/\sqrt{\text{eV}}$. At the beginning of the experiment, barrier heights of 0.6–0.7 eV were found. By successive in situ cleaning procedures, barrier heights up to 3.5 eV were found, which is close to the expected value of 4–5 eV. The small values at the beginning of the experiment were related to contamination of the tungsten tip, e.g., oxide layers or hydrocarbon/water layers. The curves represent the first observation of vacuum tunneling through a controllable gap.

Tunneling Spectroscopy on Superconductors The superconducting gap can be observed by dI/dV spectra. According to the Tersoff–Hamann theory, the dI/dV spectrum represents the LDOS of the normal, unbound electrons. For voltages $|V| < \Delta$, the electrons are bound as Cooper pairs. Typical values for Δ are in the mV regime (see Fig. 2.7). Hence, the LDOS is zero within this energy range. Hess et al. [275, 276] measured local dI/dV spectra on NbSe$_2$. A spectroscopic map revealed the Abrikosov flux lattice, where the superconductive areas could be distinguished

Fig. 2.7 dI/dV versus bias
voltage V for $NbSe_2$ at 0 T
applied magnetic field, used
to determine the gap at
1.45 K. *Inset*: gap versus
temperature and the
corresponding BCS fit.
Reprinted from [275],
https://doi.org/10.1103/
PhysRevLett.62.214, with
permission from AIP
Publishing

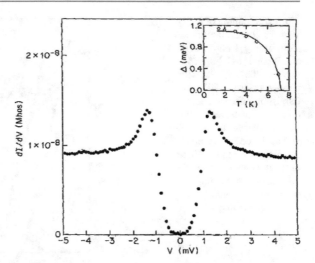

from the vortex. Tunneling into the core of the vortex revealed enhanced differential conductance (zero-bias peak). High-resolution spectroscopy data revealed a six-fold symmetry of the vortex core, which is related to the crystalline band structure. Later, tunneling spectroscopy was performed on high-T_c superconductors [416,550], where the Abrikosov flux lattice was once again imaged. Non-magnetic impurities were found to create zero-bias peaks which were related to localized low-energy excitations. The latter are typical for non-magnetic scattering in a d-wave superconductor [716]. The influence of isolated magnetic adatoms on a conventional superconductor was studied by Yazdani et al. [715]. Excitations within the superconductor energy gap, called Yu Shiba Rusinov (Shiba) bound states, were observed around these impurities. Tunneling spectroscopy on paramagnetic molecules of manganese phthalocyanine on Pb(111) showed Shiba states, which were split into triplets by magnetocrystalline anisotropy [259].

Inelastic Tunneling One of the highlights in tunneling spectroscopy was the observation of inelastic tunneling for acetylene (C_2H_2) molecules adsorbed on Cu(100) [621,622]. An increased tunneling conductance was observed at energies which corresponded well to the C–H stretch mode. An isotope effect was observed for C_2D_2 and C_2HD. The interesting aspect of vibrational spectroscopy is the availability of a database from other techniques such as infrared, Raman or electron energy loss spectroscopy. Hence, the molecules can be identified by their vibrational frequencies. The technique is also called vibrational spectroscopy. The principle of inelastic tunneling is relatively simple and is well established from experiments with planar tunnel junctions [254]. At a certain threshold $h\nu/e$ an additional current channel is opened, which corresponds to inelastic tunneling through a molecule. The molecular vibration of frequency ν is excited and the electron loses the energy $h\nu$. Figure 2.8 shows the total current I, which is the sum of elastic and inelastic tunneling. A kink

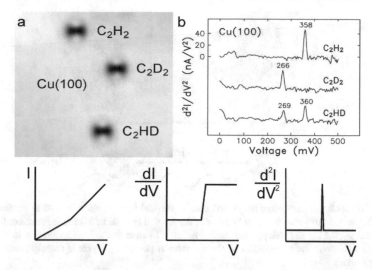

Fig. 2.8 a STM topography of acetylene molecules on Cu(100). **b** dI^2/dV^2 spectra on an acetylene molecule (C_2H_2) and on the deuterated acetylene molecules (C_2D_2 and C_2HD). The shifts of the peaks are due the isotope effect, where the different masses lead to different vibrational frequencies. Lower half: Current versus voltage curves with elastic and inelastic tunneling. A kink is observed when the inelastic electron tunneling current channel opens up. This kink becomes a step in the first derivative and a peak in the second derivative. Reproduced from [622], https://link.aps.org/doi/10.1103/PhysRevLett.82.1724, with permission from AIP Publishers. and from https://commons.wikimedia.org/wiki/File:Second_derivative.gif, Erwin Rossen, Public domain

at $V = h\nu/e$ is observed, which becomes a step in the dI/dV curve and a peak in the d^2I/dV^2 curve.

2.2.4 Manipulation Modes

STM can be used to move individual atoms (see Fig. 2.9) [52,161,627]. Vertical manipulation and lateral manipulation are the two modes of manipulation. Vertical manipulation is started by a transfer of the surface atom to the tip. The tip is retracted and moved to the desired position. The adsorbate atom is then deposited. The transfer of the adsorbate atom from the surface to the tip, or vice versa, is achieved by bringing the tip into contact or near-contact. The application of voltage pulses can be used to set the direction. Eigler et al. [162] found that Xe atoms moved in the same direction as the tunneling electrons, which was related to heat-assisted electromigration.

In the lateral manipulation mode, the adsorbate is kept adsorbed to the surface and only moved laterally along the surface. The tip is moved to the initial point and the set point is increased by about 2 orders of magnitude, which corresponds to a decrease in the distance by several angstrom units. The tip forms a weak bond with the adsorbate atom or molecule. The tip is then moved along the line of manipulation. Typical threshold resistances to slide an adsorbate are $5\,M\Omega$ to slide Xe along rows of Ni(110), $200\,k\Omega$ to slide CO along Pt(111), and $20\,k\Omega$ to move Pt adatoms along Pt(111) [627].

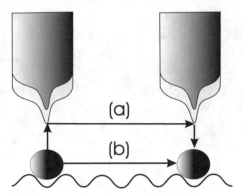

Fig. 2.9 Modes for manipulating single adatoms. **a** Vertical manipulation. **b** Lateral manipulation. In case **a**, the adatom forms a strong bond with the tip and is detached from the surface. It is then transported by the tip and redeposited on the surface. In case **b**, the interaction has to be strong enough to overcome the corrugation of the interaction potential, which is closely related to the diffusion barrier

Fig. 2.10 Tip height curves during manipulation of a Cu atom (**a**), a Pb atom (**b**, **c**), a CO molecule (**d**), and a Pb dimer (**e**)–(**g**) along [1$\bar{1}$0] on the Cu(211) surface. The tip movement is from left to right and the tunneling resistances are indicated. *Vertical dotted lines* correspond to fcc sites next to the step edge. The initial sites of the manipulated species are indicated by small sphere models. STM images of the different adparticles are shown on the *right-hand side. Arrows* indicate the direction of tip movement. Reprinted from [52], https://doi.org/10.1103/PhysRevLett.79.697, with permission from AIP Publishing

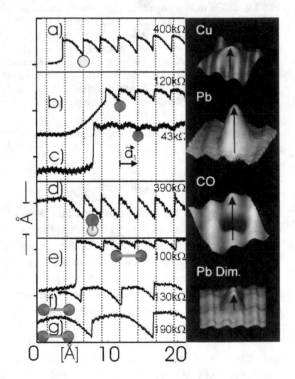

Fig. 2.11 A $40 \times 40\,\text{nm}^2$ image of a Si(100) (2 × 1):H surface imaged at 11 K after threshold STM patterning of a series of parallel lines. Reprinted from [183], https://doi.org/110.1103/ PhysRevLett.80.1336, with permission from AIP Publishing

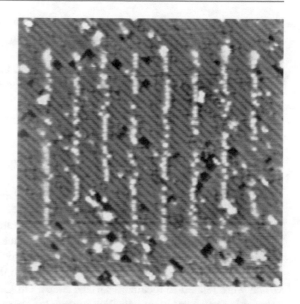

As shown in Fig. 2.10, the tip height during manipulation can be recorded, which gives some insight about the manipulation process. Alternatively, the variations in the tunneling current can be used to monitor the manipulation process. The tunneling current signal can be transformed into an acoustic signal, which can be very helpful for the experimentalist [161]. Several modes have been identified [52]:

- pulling, where the adparticle performs hops from one equilibrium position to the next to follow the attractive tip discontinuously,
- sliding, observed at low tunneling resistance, where the tip–adsorbate distance remains nearly constant, and a continuous movement of the adsorbate is observed,
- pushing, where the adsorbate is repelled by the tip and the adsorbate particle performs hops from one adsorption site to the next in front of the tip.

The hop distances depend on the size of the particles. Single atoms normally perform hops from one adsorption site to the nearest site. Dimers of Pb were observed to perform hops over multiple adsorption site distances. The presence of defects affects hopping barriers.

Alternative modes of manipulation were described by Lyding et al. [183,413]. The hydrogen passivated Si(100) surface was exposed to electronic currents with bias voltages above 5 V, which leads to the desorption of a few hydrogen atoms in the vicinity of the tip. Line widths as small as one nanometer were achieved (see Fig. 2.11). The authors related the manipulation to inelastic tunneling, which may increase the local vibrations of the adsorbate and lead to desorption. Hydrogen plays a special role, since its vibrational frequencies are high due to its small mass. Corresponding decay times of vibrational oscillations as large as 10^{-8} s were observed. Therefore, local heating and consequent desorption appears possible.

Another way to modify surfaces locally on the nanometer scale is to utilize large electric fields [627]. The field is inhomogeneous and is concentrated in the vicinity of the tip. These large field gradients parallel to the surface can lead to field-assisted directional diffusion. This method can also be used to measure dipole moments and polarizabilities of adatoms, as has been demonstrated by field ion microscopy (FIM).

2.3   Resolution Limits

2.3.1   Imaging of Semiconductors

In 1983, Binnig and Rohrer presented the Si(111)7 × 7 surface with atomic resolution [85]. This is one of the most spectacular surface reconstructions (see Fig. 2.12). In terms of resolution, the spacing between the adatoms is about 9 Å. Later, Feenstra and Stroscio presented the Si(111) 2 × 1 surface with spacings of 2–3 Å [176]. Voltage-dependent imaging showed that the protrusions depended on the polarity and did not represent the positions of the atoms, but rather maxima of the LDOS, which were related to localized surface states. On GaAs(001), Feenstra et al. [177] were able to present atom-selective images by changing the bias voltage polarity.[4] Positive polarity of the bias voltage showed the unfilled states of the surface, where the maxima coincided with the position of the Ga^+ ions, whilst negative polarity showed the filled states, where the maxima were at the sites of the As^- ions. The lateral resolution of the instrument on these semiconductors is in the range of 1 Å. The corrugation height in the z-direction is typically a few tenths of an angstrom unit, which is consistent with expectations from local variations of the LDOS. Furthermore, spectroscopic measurements (dI/dV vs. V) are in good agreement with the Tersoff–Hamann model.

2.3.2   Imaging of Metals

In 1982, the first metal surfaces, the Au(110)2 × 1 and Au(110)3 × 1 reconstructions, were resolved by Binnig and Rohrer [82]. Individual atoms were not observed at this early stage. In 1987, Hallmark et al. [248] presented the first atomic resolution image of a close-packed metal, Au(111)22 × $\sqrt{3}$. Later, Wintterlin et al. were able to present atomic resolution on Al(111) [701]. Today, a large number of clean metal surfaces can be resolved, such as Cu(111) (see Fig. 2.13), Cu(110), Cu(001), Pt(111), Pt(001), Ru(0001), Ni(001) and Ni(110) [56]. The spacing between the atoms of the close-packed surfaces is 2–3 Å. The corrugation heights are found to be rather large, of the order of a few tenths of an angstrom unit. This contradicts helium scattering experiments and calculations, which predict corrugations heights

---

[4]The bias voltage is applied to the sample side. Therefore, positive bias voltage means positive voltage on the sample and 0 V on the tip side. The I–V converter keeps the tip virtually grounded. The high gain of the operational amplifier keeps the potential of the positive and negative input at the same voltage.

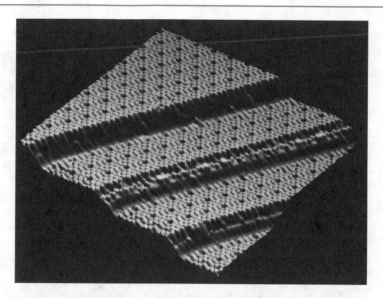

Fig. 2.12 STM image of the Si(111)7 × 7 surface in the constant current mode. Terraces are separated by steps of height 1.25 nm. The protrusions on the terraces correspond to the arrangement of the adatoms. Reprinted with permission from [635], https://doi.org/10.1116/1.585530, Copyright 1991 American Vacuum Society

of the order of 0.01 Å. (Here the corrugation height corresponds to the difference between the maximum and minimum of contours of constant LDOS.) Wintterlin et al. [701] proposed that small clusters of the sample material are important for the rather large corrugation heights. These clusters are deformed due to the rather large tip–sample interaction, which may lead to large corrugations. Alternatively, Chen et al. [109] proposed that d-states of the tip may contribute significantly to the tunneling current. These states are more localized than the s-states and may explain the high resolution. An impressive example of resolution capabilities is given by Schmid et al. [580], where Pt atoms and Ni atoms were distinguished on a $Pt_{25}Ni_{75}(111)$ surface (see Fig. 2.14). The best resolution was observed with small tunnel resistances of (50–300 kΩ). This was attributed to the interaction between adsorbates at the tunneling tip and surface atoms.

2.3.3   Imaging of Layered Materials

A number of layered materials, such as graphite or MoS_2, have been resolved. In particular, graphite has attracted a lot of attention because of the giant corrugation heights. Forces between tip and sample are considered to play an important role [614]. Other interesting examples are transition metal dichalcogenides, such as 1T-TaS_2 or 1T-$TaSe_2$, which exhibit both atomic structure and charge density waves (CDW). A review of STM work on layered materials with CDW has been published by Coleman et al. [127].

Fig. 2.13 STM images of the Cu(111) surface in constant current mode. **a** Overview image with monatomic steps. **b** Atomic resolution on Cu(111). The spacing between the protrusions is 2.5 Å

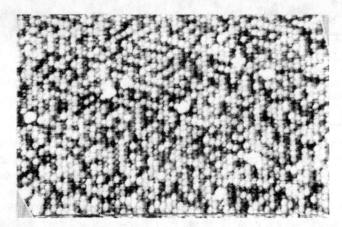

Fig. 2.14 STM image of the (111) surface of a $Pt_{25}Ni_{75}$ single crystal. A voltage of 5 mV and a current of 16 nA were applied. A rather strong 'chemical' contrast is observed, where the dark species is attributed to Pt and the bright features to Ni. The contrast is related to the interaction between tip adsorbates and surface. The image size is 125 Å × 100 Å. Reprinted from [580], https://doi.org/0.1103/PhysRevLett.70.1441, with permission from AIP Publishing

2.3.4   Imaging of Molecules

One of the most exciting applications is the imaging of adsorbates and single molecules on surfaces. Foster and Frommer presented images of single molecules [188]. Later, liquid crystals [611] and hydrocarbons were imaged.[5] Chiang and coworkers presented a systematic STM study of molecules in UHV [114,115]. Their presentation of benzene coadsorbed with CO, showing inequivalent sites of the benzene ring, is an impressive example of the resolution capabilities of STM.

---

[5]Both liquid crystals and hydrocarbons are oriented flat on the surface.

Fig. 2.15 Constant current
STM image of Cu-tetra 3, 5
di-t-butylphenyl porphyrin
(Cu-TBPP) molecules on
Ag(100) at low molecular
coverage. The monatomic
Ag substrate steps are
decorated by the molecules.
Image size $68 \times 68\,nm^2$.
$I_t = 150\,pA$ and $U = 1\,V$.
Courtesy of M. de Wild, S.
Berner, H. Suzuki and T.
Jung

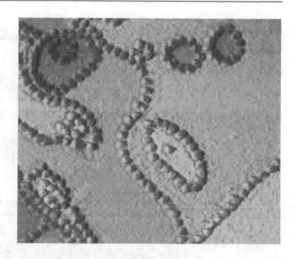

Porphyrin-like molecules have attracted a lot of attention, because of their potential
for molecular electronics (see Fig. 2.15) [215,217].

Generally, mechanisms for tunneling through molecules are rather complicated,
e.g., resonant tunneling may occur. Adatoms or small molecules (e.g., benzene
molecules) which lie flat on a surface can be imaged in most cases. Larger molecules
can be more difficult to image, because of poor conductivity. In particular, molecules
which are oriented perpendicular to the surface may cause problems. For example,
alkylthiols on Au(111) are best imaged with small currents of a few pA [140,529].
Larger currents lead to strong distortions of the film. Heim et al. [267] showed that
DNA and other biomolecules can be imaged with very small currents ($<1\,pA$) in high
humidity, but are distorted when higher currents are applied. An overview of SPM
work on organic materials, including STM measurements on molecules, has been
published by Jane Frommer [191] and preparation procedures are given in [129].
The prospect of imaging and manipulating single molecules is probably one of the
most interesting future aspects of STM. Manipulation experiments by Eigler et al.
[161] with single adatoms at low temperature show interesting aspects for fundamental science, e.g., the study of particles in a box. Manipulation experiments with
individual molecules at room temperature have been achieved by Gimzewski et al.
[215,217,218].

An important step for STM imaging of molecules was to deposit thin insulating
films, such as a few layers of NaCl, on the metallic substrate. Subsequently, molecules
can be deposited on these films. Repp et al. [551] observed that the molecular orbitals
are less disturbed and STM can be used to image the lowest unoccupied orbitals
(LUMO) and the highest occupied orbital (HOMO) by suitable choice of the applied
voltage. As seen in Fig. 2.16 the molecular orbitals of pentacene are observed and
compare well with the contours of orbital probability distribution of theoretical DFT
calculations. The agreement is even better for more realistic simulations, which
include the substrate of the molecule. Images with voltages in the gap do not show
these features. The observed resolution is improved if molecule-covered tips are used

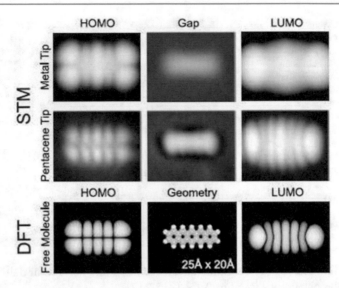

Fig. 2.16 The images are STM images with a metallic and a pentacene covered tip (first two rows). The third row shows contours of the orbital probability distribution of the free molecule. The positions of the HOMO level is observed at negative voltages of -2.4 V and the LUMO level at voltages of 1.7 V. Reprinted from [551], https://doi.org/10.1103/PhysRevLett.94.026803, with permission from AIP Publishing

Fig. 2.17 dI/dV versus voltage and I(V) curves of a pentacene molecule deposited on a thin NaCl-film on Cu(111). The HOMO and LUMO orbitals are observed as maxima in the dI/dV curve. Reprinted from [551], https://doi.org/0.1103/PhysRevLett.94.026803, with permission from AIP Publishing

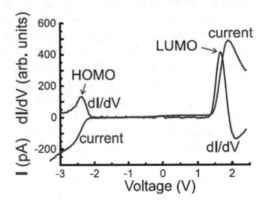

compared to metallic tips. The position of the HOMO level at -2.4 V and the LUMO level at 1.7 V is well observed in the dI/dV versus voltage curve as seen in Fig. 2.17.

2.3.5   Imaging of Insulators

Although in principle STM measurements are not possible on insulating materials, STM can be performed on thin insulating films grown on conductive substrates. Examples are STM studies of thin metal oxide films, such as Al_2O_3 [18,252], CoO [595], NiO [76], FeO [350], or thin films of ionic materials, such as CaF_2 [36] or NaCl

[264]. At low sample voltages tunneling is accomplished by the substrate electrons. For higher voltages, the electronic states of the film contribute to the tunneling. Therefore, the height of islands is sensitive to the applied bias voltage. In some cases (low bias voltage), islands can even appear as depressions. Hebenstreit et al. [264] compared the experimental results with calculations of the LDOS and found that the STM results for NaCl films on Al(100) agree with the Tersoff–Hamann theory. This is in contrast to metal surfaces, where calculated LDOS corrugations are significantly smaller than experimental values, and additional mechanisms, such as tip–sample interactions are needed to explain the results.

2.3.6   Theoretical Estimates of Resolution Limits

Right at the outset, Binnig and Rohrer estimated the resolution limits of STM [540]. With tip radii of $1000\,\text{Å}$, a relevant tunneling current region with diameter about $45\,\text{Å}$ was estimated. Stoll derived a formula for the corrugation height Δz on simple metals, depending on the atomic periodicity a [624]:

$$\Delta z = h_s \exp\left[\frac{-\pi^2(R+s)}{\kappa a^2}\right],\tag{2.9}$$

where $\kappa = \sqrt{2m(U-E)}/\hbar$ is the decay constant of the tunneling current, typically $1\,\text{Å}^{-1}$, h_s is the corrugation height of the metal surfaces, d is the distance between the sample surface and the apex of the tip (gap spacing) and R is the radius of curvature of the tip. The relevant distance is $R+s$, which is the distance between the surface and the center of curvature of the tip. The Stoll formula was found to be in good agreement with the results from Binnig and Rohrer [82] on Au(110)2 × 1 and Au(110)3 × 1. However, deviations were found for small separations s. The high resolution on close-packed metal surfaces could not be explained by this model. Other effects, such as tip–sample interaction forces, influence of the tip on the electronic structure of the sample, and more complex tip wave functions have to be taken into account (see Sect. 2.3.2). Nevertheless, the Stoll formula, which was confirmed by the Tersoff–Hamann theory up to a constant factor [641], gives a good estimate for corrugations with large gap resistance.

2.4   Observation of Confined Electrons

Since STM probes the local density of states at the Fermi energy, it is an ideal tool for studying confined electrons, which interact with surface steps or adatoms. The electrons occupying surface states on the closed-packed faces of noble metals, such as Cu(111), Ag(111) and Au(111), form a two-dimensional nearly-free electron gas [721]. These surface states were first observed by photoemission studies [345] and later by Everson et al. [137, 172] with STM on Au(111) at room temperature. A direct visualization of these surface states on Cu(111) forming standing wave patterns in the vicinity of steps and defects was acquired with a low temperature STM by Crommie et al. [131]. The wavelength λ is related to the length of the k-vector, $k = 2\pi/\lambda$, and to the momentum p by the de Broglie equation $p = \hbar k = h/\lambda$. Therefore, this experiment is an instructive example of the wavelike nature of electrons, as predicted by elementary quantum mechanics. In analogy to water waves, the confined electrons form circular patterns in the vicinity of point defects and extended 1D waves in the vicinity of step edges. A more quantitative analysis of these standing waves will be given in subsequent sections.

2.4.1 Scattering of Surface State Electrons at Steps

The energy of electrons confined in 2 dimensions is given by

$$E = E_0 + \frac{\hbar^2 k_\parallel^2}{2m^*} , \tag{2.10}$$

where $k_\parallel = \sqrt{k_x^2 + k_y^2}$. The wave function of an unscattered, delocalized state is described by

$$\psi_{k_\parallel} \propto e^{ik_\parallel r} f(z) , \tag{2.11}$$

where $\mathbf{r} = (x, y, 0)$ and $f(z)$ is a function. This indicates that the state decays exponentially perpendicular to the surface. Next, the scattering by a potential barrier, such as a surface step or an array of adatoms, is considered. The incident wave will encounter this barrier with reflectivity $R = |R|e^{i\eta}$. The reflected wave interferes with the incident wave, creating a standing wave. In one dimension, the density of states is given by

$$\text{LDOS}(E, x) \propto 1 + |R|^2 + 2|R| \cos(2k_x x - \eta) . \tag{2.12}$$

In two dimensions, the calculation is more complicated [37,131]. Assuming that the step edge extends infinitely along the y-axis and behaves like a hard wall, the standing wave function is given by

$$\psi_{k_\parallel} = C \sin(k_x x) e^{ik_y y} f(z) . \tag{2.13}$$

Then we calculate the total number of states $N(E, x)$ with energy $\leq E$ at distance x from the step:

$$N(E, x) \propto \int_{\text{disk}} |\psi_{k_\parallel}|^2 dk_x dk_y \tag{2.14}$$

$$\propto \int_{\text{disk}} \sin^2(k_x x) dk_x dk_y \propto \frac{\pi k_\parallel^2}{2} \left[1 - \frac{J_1(2k_\parallel x)}{k_\parallel x} \right] ,$$

where the disk is defined by the condition $k_\parallel \leq \sqrt{2m^*(E - E_0)}/\hbar$. The LDOS is then proportional to dN/dE:

$$\text{LDOS}(E, x) = L_0\left[1 - J_0(2k_\parallel x)\right]. \tag{2.15}$$

In these formulas, J_0 and J_1 are 0th and 1st order Bessel functions and $L_0 = m^*/\pi\hbar^2$ is the LDOS of a 2D electron gas in the absence of scattering. At large distances x, Bessel functions have the asymptotic form

$$J_m(2k_\parallel x) \propto \sqrt{\frac{1}{\pi k_\parallel x}} \cos\left(2k_\parallel x - \frac{2m + 1}{4}\pi\right). \tag{2.16}$$

Consequently, $\text{LDOS}(E, x)$ decays as $1/\sqrt{x}$, or more explicitly,

$$\text{LDOS}(E, x) = L_0\left[1 - \sqrt{\frac{1}{\pi k_\parallel x}} \cos\left(2k_\parallel x - \frac{2m + 1}{4}\pi\right)\right], \tag{2.17}$$

where

$$k_\parallel = \frac{\sqrt{2m^*(E - E_0)}}{\hbar}. \tag{2.18}$$

The oscillation of the LDOS decays, unlike the 1D case, because of the integration in k-space, as can be seen from (2.14).

Figure 2.18 shows fits of (2.15) to the dI/dV line scans in the vicinity of a step. Apart from small discrepancies very close to the step, the fit is accurate and yields values of $k_\parallel$ for different voltages and different energies E relative to the Fermi energy E_F. The $k_\parallel$ data for different energies $E - E_F$ (inset of Fig. 2.18) are fitted with (2.10), which allows one to determine the effective mass $m^* = 0.38 \pm 0.02\,m_e$, where m_e is the mass of the electron, and a band edge $E_0 = -0.44 \pm 0.01$ eV below E_F.

Hasegawa and Avouris have studied the scattering of surface states on the Au(111) surface at room temperature [258]. They observed an energy dispersion $E(k_\parallel)$ which deviates from photoemission data. A value of $m^* = 0.15\,m_e$ was determined by STM compared to a photoemission value of $m^* = 0.28m_e$. The discrepancy is not yet understood, but may be related to the fact that the total charge density near the step exhibits an oscillatory behaviour, the so-called Friedel oscillation, with a period determined by the Fermi wave number k_F: $\Delta n \propto \cos(2k_F x)$. According to Hasegawa and Avouris, the effective potential of the confined electron should also have a component $U \cos 2k_F x$, which then yields a modified dispersion relation with a wave number $k \approx (k_F + k_\parallel)/2$. This modified dispersion is found to be in good agreement with the data of Hasegawa and Avouris. Furthermore, the authors determined phase shifts along steps with different crystallographic orientations. For steps perpendicular to the $\langle\bar{1}\bar{1}2\rangle$ directions they found a phase shift of $-\pi$ which corresponds to a hard wall ($R = 1$). Along steps perpendicular to the $\langle11\bar{2}\rangle$ directions a phase shift of $-\pi/2$ was measured, which corresponds to a reflection coefficient of $R = 0.9$. The reason might be related to the open structure of the $\langle11\bar{2}\rangle$ steps, which has (001) facets.

Fig. 2.18 Spatial
dependence of dI/dV across
a step edge on Cu(111) at
4 K. For details see text.
Reprinted from [131],
https://doi.org/10.1038/
363524a0, with permission
from Springer Nature

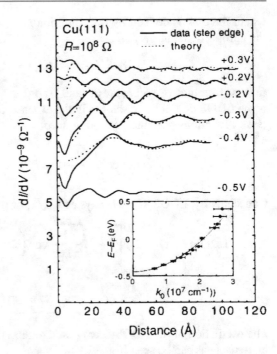

2.4.2 Scattering of Surface State Electrons at Point Defects

The scattering of surface state electrons at point defects, such as adsorbed atoms
or vacancies, can be calculated in close analogy to the previous section. Since the
wavelength of the electrons is much larger than the dimensions of the scatterer,
isotropic scattering is found. The wave function in the asymptotic form is given by

$$\psi_{m,k_{\parallel}} = \sqrt{\frac{2}{\pi k_{\parallel} r}} e^{im\Phi} \cos\left[k_{\parallel} r - \frac{2m+1}{4}\pi + \eta_m(k_{\parallel})\right], \qquad (2.19)$$

where m is the angular momentum quantum number and $\eta_m(k_{\parallel})$ is the k-dependent
phase shift. The LDOS in the vicinity of a point defect is then given by

$$\text{LDOS}(E, r) \propto \frac{2}{\pi k_{\parallel} r} \sum_{m=-\infty}^{+\infty} \cos^2\left[k_{\parallel} r - \frac{2m+1}{4}\pi + \eta_m(k_{\parallel})\right]. \qquad (2.20)$$

The phase shifts are characteristic of the point defect. Ideally, the phase shifts would
fulfill the Friedel sum rule:

$$\sum_{m=-\infty}^{+\infty} \eta_m(k_{\parallel}) = \frac{\pi Z}{2}, \qquad (2.21)$$

where Z is the charge of the point defect.

Equation (2.20) can be simplified for long wavelengths compared to the dimensions of the defect:

$$\text{LDOS}(E, r) \propto 1 + \frac{2}{\pi k_\parallel r} \left[\cos^2 \left(k_\parallel r - \frac{\pi}{4} + \eta_0 \right) - \cos^2 \left(k_\parallel r - \frac{\pi}{4} \right) \right] . \quad (2.22)$$

Only the $m = 0$ component is relevant, and this is called s-wave scattering. Crommie and Eigler found a phase shift of $\eta = -66°$ for point defects on Cu(111) [131]. This value corresponds to a repulsive potential.

Avouris et al. [37] point out that imaging of the LDOS around point defects at positive and negative bias voltage visualizes the screening of the charge of the defect. When occupied states (negative bias voltage) are imaged, the LDOS at a defect site has a sharp peak and the surrounding waves are relatively weak. Imaging of unoccupied states (positive bias voltage) shows a hole at the defect site, but the surrounding waves are more pronounced, which is consistent with screening of a negatively charged defect such as an adsorbed sulfur atom.

2.4.3   Electron Confinement to Nanoscale Boxes

If electrons are confined in a potential well, discrete energy levels are formed. The simplest case is a one-dimensional square well, as shown in Fig. 2.19. According to elementary quantum mechanics, the energy levels of an infinitely high well ($V_0 \rightarrow \infty$) are

$$E_n = \frac{h^2}{8mL^2} n^2 , \quad (2.23)$$

where $n = 1, 2, 3, \ldots$ and L is the width of the well. The corresponding wave functions are

$$\psi_n = \sqrt{\frac{2}{L}} \sin \frac{n\pi x}{L} \quad (2.24)$$

for the region within the well ($0 \leq x \leq L$). The wave function is equal to zero in the outer regions. A well of finite height has a finite number of energy levels. In addition, the wave function has non-zero, exponentially decreasing values in the outer region.

The most spectacular observation is the confinement of electrons within artificial nanoscale structures [132, 272]. Crommie et al. [132] positioned 60 Fe atoms on

Fig. 2.19 A one-dimensional square well

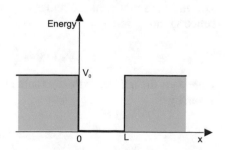

Fig. 2.20 Experimental
(*solid line*) and theoretical
(*dashed line*) voltage
dependence of dI/dV, with
the top of the STM located at
the center of an 88.7 Å
diameter, 60-atom circle of
Fe atoms on a Cu(111)
surface. Reprinted from
[272], https://doi.org/10.
1038/369464a0, with
permission from Springer
Nature

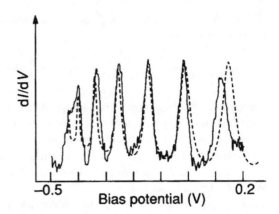

Cu(111) in the form of a circle, forming a so-called quantum corral, which confines
electrons in the inner area. A standing wave pattern, which resembles water waves
in a circular pool, was observed within the corral. dI/dV versus bias voltage curves
showed that the energy levels of the confined electrons are discrete. Good agreement
with quantum mechanical calculations was found, as shown in Fig. 2.20. The finite
width of the peaks implies a finite lifetime for the confined electrons, viz., $\Delta t_n \approx$
$\hbar/\Delta E_n$, dominated by scattering into bulk states.

In a second experiment, a 76-atom stadium was created by Heller et al. (see
Fig. 2.21) [272]. Apart from the beauty of these structures, additional information
about the scattering mechanism was found. By comparison with theory, it was determined that the Fe atoms reflect only about 25% of the incident wave, whilst 25% are
transmitted and 50% absorbed ('black dots'). Absorption is related to scattering into
bulk states. For future investigations of phenomena such as quantum billiards or for
applications such as semiconductors with tailored band gap, it will be important to
find a substitute for Fe that yields higher reflection rates.

Other examples of confinement of surface state electrons were given by Avouris
and Lyo [37]. $n = 1$, $n = 2$ and $n = 3$ states were observed on a 36 Å wide terrace
of Au(111). It was found that reflection of electrons from the side of the upper
terrace is stronger than reflection of electrons incident on the step from the lower
terrace [258]. The reflection was found to be sufficient to detect confined electrons
on these terraces. A second experiment confined electrons on 45 Å wide Ag islands
on Au(111). In this case, a standing wave pattern was found which resembles the
mechanical oscillation pattern of a clamped membrane, e.g., a drum. Avouris and
Lyo used the model of a circular box, where the analytical solutions of the wave
functions are given by

$$\psi_{n,l} \propto J_n\left(u_l\frac{r}{a}\right) \begin{Bmatrix} \cos n\Theta \\ \sin n\Theta \end{Bmatrix} . \tag{2.25}$$

Here J_n is the nth order Bessel function. The energies of the circular box states are
given by

$$E_{n,l} = E_0 + \frac{\hbar^2 u_{n,l}^2}{2m^* a^2} , \tag{2.26}$$

Fig. 2.21 Local density of states (LDOS) near E_F for a 76 Fe atom 'stadium'. Courtesy of D. Eigler

where $u_{n,l}$ is the l th root of the n th order Bessel function and a is the radius of the box. With an energy of 110 meV above the bottom of the surface state (E_0), they found a mixture of 80% of the $n = 2, l = 1$ state and 20% of the $n = 0, l = 2$ state.

A systematic study of variously-sized hexagonal Ag islands on Ag(111) has been presented by Li et al. [392]. Numerical calculations were performed to calculate the energy levels and spatial distribution of the modes. Eigenvalues scale with the inverse of the area of the hexagon. Fits of the energy levels for different areas show that the effective boundary of bound states is about 0.5 nm beyond the edge atoms. $n = 1$ to $n = 5$ modes were identified and compared to theoretical calculations for a 94 nm^2 island.

Other examples of confined electrons have been observed on aluminium, where Ar bubbles were surrounded (a 3D case), and for semiconductors, where Si dopants were screened [581,660]. An example of giant oscillations has been given by Sprunger et al. on Beryllium (0001) [617]. In this case the wavelength was found to be much smaller ($\lambda \approx 3.2$ Å) and the coupling to the bulk appears to be much weaker, representing a nearly ideal 2D system which may be of interest for future confinement experiments.

2.4.4   Summary of Dispersion Relations for Noble-Metal (111) Surfaces

The results of the effective mass from STM and angle-resolved photoemission spectroscopy (ARPES) studies seem to differ systematically, but show the same tendency: small for Au(111) and larger for Ag(111) and Cu(111) (see Table 2.1). Some agreement between STM and LDA calculations is found, with the exception of Ag(111). The energy minima of the surface states determined by STM and ARPES are quite

Table 2.1 Summary of the results from STM, angle-resolved photoemission (ARP) studies and LDA calculations for noble-metal (111) surfaces

	Cu(111)	Ag(111)	Au(111)
E_0 [eV] (ARP) [345]	0.39(1)	0.12(1)	0.41(1)
E_0 [eV] (STM)	0.44±0.01 [131]	0.072 [392]	0.45 [258]
E_0 [eV] (Calc.) [345]	0.58	0.19	0.26
m^* [$\times m_e$] (ARP) [345]	0.46(1)	0.53(1)	0.28(1)
m^* [$\times m_e$] (STM)	0.38±0.02 [131]	0.37 [392]	0.15 [258]
m^* [$\times m_e$] (Calc.) [345]	0.37	0.24	0.17

close: 0.1 eV for Ag(111) and about 0.4 eV for Au(111) and Cu(111), but the differences are not within the estimated errors. ARPES and STM may not yield the same results, since the potential of the electrons in the vicinity of steps may be modified by Friedel oscillations, as pointed out by Hasegawa et al. [258]. However, the step density is rather high, which should also affect the averaging ARPES method. Thus, there may exist other unknown reasons for the discrepancy.

2.4.5   Electron Confinement in Interfacial States of Molecular Layers on Metals

Wave functions of isolated molecules are often rather localized and do not show much dispersion. However, scanning tunneling spectroscopy on an organic monolayer film on a metallic substrate has revealed that delocalized two-dimensional band states can be observed [637]. In this case, metal-like parabolic dispersion relation with an effective mass of $m^* = 0.47 m_e$ was determined for a monolayer of PTCDA (3, 4, 9, 10-perylenetetracarboxylic-acid-dianhydride) on Ag(111), which indicates strong substrate-mediated coupling between molecules. Molecular islands of different sizes were spectroscopically analyzed. Electron confinement in these islands selects discrete wave vectors, which leads to a standing wave within the island. Local spectroscopy yields the corresponding electronic state energy. Finally, the dispersion curve of the interfacial stated can be determined (cf. Fig. 2.22).

2.5   Spin-Polarized Tunneling

STM can yield information about magnetic properties by the use of spin-polarized tunneling. Magnetic tunneling tips, such as Fe tips, are used. The spin-valve effect [321,610], which was investigated for planar tunneling junctions, predicts that the tunneling current depends on the relative orientation of the magnetic moments of the tunneling electrodes. The consequence for STM is that the tunneling current has a contribution which depends on the relative orientation of the magnetic moment of

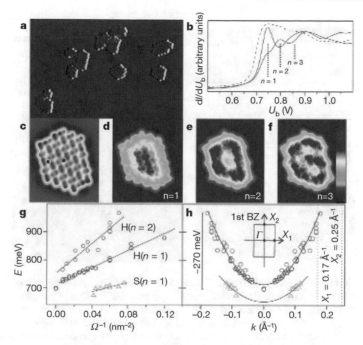

Fig. 2.22 Free-electron-like two dimensional band state of PTCDA/Ag(111). **a** STM-image (64 × 42 nm^2) of small ordered PTCDA islands on Ag(111). **b** Solid lines are STS spectra on the PTCDA island presented in c, with vertical dotted lines indicating its first three confined states. Dashed line is the spectrum from an extended monolayer. **c** STM image of 39-molecule island. **d–f** coresponding STS-images recorded at voltages of 750 mV (n = 1), 800 mV (n = 2) and 860 mV (n = 3). **g** Energies of the n = 1 (green for the square and red for the herringbone phase) and n = 2 confined state versus inverse island area. **g** Dispersion curves for two confined states. Reprinted from [637], https://doi.org/10.1038/nature05270, with permission from Springer Nature

the probing tip and the orientation of the magnetic moment of the imaged sample area.

Wiesendanger et al. [693] have explored this effect with CrO_2 tips on an antiferromagnetic Cr(001) surface. They operated the STM in constant current mode and found alternating step heights, which were related to the antiferromagnetic structure. These early experiments represented a mixture of topography (constant LDOS) and spin dependent contrast. Later, Bode et al. [90] studied the Gd(0001) surface with Fe tips by spectroscopy. Gd(0001) has an exchange-split surface state and the Fe tip was assumed to have a constant spin polarization. Using local scanning tunneling spectroscopy, the surface states were clearly identified and showed an asymmetry when the magnetization of the sample was switched by an external field. Figure 2.23d and e show a dI/dV image of the Gd film at bias voltages of -0.2 V and $+0.45$ V, respectively. Figure 2.23f shows the asymmetry image (difference between the signals from Fig. 2.23d and e divided by the sum), where domain walls are observed on the Gd islands. The resolution was estimated to be about 10 nm. A quantitative analysis indicates that the spin polarization of the majority (minority) part was $P \approx 0.45$ (-0.24).

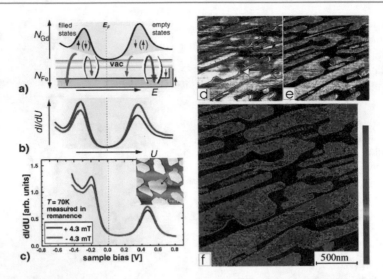

Fig. 2.23 a Principle of spin-polarized STS using a sample with an exchange split surface state, such as Gd(001) and a magnetic Fe tip with constant spin polarization. Due to the spin-valve effect, the tunneling current of the surface state spin component parallel to the tip is enhanced at the expense of its spin counterpart. **b** Expected reversal of contrast of the dI/dU spectra at the surface state peak position upon switching of the sample magnetization. **c** Experimental tunneling spectra above a Gd island (see *arrow* in the *inset*). **d–f** dI/dU images: at the majority $U = -0.2$ V (**d**) and minority $U = 0.45$ V (**e**) surface state peak position. The *arrow* indicates a tip change. **f** In the asymmetry image, the contrast is enhanced and tip changes are less visible. Reprinted from [90], https://doi.org/10.1103/PhysRevLett.81.4256, with permission from AIP Publishing

Bode et al. also observed [90] that Fe tips with film thickness <10 ML do not influence the sample magnetization, whereas Fe tips with film thicknesses >100 ML do modify the sample magnetization of the Gd film. Recently, Heinze et al. [269] have presented spin-polarized tunneling experiments on a Mn monolayer. These indicate that the antiferromagnetic structure can be observed on the atomic scale. A nonmagnetic tip shows all Mn atoms in an equivalent way, whereas a ferromagnetic tip reveals a superstructure, which is related to the orientation of the spins of the Mn atoms.

An alternative approach is to combine spin-polarized tunneling with optical techniques, where the polarization of the circularly polarized light is an additional parameter to control the experiments. Alvarado and Renaud [29] performed experiments where the light emission of a tunneling junction was analyzed. Spin-polarized electrons were ejected from a Ni tip into a GaAs sample. The recombination of these electrons led to the emission of polarized light, which was analyzed with a linear polarizer. A spin polarization of about 30% was deduced for the electrons extracted from the Ni tip. Spin-polarized tunneling from a ferromagnetic Ni sample to a nonmagnetic STM tip (Pt–Ir) was observed by Wu et al. [707] by detecting polarized light emission from the tunneling junction. An elegant way to select the spin orientation of the electron source is to use optically pumped GaAs tips. Here, the spin orientation is selected by the polarization of the light. Prins et al. [534] performed experiments on Pt/Co multilayers, which revealed spin-polarized contributions to the tunneling current. Suzuki et al. [631] presented images of Co films which indicate that the magnetic structure was observed. However, low contrast and unintended optical contrast appears to be a problem so far.

Another approach for deconvoluting topography from magnetic information was introduced by Johnson and Clarke [319]. Both the sample and the tip magnetization were modulated. Preliminary experiments in air indicated the existence of a modulated tunneling current, which could be related to spin-polarized tunneling. Later, Wulfhekel [708] performed spin-polarized tunneling experiments using tips with low coercivity, vanishing magnetostriction and low saturation magnetization under ultrahigh vacuum conditions. They changed the magnetization of these tips rapidly (40–80 kHz) and detected the modulated tunneling current. Magnetostriction effects were excluded by measurement on non-magnetic surfaces.

In summary, spin-polarized STM has been established under ultrahigh vacuum conditions. Tunneling spectroscopy, modulation of the magnetization or the combination of tunneling with optical techniques are methods for disentangling topography from magnetic information. So far, spin-polarized STM is still a laboratory technique, which has not yet been applied to technological studies. The reader is also referred to Chap. 4, where magnetic force microscopy (MFM) will be discussed. MFM has already become a valuable tool for applied science and technology problems related to magnetism. In future, spin-polarized STM might become an important tool to study the ultimate limits of magnetic storage devices or the optimization of giant-magnetoresistance. Experiments in the field of quantum computing, where individual spins are to be manipulated, have been proposed recently [166].

2.6 Observation of the Kondo Effect and Quantum Mirage

Impurities with a magnetic moment are screened by conduction electrons. The resulting many-body singlet state is a low energy excitation, called the Kondo resonance, with a characteristic width $\delta = k_B T_K$, where T_K is the characteristic Kondo temperature and k_B is the Boltzmann constant. The Kondo effect was observed macroscopically by an increased electrical resistance at low temperature. On the microscopic level, local STM spectroscopy can be used to resolve these electronic resonances at the Fermi energy. Li et al. [392] observed Ce impurities on Ag(111), where characteristic differences in the line shape were found for single Ce atoms compared to Ce clusters and continuous Ce films. A similar study was performed by Madhavan et al. on Co impurities on Au(111) [414]. Unexpected structures of the Kondo resonance were found, which were related to quantum mechanical interference of tunneling between the d-orbital and conduction electron channels.

A spatial map of the Kondo screening of Co atoms on Cu(111) was measured by Manoharan, Lutz and Eigler [427]. The Co atoms were arranged within an elliptical quantum corral. In the case where the Co atom was positioned at one focus of the ellipse, a quantum mirage was observed in the other empty focus (see Fig. 2.24). Ellipses have eigenmodes which have strongly peaked probability amplitudes near

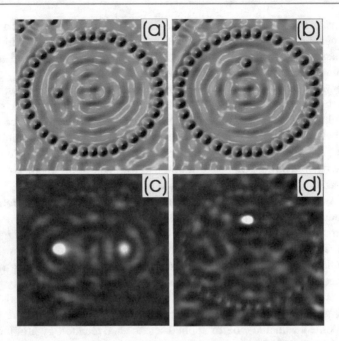

Fig. 2.24 Quantum mirage. STM topographs of an elliptical corral with a Co atom at the focus (**a**) and a Co atom moved off the focus (**b**). **c** Corresponding dI/dV difference maps showing the Kondo effect projected to the empty right focus, resulting in a Co atom mirage. **d** Corresponding dI/dV without mirage. Reprinted from [427], https://doi.org/10.1038/35000508, with permission from Springer Nature

the classical foci. The experiments reveal that the strongest mirages are observed for states close to the Fermi energy, such as the Kondo resonance. In future, the observation of quantum mirages may become a useful spectrocopic tool for probing atoms and molecules remotely, while the perturbing influence is minimized.

2.7   Observation of Majorana Bound States

In 1937, Italian physicist Ettorre Majorana predicted the existence of particles that are their own anti-particles. Attempts were made to detect these Majorana particles in elementary particle physics experiments. More recently, condensed matter systems were explored in this respect. Magnetic wires were deposited on a superconductive substrate and it was predicted theoretically, that at the borders of the wires, Majorana bound states might exist. One specific system, Fe wires on Pb, turned out to be the system of choice. Fe is deposited under ultrahigh vacuum conditions onto Pb(110), where moderate annealing leads to the self-assembly of Fe wires along the Pb(110) rows. Tunneling spectroscopy reveals zero bias peaks at the end of the wires, which can be related to the existence of Majorana bound states [477]. In addition, the Majorana wave function was mapped by LDOS imaging arond zero bias voltage [508]. Additional information was gained by combined STM and AFM imaging,

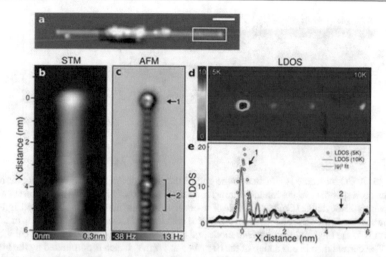

Fig. 2.25 Zero-bias electronic and structural characterisation of the Fe chain. **a** STM overview image of a Fe chain with a scale bar of 5 nm. **b** Fe chain end imaged by STM **c** Fe chain end imaged by AFM, where individual Fe atoms can be seen with separations of 0.37 nm. **d** Zero bias conductance map acquired at 5 K shows evidence for a Majorana bound state (MBS). At 10 K the MBS disappears due to the suppression of the superconductivity. Figure from [508], https://doi.org/10.1038/npjqi.2016.35, distributed under the terms of the Creative Commons Attribution License (http://creativecommons.org/licenses/by/4.0)0

which showed that the Fe wire consists of single Fe atoms at a separation of 0.37 nm (Fig. 2.25).

2.8   Single-Molecule Inelastic Tunneling Probe

STM gives atomic resolution on semiconductors, metals and even on thin insulating films (cf. Sect. 2.3.5). The imaging of molecules gives information about molecular orbitals related to the HOMO, LUMO levels (cf. Sect. 2.3.4). Though this information is very valuable, it is not directly related to the structural information. One way to overcome this limitation, is to use inelastic tunneling through a CO-terminated tip [655] (see also Sect. 2.2.3). In this case, the hindered translational vibration mode of the attached CO molecule can be observed at different positions above the molecule under inverstigation. The CO vibration senses the landscape of the molecule. By mapping the variations of current variations at a voltage, corresponding to the energy of the vibrational mode, the sceleton of the molecule can be nicely observed, representing the molecular structure in a relatively straightforward way. There is also evidence that the local hydrogen bonds are visualized as well. Thus, this imaging mode represents local bonding between atoms within the adsorbed molecule. It is called itProbe, which stands for inelastic tunneling probe. Measurements were performed at 600 mK and with voltage modulations of 1 mV to achieve the high spectroscopic resolution (Fig. 2.26).

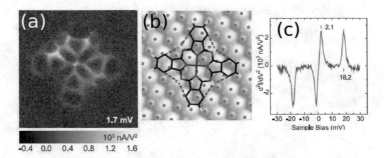

Fig. 2.26 A CO molecule is attached to the end of the probing tip. Inelastic tunneling currents at voltages corresponding to the hindered translation mode of the CO molecule show the sceleton of the cobalt phthalocyanine (CoPc) molecule. The observed current mapping (**a**) is in good agreement with the structure (**b**). **c** dI^2/dV^2 versus voltage of the attached CO molecule. The peak at 2.1 meV corrsponds to the hindered translational mode and at 18.2 meV to the hindered rotational vibration mode. The current map in **a** is a map of the dI^2/dV^2 at 1.7 mV, which is dominated by the hindered translational mode. Interactions with the atoms of the molecule lead to small shifts of the peak in the range of 1–3 meV. Figure adapted from [717], https://doi.org/10.1080/23746149.2017.1372215, distributed under the terms of the Creative Commons Attribution License (http://creativecommons.org/licenses/by/4.0)

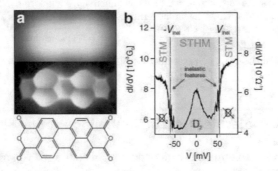

Fig. 2.27 Molecular hydrogen is dosed into the UHV system and tunneling junction with incorporated hydrogen is formed. This type of tip conformation gives very high resolution on molecules. It is related to Pauli repulsion forces, which influence the distance dependence of tunneling and yield submolecular resolution comparable to submolecular resolution observed by nc-AFM. A PTCDA (3, 4, 9, 10-perylenetetracarboxylic-dianhydride) molecule is imaged by scanning tunneling hydrogen microscopy (STHM). Figure from [688], https://doi.org/10.1103/PhysRevLett.105.086103, distributed under the terms of the Creative Commons Attribution License (http://creativecommons.org/licenses/by/3.0)

2.9   Scanning Tunneling Hydrogen Microscopy

The attachment of a hydrogen molecule to the tip of a STM and the choice of suitable bias voltage leads to a different operation mode of STM, where submolecular resolution in close analogy to nc-AFM can be observed [638,688]. At bias voltages below 50 mV inelastic tunneling raises and the images cannot be related to LDOS variations anymore, but resemble to the high resolution AFM imaging. Detailed analysis show that this STM contrast is dominated by Pauli repulsion forces. Similarly, Xe, CH_4 or CO can be attached to the probing tip and give atomic contrast closely related to the contours of total charge density [347]. Some distortions due to tilting of the attached molecule are observed (Fig. 2.27).

Force Microscopy

3

Abstract

An introduction is given into atomic force microscopy (AFM), where a force is detected between probing tip and sample. Force detections schemes as well as operation modes are discussed. Various force interactions can be distinguished, such as repulsive contact forces, chemical forces, electrostatic forces or magnetic forces. Application examples will be shown for the various sample categories ranging from metals to insulators.

3.1 Concept and Instrumental Aspects

The basic concept of force microscopy is the measurement of forces between a sharp tip and a sample surface. Most commonly, the tip is mounted on the end of a cantilever which serves as a force sensor. Either the static deflection of the cantilever or the change in its dynamic properties due to tip–sample forces can be exploited. The limit of force detection is far lower than the force between atoms at lattice distances, explaining the widely used term atomic force microscope. In analogy to scanning tunneling microscopy, we will refer to the method as scanning force microscopy (SFM).

3.1.1 Deflection Sensors: Techniques to Measure Small Cantilever Deflections

There are different techniques to detect the small bending of the cantilever due to tip–sample forces (see Fig. 3.1). Most instruments use the beam-deflection method [24,432,453]. A light beam is reflected at the rear side of the cantilever and the deflection is monitored by a position-sensitive photodiode. A schematic drawing of the setup is shown in Fig. 3.2. A four-segment photodiode allows one to detect not

© Springer Nature Switzerland AG 2021

E. Meyer et al., *Scanning Probe Microscopy*, Graduate Texts in Physics,
https://doi.org/10.1007/978-3-030-37089-3_3

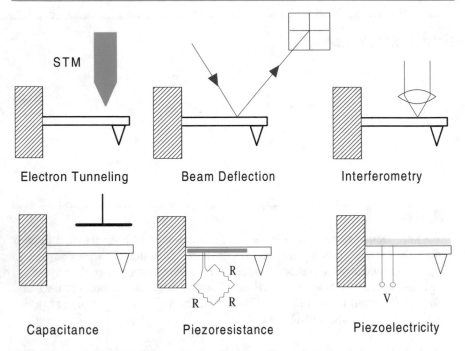

Fig. 3.1 Deflection sensors for scanning force microscopy

only the normal bending, but also the torsion of the cantilever caused by lateral forces acting on the tip.

An alternative deflection sensor can be implemented by using the cantilever as one mirror of an optical laser interferometer [560]. This technique has the advantage of easy calibration by the wavelength of the light. Furthermore, it can be implemented if space is limited, as in low-temperature experiments. Both optical methods have achieved a sensitivity which is limited by the thermal noise of the cantilever. An intriguingly simple version of an interferometer has been presented by Sarid et al., where the reflected light from the cantilever is fed back into the laser diode cavity [573].

The cantilever deflection can also be detected by a change in the capacitance between the cantilever and a counterelectrode [443]. This technique is capable of a very fast measurement, and the whole force sensor can be produced by microfabrication techniques [88]. A drawback with this sensor is the force between cantilever and counterelectrode, which cannot be neglected. On the other hand, the electrostatic force can also be used to control the deflection in a force feedback scheme [168]. In the original force microscope of Binnig et al., the cantilever deflection was detected by means of a tunneling current from the cantilever to an STM tip positioned at the rear side of the cantilever [86]. This otherwise very sensitive setup is also complicated by the force between the STM tip and the cantilever.

Self-sensing cantilevers form a very elegant class of deflection sensors. Most are realized by producing a piezoresistive layer on a silicon cantilever [650]. Although

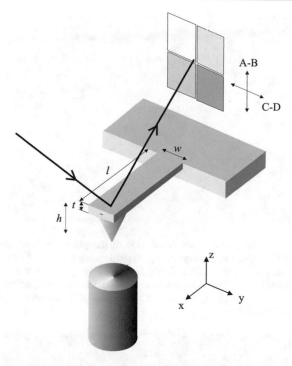

Fig. 3.2 Schematic diagram of the beam-deflection SFM. The relevant dimensions of the rectangular cantilever are indicated: length l, width w, thickness t and height of the tip h. Note that $h = h_{tip} + t/2$. Normal and lateral forces acting on the tip are measured via normal and torsional motions of the cantilever. A light beam is reflected off the rear side of the cantilever. Angular deflections of the laser beam are measured with a position-sensitive detector (4-quadrant photodiode). The A-B signal is proportional to the normal force and the C-D signal is proportional to the torsional force

the signal-to-noise ratio is comparable to beam-deflection schemes, the first dynamic measurement showing atomic resolution was performed with piezoresistive cantilevers [203]. Other dedicated designs allow the piezoresistive detection of lateral forces [119,225]. Piezoelectric cantilevers have the advantage of being sensor and actuator for dynamic measurements at the same time. This ability has been exploited in the compact design of a high-vacuum dynamic force microscope [118] and for a significant enhancement of the speed of tapping mode force microscopy [630]. Commercially available quartz tuning forks are cheap piezoelectric sensors with high frequency and spring constant. High-resolution images in dynamic mode have been achieved by attaching sharp probing tips to such quartz tuning forks [123,205,209,210]. The setup is ideal for low temperatures in combination with STM. In this case, a tungsten wire is attached to one the prongs and the second prong is fixed, also called Q-Plus setup [210]. It is also found that the Q-factors at low temperatures can reach 30'000–50'000, which gives excellent force (gradient) sensitivity. Under these conditions submolecular resolution was achieved on planar molecules [231] (Fig. 3.3).

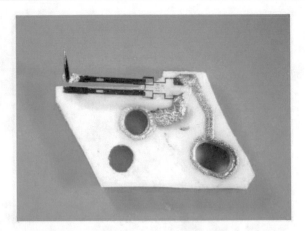

Fig. 3.3 Photograph from a tuning fork AFM. One of the prongs is fixed, also called Q-Plus [210]. The probing tip is glued to the other prong. Typical spring constant is 1800 N/m. Oscillation amplitudes of 10–100 pm can be selected. Especially, at low temperatures the internal damping is very low and Q-factor of 30'000–50'000 are measured. Therefore, excellent force sensitivity can be achieved. Reprinted from [210], https://link.aps.org/doi/10.1103/RevModPhys.75.949, with permission from AIP Publishing

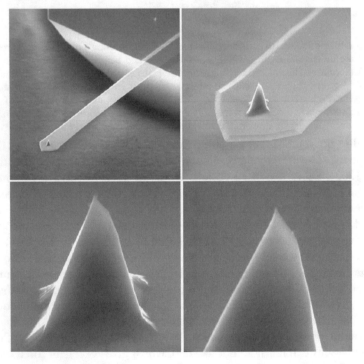

Fig. 3.4 Scanning electron microscopy images of a rectangular silicon cantilever with integrated probing tip, manufactured by Nanosensors. The height of the probing tip is 12.5 μm [10]

3.1.2 Spring Constants of Rectangular Cantilevers

The spring constant of a rectangular cantilever (Fig. 3.4) can be calculated from its geometry. The spring constant k_N for normal bending is given by

$$k_N = \frac{Ewt^3}{4l^3},$$ (3.1)

where w is the width, l the length, t the thickness of the cantilever and E the Young's modulus of the material. Width and length can be measured by means of scanning electron microscopy. The thickness t can be more precisely determined from the resonance frequency f of the cantilever:

$$t = \frac{2\sqrt{12}\pi}{1.875^2} \sqrt{\frac{\rho}{E}} fl^2,$$ (3.2)

where ρ is the mass density. In the case of silicon, $\rho = 2330\,\text{kg/m}^3$ and $E = 1.69 \times 10^{11}\,\text{N/m}^2$, and the thickness is

$$t = 7.23 \times 10^{-4}\,\text{s/m} \times fl^2.$$ (3.3)

The relation between the signal from the deflection sensor and the actual normal force can be calibrated by recording deflection versus distance curves in the contact mode (see Sect. 3.4.1).

The torsional spring constant k_T, which is needed for a quantitative analysis of lateral force measurements, is given by

$$k_T = \frac{Gwt^3}{3h^2l},$$ (3.4)

where $G = 0.68 \times 10^{11}\,\text{N/m}$ is the shear modulus of silicon. The resonance frequencies and corresponding spring constants of higher oscillation modes will be considered when the noise of the cantilever is analyzed in Sect. 3.9. Triangular cantilever geometries are widely used for topographic imaging since their high torsional stiffness reduces the influence of lateral forces on the measurements. The calculation of spring constants for triangular cantilevers (Fig. 3.5) has been discussed in [567].

Several methods have been suggested for calibrating the force constants of cantilevers experimentally (see references in [566]). The most elegant has been proposed by J. E. Sader. It is based on the viscous damping of the cantilever motion in air. From measurements of length, width, resonance frequency, and quality factor, the spring constant can be evaluated using some sophisticated mathematics. Note that the difficult measurement of the cantilever thickness is not necessary in this method [566].

There are some simple criteria to be considered when cantilevers are fabricated. The resonance frequency f_R should be selected above building vibrations (1–100 Hz)

Fig. 3.5 Scanning electron microscopy images of a triangular silicon nitride cantilever with integrated probing tip, manufactured by Park Scientific. The tip height is about 2 μm [9]

and sound waves (1–10 kHz). If high resolution is required, the spring constant k_N of the cantilever should be comparable to the spring constants of atomic springs. A simple estimate of atomic springs k_{atom}, is given by the formula $k_{atom} = m\omega^2$, where $m = 5 \times 10^{-26}$ kg is a typical mass of an atom and $\omega = 10^{13}$ Hz is a typical phonon frequency. Thus, an atomic spring has a value of about 5 N/m, which gives us an order of magnitude estimate of the spring constant of the cantilever $k_N \approx 0.1$–100 N/m. Thermal vibrations of the cantilever should be limited to less than 0.1 nm. With $k_N \langle \Delta z^2 \rangle = k_B T$, this limit corresponds at $T = 300$ K to $k_N \geq 0.4$ N/m. It can be shown that only cantilevers with dimensions in the micrometer range fulfill these design criteria [452]. Generally, higher resonance frequencies require smaller cantilevers.

3.1.3 Cantilever and Tip Preparation

Nowadays most cantilevers are produced with integrated tips in a microfabrication process using methods established in the semiconductor industries. The preferred materials are silicon and silicon nitride. Typical silicon cantilevers are made from single crystalline material. The pyramidal tip points along the ⟨100⟩ direction and has a macroscopic cone angle of 50°. At the apex, the cone angle is reduced and tip radii down to about 10 nm can be realized. Highly doped silicon is used to avoid charging and to allow for combined tunneling and force microscopy experiments. Silicon tips taken from the wafer are covered by a non-conducting layer of native oxide, which can be removed by either dipping into hydrofluoric acid or sputtering in vacuum. Silicon nitride cantilevers are commercially available in triangular form, often with several cantilevers of different spring constant attached to one chip. Tip radii of 20–50 nm can be achieved. Gold or aluminum coating of the cantilevers provides high reflectivity for beam deflection instruments. There is a growing number of manufacturers. For a list see [7].

The tip apex radius and aspect ratio are crucial parameters for lateral resolution on rough surfaces, and also for the relation between short-range and long-range forces. Several expensive improvements are offered by the manufacturers, which are based on sophisticated etching techniques (see Fig. 3.6a). Another possibility is the selective deposition of contaminants on the tip in a scanning electron microscope. Alternatively, the use of carbon nanotubes as chemically inert and mechanically

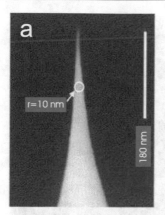 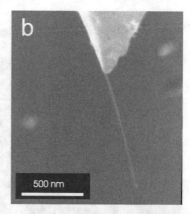

Fig. 3.6 Scanning electron microscopy images of modified silicon tips. **a** Ultrasharp silicon tip (Nanosensors). **b** Carbon nanotube grown by chemical vapor deposition on a silicon tip

stable tips has been demonstrated. They have been transferred to the tip under optical microscopes using adhesives [134] or grown directly on the tip by chemical vapor deposition techniques (see Fig. 3.6b) [247]. For applications based on electrostatic interactions, a metal coating on the tip or even a metallic tip is advantageous [652]. Diamond coatings on tips have also been realized [77]. In order to reduce long-range forces, it is found to be useful to use the focussed ion beam to sharpen the tips as shown in Fig. 3.7. In this case, the diameter is reduced by a factor of ten, which leads to a reduction of long-range forces (van der Waals and electrostatic forces) by a similar factor. This is useful for high resolution force microscopy, where tips are terminated with single atoms and operated in the repulsive regime. Under these conditions, the long-range forces have to be counter-balanced by the front-end atom. Therefore, it is important to minimize this long-range part by reduction of the radius of curvature and application of minimum electrostatic fields.

3.1.4   Implementations of Force Microscopy

Force microscopy has been implemented in a multitude of instruments. They can be sorted by mechanical design, environment, or operational temperatures. The most widely used setup is the beam-deflection type with a scanned sample, which has also been implemented in ultrahigh vacuum [287,454]. The necessary alignment of the light beam makes a fixed position of the tip preferable. On the other hand, a scanned-tip design allows the construction of a stand-alone microscope where samples of any desired size can be studied [277]. Furthermore, the sample can be cooled and heated without risking a misalignment of the light beam [135]. Stand-alone force microscopes have been integrated into optical microscopes, allowing the user to switch easily between the objective and the force microscope head. If no lateral force measurements are required, the use of interferometric deflection sensors is more suitable for scanned-tip designs, due to the compact optical setup. This is of

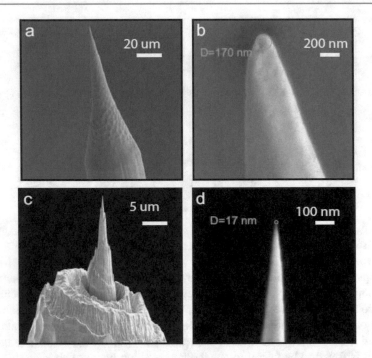

Fig. 3.7 Scanning electron microscopy images of modified tungsten tips, which were attached to tuning fork sensors. **a** Tungsten tip after electrochemical etching. **b** The diameter is 170 nm. **c** Tungsten tip modified by the focussed ion beam. **d** The diameter of 17 nm is reduced by a factor of 10. Therefore, long range forces are drastically reduced. Courtesy of Shigeki Kawai

particular advantage in low-temperature force microscopes [27,297]. The adaption of force microscopy to liquid cells should be mentioned. It allows measurements with low forces in biological buffer solution as well as in electrochemical experiments [406]. Finally, an SFM has been developed for planetary missions [20].

3.2   Relevant Forces

In this section, we give a short review of the relevant forces acting between tip and sample. The interaction range of the different types of force is of great importance for force microscopy, since different parts of the tip and cantilever contribute differently to the total force which is measured. Van der Waals forces act between the mesoscopic tip end and the surface, while the electrostatic force can originate from the whole cantilever. This is in contrast to STM, where contributions of the outermost atoms of the tip dominate the tunneling current due to the exponential decay with distance. The tunneling current corresponds to short-range chemical forces, which can give atomic resolution in force microscopy. However, long-range van der Waals and electrostatic forces are always to be taken into account in force microscopy, even though they may not be of any interest for the actual measurement. For example, the typical situation in contact mode is determined by an equilibrium between the attractive long-range forces and the repulsive short range forces, where only the latter provide good resolution.

3.2.1 Short-Range Forces

Short-range chemical forces arise from the overlap of electron wave functions and from the repulsion of the ion cores. Therefore, the range of these forces is comparable to the extension of the electron wave functions, i.e., less than one nanometer. Short-range forces can be both attractive or repulsive. Forces are attractive when the overlap of electron waves reduces the total energy. These situations are comparable to molecular binding. On the other hand, the Pauli exclusion principle can lead to repulsive forces due to strong electron wave overlap. These forces are directly connected to the total electron density. The ionic repulsion acts over small distances, where the screening of the ion cores by the electrons falls away. Model potentials like the Lennard–Jones or the Morse potential have been introduced for the description of short-range forces. However, their application is essentially limited to the pairwise interaction of atoms. For the tip–sample interaction, at least the interaction with the nearest neighbor atoms has to be included and, furthermore, the displacement of atoms by the action of the short-range force has to be taken into account.

Attractive short-range forces are of the order of 0.5–1 nN per interacting atom at tip–sample distances typical for STM operation. The decay length of such forces is expected to be of the order of atomic units, i.e., 0.05 nm for metallic adhesion, but around 0.2 nm for covalent bonding [514]. The forces have been treated in theoretical studies of different tip/sample systems. For example, metallic binding of an Al/Al system was calculated from first principles [120], covalent binding forces of a silicon tip over a Si(111) reconstructed surface were determined, including the displacement of atoms [514], and the forces between a MgO tip and surfaces of ionic crystals have been studied in atomistic simulations [399]. Imaging of compound semiconductors has been modelled [618], as has chemical bond formation between a silicon tip and a metal surface [619].

Experimentally, attractive short-range forces around 1 nN between metallic tips and surfaces have been found at distances where STM operation was possible [152, 404]. The decay length of these metallic adhesion forces is sometimes larger (up to 0.3 nm) than the decay length of the tunneling current [153,404,576]. On the Si(111) surface, short-range forces of the order of 0.5 nN have been measured in a regime where atomic resolution was obtained [167]. Low-temperature experiments even allow site-specific studies of the short-range forces, which confirm the origin of the force in dangling bonds of the reconstructed surface [385]. In this case, a decay length close to the theoretical predictions was obtained. When comparing experimental force–distance curves with the range of interatomic potentials, one has to consider the displacement of tip and surface atoms which distort the distance axis.

The variation of short-range forces on the atomic scale makes atomic resolution possible in force microscopy. The above-mentioned studies allow the conclusion

that, at distances of 0.5 nm from the surface, short-range forces between tip apex
and surface atoms become comparable to long-range forces between tip and sample.
Consequently, this is a characteristic distance at which atomic resolution in non-
contact modes is obtained. The limits of resolution using repulsive forces in contact
modes are discussed in Sect. 3.4.

3.2.2   Van der Waals Forces

Van der Waals forces are dipole–dipole forces. The most important forces are not
those between permanent dipoles but the so-called dispersion forces. These act
between dipoles that arise from fluctuations and dipoles induced in their electric
field. They are always present and attract even chemically inert noble gas atoms. The
range of dispersion forces is limited. When the distance between molecules is larger
than the distance light can travel during the characteristic lifetime of the fluctuations, the dispersion forces are weakened. This effect is called retardation. The van
der Waals force at short distances decays as $F \propto 1/r^7$, whereas beyond $r \approx 5$ nm
this power law reduces to $F \propto 1/r^8$.

The van der Waals force between macroscopic bodies can be calculated in two
ways. First, the molecular forces can be summed up for each geometry, assuming that
they are pairwise additive and non-retarded. However, the validity of the additivity
assumption is limited by the fact that the mutual polarization of atoms is influenced by
the presence of third atoms. A more rigorous approach for the calculation of the van
der Waals forces between macroscopic bodies is the Lifshits theory, which treats the
bodies as a continuum. The distance dependence found by both approaches has the
same form. Only the Hamaker constant, which accounts for the materials involved,
is calculated in a different way. Since the range of van der Waals forces is limited, the
tip–sample geometry of the force microscope can be well approximated as a sphere
approaching a semi-infinite body. For this configuration, the van der Waals force is

$$F_{\text{VdW}} = \frac{HR}{6D^2} \,, \tag{3.5}$$

where H denotes the Hamaker constant, R the tip radius, and D the distance between
tip and sample surface. The value of the Hamaker constant H is of the order of
10^{-19} J. For materials with high dielectric constant like metals, it is about a factor
of ten higher than for insulators. For a tip of radius $R = 30$ nm, the van der Waals
force in vacuum at a distance $D = 0.5$ nm is of the order of $F_{\text{VdW}} = 2$ nN.

The van der Waals force can also be determined for more complex tip–sample
geometries like a half-sphere at the end of a truncated cone [570]. However, the term
given above dominates for close tip–sample distances in all models.

It is important to note that the medium between tip and sample has a great influence
on the van der Waals force. Very much simplified, the Lifshits theory predicts that
the force is proportional to $(\varepsilon_1 - \varepsilon_3)(\varepsilon_2 - \varepsilon_3)$ and $(n_1^2 - n_3^2)(n_2^2 - n_3^2)$, where ε and
n denote the dielectric constant and the refractive index of the tip (1), the sample (2),
and the medium in-between (3). It is clear that a medium with ε and n close to the

respective values of the tip and sample will greatly reduce the van der Waals forces compared to the vacuum. For most solid materials, this is the case when immersing tip and sample in water. A clever choice of the immersing liquid can even lead to a negative Hamaker constant and, consequently, to repulsive van der Waals forces [302].

3.2.3 Electrostatic Forces

Electrostatic forces act between localized charges on insulating tips and samples. Their strength and distance dependence obey Coulomb's law. Charges can easily be trapped at sample surfaces in the course of surface preparation, for example by sample cleavage or by UHV techniques like ion sputtering. Furthermore, contact electrification can charge tip and sample after their contact is broken. Even in air, such charge can persist for hours particularly on polymers, and in vacuum for days. The decay of the surface charge can be measured by the electrostatic force down to single charges [587].

Charges on the surface also attract conductive tips. The method for determining the force is to calculate the interaction between the charges and their mirror image in the tip. Even neutral but polar surfaces interact with a conductive tip via image forces [326]. Likewise, a polar group at the tip apex causes an attractive force towards a conductive sample.

Electrostatic forces also act between conductive tips and conductive samples when they are at a different potential. If one considers the tip–sample system as a capacitor with distance-dependent capacitance C, the force is given by

$$F_{el} = \frac{\partial C}{\partial z}(U_{bias} - U_{cpd})^2 \,, \tag{3.6}$$

where U_{bias} is the bias voltage applied between tip and sample and U_{cpd} is the contact potential difference caused by the different work functions of tip and sample. Note that a zero bias voltage does not normally correspond to a minimal electrostatic force but that the contact potential difference has to be compensated. Unfortunately, the work function is very sensitive to perturbations at the surface. The irregular shape of the tip gives rise to patch charges, which cannot be completely compensated by a bias voltage [94]. A minimum of the electrostatic force at non-zero bias voltages has also been found for insulating surfaces, which have no well-defined work function. For such materials, the electrostatic force follows changes in the bias voltage with a delay proportional to the resistivity of the sample [62].

The term $\partial C/\partial z$ depends on the geometry of the tip, which can be modeled as a truncated cone and a half sphere [570]. The dominating term for small tip–sample distances is

$$F_{el} = \pi \varepsilon_0 \frac{R}{z}(U_{bias} - U_{cpd})^2 \,. \tag{3.7}$$

For a tip of radius $R = 20\,\text{nm}$ at a tip–sample distance of $z = 0.5\,\text{nm}$ and a potential difference $U_{bias} - U_{cpd} = 1\,\text{V}$ the electrostatic force is about $F_{el} = 0.5\,\text{nN}$.

For increasing tip–sample separation, the distance dependence changes its form as

$$-\frac{1}{z} \rightarrow -\frac{1}{z^2} \rightarrow \ln(z/H) \,, \tag{3.8}$$

where H is an effective height with $z < H$ [386]. The exact form depends critically on the tip shape. Consequently, it is difficult to quantify the electrostatic contribution to the total force. Due to their long-range character, one even has to consider electrostatic interactions between the sample and the cantilever.

3.2.4 Magnetic Forces

The forces acting on magnetic dipoles located in a magnetic field are called magnetic forces. In force microscopy experiments, the magnetic dipoles are usually contained in the ferromagnetic material on/of the tip of the cantilever and the magnetic field is produced by a ferromagnetic sample or a current distribution located in close proximity to the tip. The current distribution may be that of a current-carrying device or it may arise in a superconductor from the tip's stray field.

The calculation of the magnetic forces acting on the tip can become challenging, because the magnetic stray field of the tip or sample does generally change the magnetic moment distribution of the sample or tip, respectively. Moreover, these changes can be reversible or irreversible. In some cases, an experimentalist can deliberately select between the two cases. The tip-sample distance can, for example, be decreased to modify the magnetic sample structure locally, and then again increased for imaging. During imaging the goal is to keep the inherent changes of the sample or tip magnetic moment distributions reversible such that the imaging process can be repeated with the same result. This can be achieved by minimizing the tip-sample interaction force, i.e. by lowering the magnetic moment of the tip or by increasing the tip-sample distance. This however lowers the measured signal such that care has to be taken to minimize measurement noise. In the best case, the modification of the sample and tip magnetic moment distribution by the imaging process becomes negligible. In this case, the magnetic interaction between the tip and the sample can be described by the magnetostatic energy of the tip with a magnetization distribution $\mathbf{M}_{\text{tip}}$ located at the position $(\mathbf{r}, z)$ in a stray field emanating from a sample $\mathbf{H}_{\text{sample}}$:

$$E_m(\mathbf{r}, z) = -\mu_0 \int \mathbf{M}_{\text{tip}}(\mathbf{r}', z') \cdot \mathbf{H}_{\text{sample}}(\mathbf{r} + \mathbf{r}', z + z') \, d\mathbf{r}' \, dz', \tag{3.9}$$

where the integration is performed in the dashed coordinate system that is attached to the tip. The force acting on the tip is then given by

$$\mathbf{F}(\mathbf{r}, z) = \mu_0 \int \nabla \left[\mathbf{M}_{\text{tip}}(\mathbf{r}', z') \cdot \mathbf{H}_{\text{sample}}(\mathbf{r} + \mathbf{r}', z + z')\right] d\mathbf{r}' \, dz'$$

$$= \mu_0 \int \left[\mathbf{M}_{\text{tip}}(\mathbf{r}', z') \cdot \nabla\right] \mathbf{H}_{\text{sample}}(\mathbf{r} + \mathbf{r}', z + z') \, d\mathbf{r}' \, dz'. \tag{3.10}$$

Note that the magnetization field of the tip, $M_{tip}(r', z')$, is generally not known. Models of the tip magnetic moment distribution are then often employed to understand the magnetic tip-sample interaction. The simplest of these are magnetic point-pole models, where a magnetic point charge q_{tip} or a magnetic point dipole m_{tip} is used to describe the tip-sample interaction. The force given in (3.10) then simplifies to

$$F(r, z) = \mu_0 q_{tip} \cdot H_{sample}(r, z) \tag{3.11}$$

$$= \mu_0 \left[m_{tip} \cdot \nabla \right] H_{sample}(r, z) . \tag{3.12}$$

Alternatively, the interaction between the tip magnetization distribution and the sample stray field can be described in a two dimensional Fourier space where the coordinates (r, z) are transformed to (k, z). The force on the tip then becomes

$$F(r, z) = \mu_0 \sigma_{tip}^*(k) \cdot H(k, z) , \tag{3.13}$$

where $\sigma_{tip}^*(k)$ is the complex conjugate of the tip equivalent magnetic surface charge density $\sigma_{tip}(r)$ in an xy-plane running through the apex of the tip. It is equivalent to z-component of the stray field of the tip in the latter plane, i.e.

$$H_{z,tip}(r) = \frac{1}{2}\sigma_{tip}(r) . \tag{3.14}$$

3.2.5   Capillary Forces

It is well known that microcontacts act as condensation nuclei. In air, water vapor plays the dominant role. If the radius of curvature of the microcontact is below a certain critical radius, a meniscus will be formed. This critical radius is defined approximately by the size of the Kelvin radius $r_K = 1/(1/r_1 + 1/r_2)$, where r_1 and r_2 are the radii of curvature of the meniscus. The Kelvin radius is connected with the partial pressure p_S by the equation

$$r_K = \frac{\gamma V}{RT \log (p/p_S)} , \tag{3.15}$$

where γ is the surface tension, R the gas constant, T the temperature, V the molar volume and p_S the saturation vapor pressure [17]. The surface tension of water is $\gamma = 0.074 \, N/m$ at $T = 20 \, °C$, which gives the parameter $\gamma V/RT = 0.54 \, nm$. Therefore, for $p/p_S = 0.9$, we obtain a Kelvin radius of $100 \, nm$. For smaller vapor pressures, the Kelvin radius becomes comparable to the dimensions of the molecules and the Kelvin equation is no longer applicable. In scanning force microscopy, typical tips with radii less than $100 \, nm$ are possible nuclei for condensation. If a meniscus is formed, an additional capillary force acts on the tip. A simple estimate is given by

$$F = \frac{4\pi R\gamma \cos \Theta}{1 + D/[R(1 - \cos \phi)]} , \tag{3.16}$$

where R is the radius of curvature, Θ the contact angle, D the distance between tip and sample, and ϕ the angle of the meniscus [306]. The maximum force is given by $F_{max} = 4\pi R\gamma \cos \Theta$. For a tip radius of 100 nm, we obtain a force $F_{max} = 9.3 \times 10^{-8}$ N, which is significantly stronger than the corresponding van der Waals force. Typical force versus distance curves in ambient conditions reveal adhesion forces of the order of 10^{-8}–10^{-7} N, which mainly originate from capillary forces [620]. These forces limit the minimum force which acts on the outermost tip region, and have to be equilibrated by the repulsive force in this small contact region. Consequently, capillary forces can determine the size of the contact and play an essential role in force microscopy measurements in air. Depending on the humidity, the presence of a meniscus has to be taken into account. The finite time of capillary formation can influence SFM results in the form of a modified velocity dependence [552]. A reduction of capillary forces is achieved by covering hydrophilic samples and probing tips with amphiphilic molecules that render the surfaces hydrophobic. Then the surfaces are not covered by thick water films and capillary forces are reduced. Immersing the whole tip–sample area under liquids can completely remove capillary interaction.

3.2.6   Forces in Liquids

As we have described above, the immersion of tip and sample in a liquid can cause a dramatic reduction in the tip–sample force. In particular, the van der Waals forces can be reduced and the capillary forces removed. Here we briefly describe some properties of the forces which arise only in a liquid environment.

When the tip–sample distance is of molecular dimensions, the short-range order of the liquid can cause an oscillation of the force [312]. Short-range forces which are introduced by the interaction between solvent molecules and the surfaces are usually referred to as solvation forces.

Surface charging can be produced by ionization or dissociation of a chemical group at the surface or by adsorption of ions from the liquid. Such charge is compensated by a certain density of counter ions in front of the surface, resulting in a so-called electric double layer. The double layer can interact with other double layers or with polarizable surfaces. The repulsive interaction of electric double layers has been exploited to achieve high resolution in force microscopy [613].

Finally, forces based on change of entropy should be mentioned. For example, when two surfaces with attached polymer chains approach each other, the unfavorable entropy associated with the confined geometry will create a repulsive force. For a discussion of these steric or fluctuation forces, we refer the reader to the text book by Israelachvili [306].

Fig. 3.8 Operation modes of force microscopy sorted with respect to static or dynamic detection and tip–sample contact formation

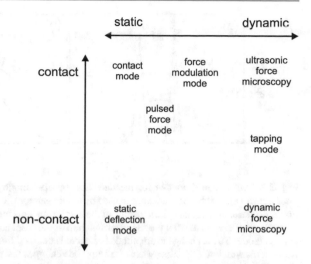

3.3   Operation Modes in Force Microscopy

Several modes of operation have been introduced in force microscopy, with a variety of names describing their characteristic features. These operational modes can be divided into static and dynamic modes, where the static bending of the cantilever or its dynamic properties are measured, respectively. Furthermore, the operation modes are often distinguished with respect to whether the tip is in contact with the surface or not. For dynamic modes, however, the tip may be temporarily in contact during each oscillation cycle. The modes are listed in Fig. 3.8.

The most important static mode is the so-called contact mode, where the tip–sample distance is controlled to maintain a constant cantilever bending. In this mode, a topographic surface of constant force is recorded. On the other hand, scanning in constant height above the surface yields maps of the force. Due to mechanical instabilities (jump-to-contact) and drift problems, this mode is mainly used to image long-range magnetic forces.

In the dynamic modes, changes in vibrational properties of the cantilever due to tip–sample interactions are measured. These properties include the eigenfrequency, the oscillation amplitude, and the phase between excitation and oscillation of the cantilever. The dynamic modes can be differentiated by the feedback parameters used for distance control. A complete review of dynamic methods has been given by Garcia and Perez [195]. Using the frequency shift of a self-exciting oscillation loop as feedback parameter opens the possibility of a reliable non-contact mode (see Fig. 3.9a). This method is often referred to as non-contact AFM or dynamic force microscopy and has been successfully applied in vacuum environments, where the cantilever oscillations exhibit high quality factors. The change in amplitude of an oscillation excited with constant frequency can be exploited as feedback parameter in gaseous or aqueous environments with lower cantilever quality factors (see Fig. 3.9b). Here, intermittent tip–sample contact is possible and can even be used to study surface elasticity. When a cantilever oscillation of several nanometers amplitude is

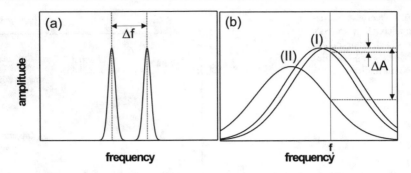

Fig. 3.9 Two dynamic modes for the detection of tip–sample interactions. In dynamic force microscopy (**a**), the shift in frequency caused by the tip–sample interaction is detected. The oscillation is driven at the actual eigenfrequency of the cantilever. This mode is mainly applied in vacuum, where high Q-factors allow a precise determination of the resonance frequency. In the intermittent contact mode (**b**), the change in amplitude is detected at a fixed frequency f_0. This mode is suitable for systems with low Q-factors, where the amplitude changes fast enough upon tip–sample interactions. The decrease in amplitude at f_0 may be due either to a frequency shift (I) or to an additional decrease in the total amplitude due to damping (II)

excited close to its resonance frequency, this mode is usually called tapping mode or intermittent contact mode.

Other approaches prefer an excitation to small amplitudes far from the resonance in order to overcome the problem of a non-linear force curve in the range of the oscillation [282]. Measurements of cantilever oscillations with the tip in permanent contact with the sample allow studies of the mechanical response of the sample. Here, the ultrasonic frequency range is of special interest and the mode is often called ultrasonic force microscopy [61]. Finally, an analysis of the thermal noise in the cantilever movement allows one to study interactions with minimal amplitudes (see also Sect. 3.7).

These modes are not limited to mapping topography and surface properties. Further information can be obtained by recording the respective signals as a function of distance or other parameters, like tip–sample voltage. These force–distance curves, also known as force spectroscopy, contribute to our understanding of the tip–sample interaction.

The typical features of non-contact and contact mode are depicted in Fig. 3.10. In the non-contact situation shown in Fig. 3.10a, all forces are attractive and the short-range force between the outermost tip atom and one surface atom makes a significant contribution to the total force. This ideal constellation allows imaging with true atomic resolution without irreversible damage to the surface. Figure 3.10b displays the more complex case when some atoms of the tip and sample are in repulsive contact. Here the total force can still be attractive depending on the relation between short- and long-range forces. In any case, the deformation of the surface and the transition between repulsive and attractive forces will establish a contact consisting of several atoms. True atomic resolution of point defects and of step edges cannot then be expected.

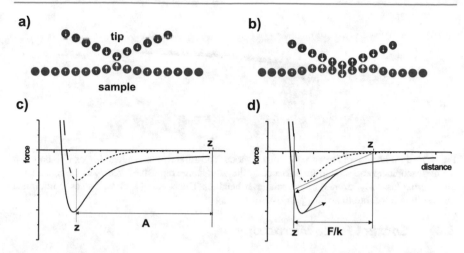

Fig. 3.10 Contrast formation in non-contact and contact mode. In **a** and **b**, *arrows* indicate forces acting on the atoms of tip and surface. In **c** and **d**, the total tip–sample force is sketched by the *solid line*, while the short-range contribution appears as a *dashed line*. The *arrow* marked A in (**c**) quantifies the oscillation amplitude in the dynamic mode. *Arrows* in (**d**) mark the jump into and out of contact, whilst the *straight dashed line* represents the value of the cantilever spring constant k

Figure 3.10c and d explain the concepts of dynamic and static force microscopy based on force versus distance curves. These two modes try to exploit the situations illustrated in Fig. 3.10a and b, the dynamic mode using stiff cantilevers and the static mode using soft cantilevers. The vertical bar at position z in Fig. 3.10c indicates a position comparable to that in Fig. 3.10a, where strong short-range forces form a significant part of the total force. The arrow bar marked A indicates a typical tip oscillation amplitude in the dynamic mode, which ensures a restoring force at the turning point to prevent the tip from being trapped at the sample surface. Note that the total force can still be attractive when the short-range force enters the repulsive regime. The arrows in Fig. 3.10d indicate the instabilities known as 'jump-to-contact' in the static mode. The straight dashed line represents the value of the spring constant. When the derivative of the total force becomes larger than the spring constant, the tip jumps into a position where repulsive short-range forces balance attractive forces. The 'jump-off' upon retraction of the tip takes place again when spring constant and force derivative become equal (see Fig. 3.12).

In the following sections, the contact mode and dynamic force microscope will be discussed in detail. On the basis of this introduction, we will also discuss the tapping mode and describe other members of the family of force microscopy modes.

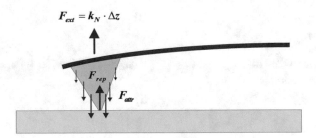

Fig. 3.11 Equilibrium of forces in the contact mode. The attractive long-range forces F_{attr} between the mesoscopic tip and sample are balanced by the short-range repulsive force F_r at the contact and the external force F_{ext} exerted by the cantilever bending. The bending is chosen so as to minimize the resulting force, but to avoid a jump out-of-contact

3.4   Contact Force Microscopy

3.4.1   Topographic Imaging

Contact mode force microscopy is based upon the static measurement of deflection of the cantilever. Topographic images are recorded by scanning the tip over the sample surface at constant cantilever deflection. The deflection corresponds to a normal force which can be calculated by multiplying by the spring constant. In this mode of operation, the probing tip is brought into the repulsive force regime. The position of the tip is given by an equilibrium of forces: the attractive force between the mesoscopic tip and sample has to be compensated by the repulsive force between the tip apex and sample, and the external force exerted by the cantilever spring has to be added (see Fig. 3.11). Consequently, the normal force calculated from the cantilever deflection is not the total force to be compensated by repulsive forces, but the attractive tip–sample force has to be added.

Repulsive forces increase very strongly with decreasing tip–sample distance. Therefore, images of constant repulsive force are often identified with topography. One has to take into account the fact that local variations in the elasticity may influence measurements. The softer a certain area of the surface is, the more deeply the tip will indent with constant normal force. This effect can be exploited in a simple extension of the contact mode: the elastic response of the tip–sample contact to a slow height modulation is measured in the force modulation mode, using a lock-in technique [150]. A second source of topographic artifacts is the strong torsion of the cantilever by frictional forces which may falsify the measurement of normal forces. This effect can be excluded by comparing forward and backward scans. If the cross-talk from torsional bending is negligible, the topography images should be identical for the forward and backward direction, while the lateral force maps are inverted as shown in Fig. 3.12. Topographic artifacts are discussed in more detail in Sect. 6.4.2.

Contact mode force microscopy is accompanied by the jump-to-contact for a soft cantilever. When the gradient of the increasing attractive force becomes larger than the spring constant of the cantilever, an instability occurs and the tip jumps to contact. The force curve is depicted in Fig. 3.10 and the resulting measurement in Fig. 3.13.

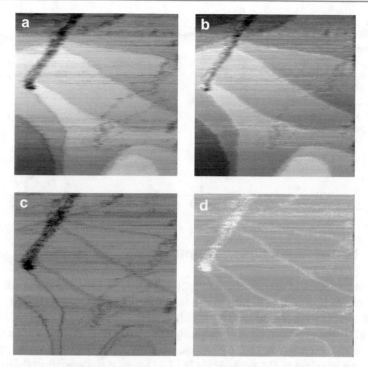

Fig. 3.12 Topography (**a**), (**b**) and lateral force (**c**), (**d**) maps recorded on a Cu(111) surface. Atomically flat terraces separated by monatomic steps and a scratch can be recognized. The frame size is 115 nm. Increased friction is found at steps and in the wide scratch. Note that the forward scan direction (**a**), (**c**) and the backward scan direction (**b**), (**d**) are identical in the topography signal but inverted in the lateral force signal

The adhesion of the tip to the sample generates a hysteresis in the deflection versus distance curve. A significant negative normal force is required to release the jump out-of-contact. Note that a measurement of deflection versus distance like the one shown in Fig. 3.13 can be used to calibrate the normal force based on the cantilever deflection. The calculation of the necessary spring constant has been introduced in Sect. 3.1.2.

Once the tip is in contact, the operator will try to minimize the forces on the tip apex by bending the cantilever. This technique is sometimes referred to as applying a negative load. The minimum force on the tip apex is achieved close to the jump out-of-contact. Experimentally, it is found that the minimum force is a significant fraction of the attractive tip–sample forces (typically 10–50%). Thus, the compensation by cantilever bending is not complete.

The interplay of forces depicted in this section and the elasticity of the materials leads to the formation of a finite contact area between tip and sample. In the following section, we will discuss the role of this contact area for the lateral resolution in contact mode force microscopy.

Fig. 3.13 Cantilever
bending versus sample
displacement recorded with
a soft cantilever. Spring
constant $k = 0.024$ N/m.
Arrows indicate approach
and retraction. The *dotted
line* represents the value of
the spring constant

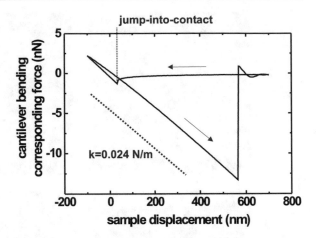

3.4.2 Lateral Resolution and Contact Area

Since the invention of atomic force microscopy in 1986 [86], the quest for atomic
resolution has been a central part of the research. The first results showing atomic
periodicity were presented in 1987 for highly oriented pyrolitic graphite (HOPG) [87]
and for a layered insulator, highly oriented boron nitride [22]. It was soon recognized
that the resolution might not be true atomic resolution, but due to the shear of flakes,
which were attached to the probing tip [517]. Due to the commensurability of the
flake and the surface, a constructive interference can occur between these finite-
area surfaces, known as the Moiré effect. Measurements on non-layered materials,
such as LiF(001) [449] or AgBr(001) [450], showed that atomic periodicity does not
depend on the existence of flakes. Even the highly-reactive Si(111)7×7 surface has
been imaged in contact mode with atomic periodicity. In this case, strong chemical
interactions between the probing silicon tip and the sample made it necessary to
cover the tips with PTFE before scanning [288].

However, the observation of atomic features in contact mode is limited to the
imaging of the lattice. Single point defects have not been observed so far. This lack
of true atomic resolution in contact mode is explained by the fact that the area of the
tip–sample contact is larger than atomic dimensions. If the contact was of atomic
dimensions, the attractive forces would cause a pressure of the order of GPa. Thus,
tip and sample would be deformed elastically or plastically in order to increase the
contact area and to reduce the pressure (see Fig. 3.10). In a very simple approach,
the Hertz model can give an estimate of the contact diameter a:

$$a = 2(DRF)^{1/3} \, , \tag{3.17}$$

where $D = (1 - v_1^2)/E_1 + (1 - v_2^2)/E_2$ and v_i and E_i are the Poisson ratios and
Young's moduli of probing tip and sample. With typical parameters ($E_1 = E_2 =
1.7 \times 10^{11}$ N/m², $v_i = 0.3$ and $R = 90$ nm), the contact diameter in ambient atmo-
sphere is 2–10 nm (1 nN $< F <$ 100 nN). In ultrahigh vacuum, contact diameters are
estimated to be 1–4 nm (0.1 nN $< F <$ 10 nN), whereas in liquids the achievement

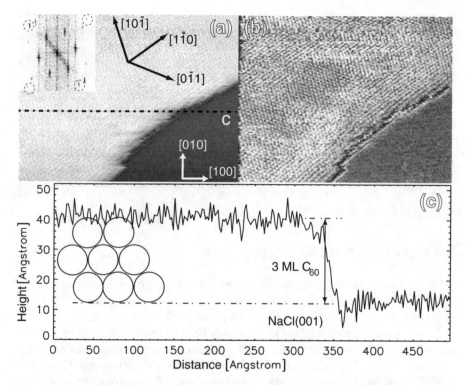

Fig. 3.14 a High-resolution constant force image of C_{60} on NaCl(001). The *inset* is the FFT image, showing the spots from both the C_{60} periodicity and the NaCl(001) lattice. **b** Corresponding friction force map. The molecular structure is visible on both the C_{60} terrace and the NaCl lattice. The observation of molecular structure at the step edge confirms that the resolution is about 1 nm, which corresponds to the distance between C_{60} molecules. **c** Profile as indicated in (**a**)

of true atomic resolution can be expected under the best conditions ($a < 1$ nm). These rough estimates are confirmed experimentally. In ambient pressure, the resolution is typically in the range 5–10 nm. The resolution in ultrahigh vacuum is demonstrated in Fig. 3.14. The atomic periodicity of both the cubic NaCl substrate and the hexagonal C_{60} film are reproduced and the step edge is resolved down to 1 nm. In liquids, the attractive forces can be reduced to 10 pN and true atomic resolution has been achieved in the static deflection mode [499].

In conclusion, typical contact diameters in contact mode force microscopy lie in the range 1–10 nm, limiting the lateral resolution. Reduced long-range forces as present in vacuum or in liquids improve the situation. A more exact calculation of the contact area requires sophisticated elasticity models for the given combination of materials. Generally, atomic-scale images have to be interpreted with care. Even when atomic-scale features are observed in normal or lateral force, the contrast originates from a multiple-atom contact. Exceptions to this rule are possible in liquids with ultralow forces (<100 pN), and by non-contact imaging in ultrahigh vacuum, where true atomic resolution can be achieved (see Sect. 3.5).

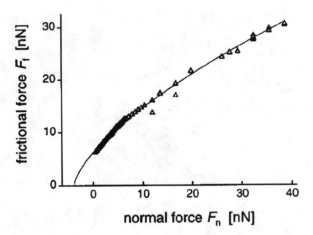

Fig. 3.15 Load dependence of the lateral force. Data were recorded on amorphous carbon under argon atmosphere using a well-defined tip with a radius of $R = 58$ nm. Reprinted from [591], https://link.aps.org/doi/10.1103/PhysRevB.56.6987, with permission from AIP Publishing

3.4.3   Friction Force Microscopy

When in contact mode force microscopy the tip is moved over the surface, friction in the tip–sample contact will produce a lateral force on the tip apex. This lateral force causes a torsional bending of the cantilever, which can be recorded in beam-deflection-type force microscopes. In this way the local variation of friction can be studied with high resolution and for various values of parameters like external load or scanning velocity. One of the great benefits of friction force microscopy (FFM) is the possibility of deciding whether wear has taken place in the course of the experiment by subsequent imaging of the relevant area.

The friction between two macroscopic bodies is composed of contributions from a multitude of small microcontacts. With the invention of the FFM, it became possible to study a single contact of relevant size. It is important to note that the tribology of a single contact differs completely from that of macroscopic bodies. For example, the well known laws of macroscopic friction state that friction is proportional to the load and independent of the geometric contact area. Figure 3.15 displays the load dependence of the lateral force as measured by FFM. The lateral force increases as $F_{\mathrm{N}}^{2/3}$, i.e., in proportion to the contact area [see (3.17)]. The discrepancy with the macroscopic laws of friction can be resolved by considering the real area of contact, which is the sum of all microscopic contact areas [452]. The real contact area depends on the load, not only by the relation given in (3.17), but also by the number of contacts formed at all. Another difference between microscopic and macroscopic laws is the velocity dependence, discussed in Sect. 3.4.6.

Lateral friction contrast can be caused by geometrical features like steps or holes. As an example, consider the increased friction at the monatomic steps of a Cu(111) surface shown in Fig. 3.12. Locally increased friction should always produce an inverted contrast in forward and backward scans, otherwise cross-talk from the normal force into the lateral force signal can be suspected. A further source of friction contrast is the local variation of tip–sample forces on inhomogeneous surfaces, usually referred to as material-specific contrast. A typical example is the friction contrast

Fig. 3.16 Material-specific
contrast in friction force
microscopy. Topography (**a**)
and lateral force (**b**) recorded
on a mixed
Langmuir–Blodgett film.
Circular areas of
hydrocarbon molecules are
surrounded by fluorocarbon
molecules. Within the
hydrocarbon islands,
material has been removed
by the action of the tip (low
scanning velocity, high load)

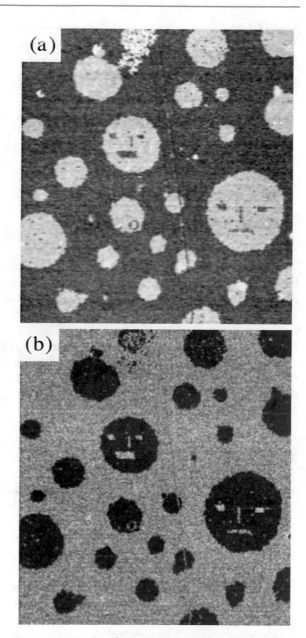

on mixed Langmuir–Blodgett films shown in Fig. 3.16. Note that the lateral force
contrast can be caused not only by friction contrast, but also by a change in the
normal force, which influences the lateral force. For example, lateral force contrast
has been found on films of molecules with varying length but the same chemical
endgroup [359].

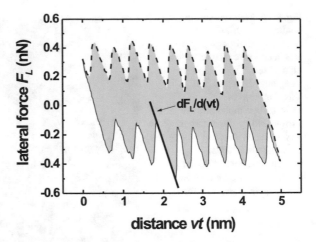

Fig. 3.17 Lateral force versus scanned distance measured in ultrahigh vacuum on a NaCl(100) surface

For a quantitative analysis of FFM images, the torsional bending of the cantilever has to be calibrated as explained in Sect. 3.1.2, or by alternative methods [497]. Furthermore, the difference between the lateral force data of forward and backward scans divided by two should be used in order to account for asymmetries arising from imperfect alignment of the light beam. As mentioned above, one has to pay attention to cross-talk from the normal bending of the cantilever in the lateral force signal. This problem can be avoided by using sophisticated cantilever designs with force feedback [311]. A further interesting extension is dynamic friction force microscopy, where the friction force on a laterally oscillating sample is detected by means of lock-in techniques [368].

3.4.4 Atomic Friction Processes

The lateral force measured by FFM on well-defined surfaces can exhibit atomic-scale features. An example is given in Fig. 3.17 for a NaCl(100) surface cleaved and studied in ultrahigh vacuum. A sawtooth-like behavior is observed with the periodicity of the surface lattice. This phenomenon is called atomic-scale stick–slip. The lateral force increases while the contact is locked onto one atomic position until it is strong enough to initiate a slip to the next atomic position on the surface. Furthermore, the lateral force exhibits hysteresis after reversing the scan direction. From the area that is enclosed by the friction force loop, one can directly calculate the energy that is dissipated.

Atomic-scale stick–slip was first observed on graphite using a tungsten tip [438]. In this pioneering work, however, rather high normal forces of up to 5×10^{-5} N were applied and lateral forces of the order of 100 nN were found. These values far exceed what is expected as a force associated with an atomic process. For the detachment of single atoms, one would expect forces of the order of 1.6 nN, estimated as the quotient of a binding energy of 1 eV and a binding length of 0.1 nm. It was suspected that the strong forces are related to a large contact area, possibly even to the dragging of

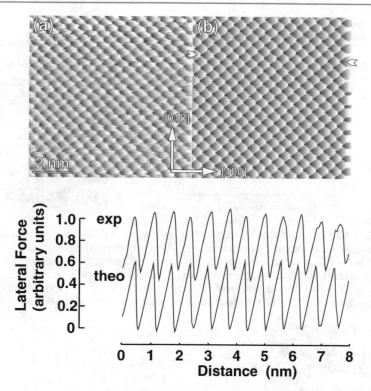

Fig. 3.18 a Experimental and **b** theoretical lateral force map. The experiment was performed in ultrahigh vacuum with a silicon tip. As shown in the profiles, the typical atomic-scale stick–slip is revealed in both images

a flake of material attached to the tip. However, studies on ionic surfaces in ultrahigh vacuum involve contacts of atomic dimensions. The data presented in Fig. 3.17, i.e., lateral forces below 1 nN and an energy dissipation of less than 1 eV per slip event, are in agreement with the assumption of an atomic friction process.

In Fig. 3.18, an experimental FFM image recorded on KBr(100) is compared with a theoretical image which has been calculated by means of the two-dimensional Tomlinson model [245, 411]. The spring potential of the cantilever is dynamically added to a two-dimensional surface potential, and the conditions for sticking and slipping are evaluated. Qualitatively, the experimental data fits well with the theory. Note that only one type of ion is imaged as a protrusion or, in other words, the unit cell of the surface lattice is represented in this map. Furthermore, no atomic-scale defect has ever been found in this type of experiment. The absence of defects could be explained either by the averaging of the coherent contributions of all atoms in contact or by a displacement of point defects in the stress field of the approaching tip [605]. These considerations evidence the need for an atomistic model beyond continuum mechanics and the Tomlinson model. An example of the modeling of FFM at surfaces of ionic crystals is given in Fig. 3.19. It is shown that, after the start of a scan, the atomic configuration of the tip–sample contact is repeatedly rearranged

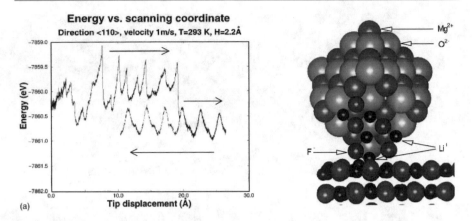

(a)

Fig. 3.19 Simulation of FFM on an ionic surface. **a** Energy of the tip–sample system after start
of scanning. At the beginning, irregular stick–slip behavior is observed corresponding to frequent
jumps of atoms between tip and sample. After a scanned distance of 19 Å, a stable configuration
with a significantly lower energy is established, which subsequently provides regular atomic stick–
slip periodicity. **b** Atomic configuration of tip and sample after scanning. Reprinted from [398],
https://link.aps.org/doi/10.1103/PhysRevB.56.12482, with permission from AIP Publishing

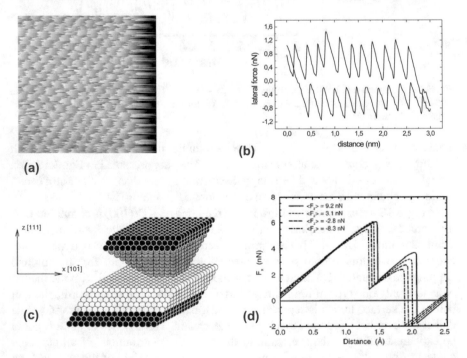

Fig. 3.20 **a** Friction force map of a Cu(111) surface showing atomic-scale stick–slip. The frame
size is 3 nm. **b** Lateral force versus scanned distance for one cross-section of (**a**). **c** Model set up for
a molecular dynamics simulation of a copper neck sliding over a Cu(111) surface. **d** Resulting lateral
force versus scanned distance. Reprinted from [616], https://link.aps.org/doi/10.1103/PhysRevB.
53.2101, with permission from AIP Publishing

until a stable configuration is reached. This configuration is associated with a lower total energy and, in the further process of scanning, provides an atomic-scale stick–slip behaviour. The rearrangement of the contact for stable friction processes has been named self-lubrication [398].

Atomic-scale stick–slip was also found on a metallic surface. The results of an experiment on a copper surface and a corresponding molecular dynamics simulation are presented in Fig. 3.20. By comparing experiment and simulation, it was possible to obtain a reliable description of the contact area, in this case, the formation of a copper neck between the tip and Cu(111) surface with a commensurate contact of about 25 atoms [63].

The preceding examples involve experiments performed on well-defined surfaces under vacuum conditions. By using low normal forces, any wear processes were avoided. However, when the normal force is carefully increased so that the onset of abrasive wear is initiated, atomic processes may still be observed. Gnecco et al. have shown that the debris of a scratch in a KBr surface exhibits perfect crystalline ordering by reducing the normal force after scratching for imaging [220].

In ambient atmosphere experiments, adsorption and desorption of contaminants by the action of the probing tip can lead to atomic-scale features in the lateral forces. Therefore, the observed friction might not be purely wearless. In experiments with surface force apparatus, shear forces are found to show characteristic oscillations as a function of the thickness of the liquid layer between curved mica sheets [306]. The confinement of the liquid causes a solid-like behavior. Analogously, confined molecules at the interface of the tip and sample may lead to shear forces that vary on an atomic scale.

3.4.5 Lateral Contact Stiffness

One of the central questions of contact mode force microscopy is the determination of the contact area between tip and sample. Imaging of well-defined structures is one way to estimate the size of the contact. Alternatively, continuum elasticity models can be used to describe the nanometer-sized contact. The lateral contact stiffness method is particularly interesting, because it is rather independent of the selected model of the contact [99].

The slope of lateral force versus distance during sticking (thick bar in Fig. 3.17) represents the effective stiffness k_{eff} of the experiment. It is composed of the torsional stiffness of the cantilever k_T, the bending stiffness of the tip k_{tip}, and the lateral stiffness of the contact k_{con}:

$$\frac{1}{k_{eff}} = \frac{1}{k_T} + \frac{1}{k_{tip}} + \frac{1}{k_{con}}. \tag{3.18}$$

The stiffness of a typical silicon tip has been calculated to be around $k_{tip} = 84 \, \text{N/m}$ [381]. The data presented in Fig. 3.18 have been recorded using a cantilever with $k_T = 35.5 \, \text{N/m}$. From the effective stiffness $k_{eff} = 10.5 \, \text{N/m}$, a contact stiffness of

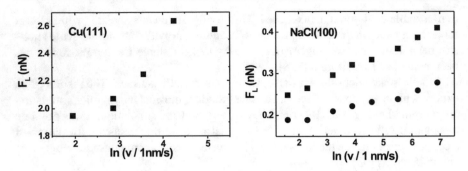

Fig. 3.21 Velocity dependence of the mean lateral force F_L in atomic stick–slip experiments on Cu(111) and NaCl(100). For the case of NaCl(100), *squares* and *circles* denote external loads of 0.65 nN and 0.44 nN, respectively

$k_{con} = 18$ N/m is determined. Now, the contact radius a can be estimated to be

$$a = \frac{k_{con}}{8G^*} , \qquad (3.19)$$

where the effective shear modulus is given by

$$\frac{1}{G^*} = \frac{2 - \nu_1^2}{G_1} + \frac{2 - \nu_2^2}{G_2} . \qquad (3.20)$$

G_1 and G_2 are the shear moduli of the sample and tip, and ν_1 and ν_2 are the Poisson ratios [126,318]. The Young's modulus E_1, shear modulus G_1, and Poisson ratio ν_1 of the sample can be calculated from the elastic coefficients [452], which can be found for a variety of materials in [263]. In combination with the shear modulus of silicon $G_2 = 6.8 \times 10^{10}$ N/m^2 and its Poisson ratio $\nu_2 = 0.22$, a contact radius of $a = 0.42$ nm is found. While continuum mechanics provides a valid approach to describe the friction of contact areas as large as tens of nanometers, atomic friction processes need atomistic modelling for a detailed understanding.

3.4.6 Velocity Dependence of Atomic Friction

The laws of macroscopic friction state that friction is independent of the sliding velocity. This law is approximately valid for unlubricated contacts of rather hard metals, such as titanium or steel, while soft materials exhibit a non-monotonic velocity dependence [542]. FFM on the micron scale has revealed no velocity dependence on carbon surfaces [733], but a logarithmic increase of the lateral force for polymers grafted on silicon [92].

The latter dependence is also found in experiments where atomic-scale stick–slip behaviour is observed. Mean lateral forces versus scan velocity are plotted in Fig. 3.21 for NaCl(100) and Cu(111). These results can be explained by including the effects of thermal activation in a one-dimensional Tomlinson model. One obtains the following dependence of the lateral force F_L on velocity v and temperature T just before a slip process (for the derivation see [219]):

$$F_L(v, T) = F_L^0 + \frac{k_B T}{\lambda} \ln\left(\frac{v/a}{f_0} \frac{\lambda k_{eff} a}{k_B T}\right) ,$$ (3.21)

where a is the lattice constant of the surface in the scan direction and λ is a length parameter related to the distance involved in the slip. The physical origin of the velocity dependence can be described as follows. The slower the scanning is done, the higher is the probability of an early thermally activated jump. Therefore, the friction force increases with increasing scanning velocity. The logarithmic dependence in (3.21) consists of two factors. The first is the ratio of the frequency of stick–slip events v/a and a characteristic frequency f_0. The second factor is the ratio of the energy barrier $\lambda k_{eff} a$ overcome in the slip process and the thermal energy $k_B T$. The increase in lateral force is, of course, limited to low scan velocities, and the maximum friction is the same as that expected at very low temperatures without thermal activation.

The theory of the velocity dependence has been more thoroughly derived and experiments have been simulated by Sang et al. [572]. Lateral force measurements in ambient atmosphere have revealed that the velocity dependence of friction on hydrophilic surfaces is governed by the capillary condensation process [552].

3.4.7   Temperature Dependence of Atomic Friction

The temperature dependence of atomic friction has been addressed in a number of experiments, where a non-monotonic temperature dependence was observed. Experiments of silicon tips on silicon show a peak-like enhancement of friction around 100 K [577]. These results were explained by a collective effect, where muliple asperities contribute to atomic friction [50]. On MoS_2 an exponential increase was observed with decreasing temperature until 220 K and a transition to athermal friction below this temperature [730]. Jansen et al. [309] have investigated temperature dependent atomic friction on graphite. As seen in Fig. 3.22 the representative friction loop shows that atomic friction increases with decreasing temperature. At all investigated temperatures friction loops were acquired and the distribution of maximum lateral forces to induce slip events were analyzed by histograms. The most frequent lateral forces, at which slip occurs, were plotted against velocity in Fig. 3.22c, where a logarithmic increase with temperature is observed as discussed in the previous section. In Fig. 3.22d the temperature dependent data are plotted $(F_c - F)^{3/2}/T$ versus $ln(f_0 T/v)$ and all data are found to collapse on one curve. This is in agreement with the model from Sang et al. [572], which corresponds to the more accurate form of the Prandtl–Tomlinson model. In this case the dependence of the frictional force versus temperature is described by

$$F_L(v, T) = F_c - \left(\beta k_B T ln(\frac{v_c}{v})\right)^{2/3}$$ (3.22)

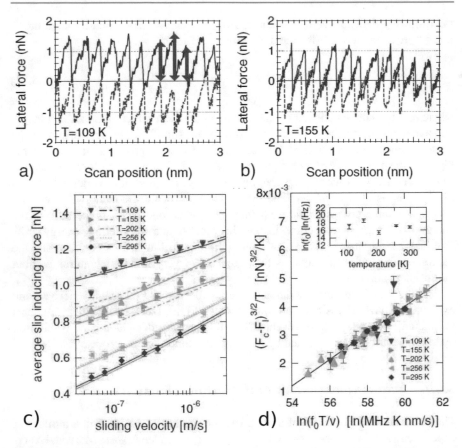

a) Scan position (nm)

b) Scan position (nm)

c) sliding velocity [m/s]

d) $ln(f_0 T/v)$ [ln(MHz K nm/s)]

Fig. 3.22 Temperature dependence of atomic friction on highly oriented pyrolitic graphite. **a** Friction force loop at 109K. **b** Friction force loop at 155K. **c** Temperature dependence of atomic friction as a function of velocity at different temperatures. **d** $(F_c - F)^{3/2}/T$ versus $ln(f_0 T/v)$ with inset $ln(f_0)$ versus T. Reprinted from [309], https://link.aps.org/doi/10.1103/PhysRevLett.104.256101, with permission from AIP Publishing

and F_c is maximum lateral force at $T = 0 \, K$ before sliding occurs. The parameter

$$v_c = (2f_0 \beta k_B T)/(3c_{eff}\sqrt{F_c}) \tag{3.23}$$

is the critical velocity, f_0 is the attempt frequency and β is a parameter to describe the curvature of the potential energy corrugation. For a sinusoidal shape it is $\beta = (3\pi\sqrt{F_c})/(2\sqrt{2}a)$. Jansen et al. find that β is increased by a factor of two, which shows that the potential landscape is non-sinusoidal. Second and more important, they observe that the attempt frequency f_0 depends on temperature (see inset of Fig. 3.22d). This is an aspect so far not expected, because f_0 was often related to the mechanical resonances of the cantilever. Therefore, some other mechanisms seem to affect the attempt frequency at different temperatures. Nonetheless, the careful work from Jansen et al. has shown that in the case of single asperity contacts, where

Fig. 3.23 Example of vertical pulling of a polymeric chain to be separated from a Au(111) surfaces. Oscillatory forces are related to the detachment of subunits from the surface. In addition, lateral forces of the remaining part of the chain on the surface are observed

atomic stick-slip is observed at all investigated temperatures, the data agree well with the Prandtl–Tomlinson model, which states that the atomic friction is reduced by thermal excitation.

3.4.8   Molecular Friction

Friction force microscopy is focussed on the dissipative interactions between the sharp probing tip and a sample surface operated in repulsive mode. An interesting extension is to attach a molecule to the end of the probing tip and to measure the interaction with the surface with this molecule attached to the tip. The tip can either be moved vertically to the surface, called vertical pulling experiment, or laterally, called lateral pulling experiment. The first case is shown in Fig. 3.23, where a polymeric chain, consisting of fluorene monomers, is pulled away from a gold surface [337]. During the manipulation the frequency shift of the cantilever is measured, showing a number of oscillations, which corresponds to the number of molecular subunits. The spacing between these maxima is approximately the length of one of the subunits (0.84 nm). These data were compared to a minimalistic model, which provided quantitative estimates of the adhesive energy to detach one subunit (0.27 eV), but also indicated that the part of the chain remaining in contact with surface is sliding, which gives raise to some additional features in the lateral forces. Therefore, the vertical pulling technique gives valuable information about the energetics of motion

Fig. 3.24 Graphene nanoribbons are pulled across a Au(111) surface. The periodic oscillations are related to the atomic spacing of gold. Beating patterns are related to the commensurability of the contact.

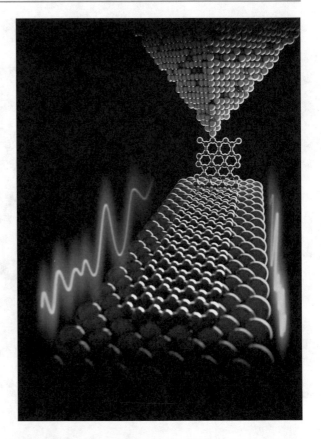

and detachment of molecular chains. As a second example, Fig. 3.24 represents a graphene nanoribbon sliding on a gold surface [338]. A nearly periodic force pattern related to the atomic spacings between gold atoms is observed. In addition, a beating pattern is observed, which was related to the motion of the graphene nanoribbon across the hearingbone reconstruction. These results clearly demonstrate that the observed friction is related to the commensurability of the contact formed. Typical forces are on the order of a few tens to hundred pico-newtons. These forces are nearly independent of the length of the ribbon, but are dominated by the edge forces. An essential ingredient for such low lateral forces is the high elastic modulus of graphene, which makes the contact rigid and suppresses the formation of conformational contacts, which is essential for structural superlubricity.

3.5   Dynamic Force Microscopy

Dynamic force microscopy (DFM), often also called non-contact AFM, is currently the only operation mode providing true atomic resolution and images of a quality comparable to STM. Progress in experimental and theoretical work in this field is

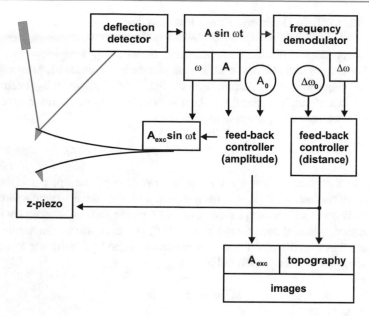

Fig. 3.25 Scheme of the instrumentation for dynamic force microscopy in the constant amplitude mode

documented in the proceedings of a series of workshops [65,469,590], and in a monograph [470].

In this mode, the cantilever is excited by a piezoactuator to oscillate at its mechanical eigenfrequency. This is done by a phase-locked loop: the electrical signal, which is proportional to the cantilever oscillation, is applied to the piezoactuator after amplification and phase shifting. This self-exciting oscillation changes its frequency due to tip–sample interactions, as schematically depicted in Fig. 3.9. In order to track the frequency shift, fast frequency demodulation schemes are employed, as first introduced by Albrecht et al. [23]. The frequency shift signal is used to control the tip–sample distance.

The oscillation amplitude should be large enough to ensure that the restoring force at the lower turning point is larger than the attractive force between tip and sample. This will avoid an instability, which would stop or at least seriously distort the oscillation. Two strategies are available for controlling oscillation amplitudes. The first is to use a constant excitation amplitude, resulting in a decrease in the oscillation amplitude upon damping close to the surface. Consequently, there is cross-talk between damping phenomena and topography signal. The second strategy is to employ a further control circuit, which maintains a constant amplitude in all experimental situations. With the latter system, stable operation without tip crashes is more difficult to obtain. However, it has the advantage that a quantitative analysis of forces becomes possible. A schematic setup for DFM is shown in Fig. 3.25.

3.5.1 Modelling Dynamic Force Microscopy

In this section we will discuss the relationship between tip–sample forces and frequency shifts as recorded in the constant amplitude mode of DFM. The deflection of an oscillating cantilever is derived from the theory of elasticity. It has been shown how the system of cantilever and tip can be described as a damped harmonic oscillator [545]. Here, we use the equation of motion

$$m\ddot{z} = -k\big[z - A_{\text{exc}} \cos{(\omega t + \varphi)}\big] - \gamma\dot{z} + F(z) , \qquad (3.24)$$

where m is the effective mass, z the vertical position of the tip, γ the damping coefficient of the internal friction of the material, and $F(z)$ the tip–sample force. The excitation is done with an amplitude A_{exc} and a phase shift φ between oscillation and excitation, which should always be $\varphi = 90°$. For the constant amplitude mode, we assume that the frictional force $-\gamma\dot{z}$ is compensated by the driving force $F_0 = kA_{\text{exc}} \cos(\omega t + \varphi)$. Then, (3.24) reduces to

$$m\ddot{z} = -kz + F(z) . \qquad (3.25)$$

For small amplitudes, a linear expansion of the force can be carried out and a frequency shift proportional to the force gradient is found:

$$\frac{\Delta f}{f} = -\frac{1}{2k}\frac{\partial F}{\partial z} . \qquad (3.26)$$

Note that this approximation holds only for amplitudes that are very small compared to a characteristic decay length of the force. Usually, the opposite situation is found in DFM. The interaction between tip and sample disturbs the harmonic potential of the cantilever spring only close to the lower turning point of the tip. Therefore, an appropriate approximation is to assume that the trajectory of the tip is harmonic. Inserting $z = z_0 + A \sin \omega t$ into (3.25) and multiplying by $\sin \omega t$ yields

$$- mA\omega^2 \sin \omega t = -kA \sin \omega t - kz_0 + F(z_0 + A \sin \omega t) . \qquad (3.27)$$

Integration over one oscillation period results in

$$kA \left(1 - \frac{\omega}{\omega_0}\right) = \frac{1}{\pi} \int_0^{2\pi/\omega} \sin \omega t\, F(z_0 + A \sin \omega t)\mathrm{d}t . \qquad (3.28)$$

For small shifts in the frequency we obtain a relation between experimental parameters and the force averaged over the oscillation cycle:

$$\frac{\Delta f}{f} kA = \frac{1}{\pi} \int_0^{2\pi/\omega} \sin \omega t\, F(z_0 + A \sin \omega t)\mathrm{d}t . \qquad (3.29)$$

Fig. 3.26 Conversion of a Δf versus distance curve into a force versus distance curve according to the iterative procedure introduced by Dürig [154]. The data are recorded with a silicon tip approaching a Si(111)7×7 surface. The long-range contribution has been subtracted

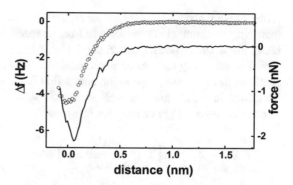

This result was first suggested by Giessibl in a mathematically more rigorous form [204]. The integral on the right-hand side of (3.29) can be evaluated for large oscillation amplitudes, assuming different force laws. If the distance between tip and sample at closest approach is smaller than the tip radius, the long-range interactions are dominated by the spherical cap of the tip. In this case one obtains simple expressions for the frequency shift Δf_V caused by electrostatic interaction and the frequency shift Δf_{VdW} caused by van der Waals interaction [241]:

$$\frac{\Delta f_V}{f_0}kA = -\frac{\pi\epsilon_0 R(V_{bias} - V_{cpd})^2}{(2\bar{s}A)^{0.5}}, \tag{3.30}$$

$$\frac{\Delta f_{VdW}}{f_0}kA = -\frac{HR}{6\bar{s}(2\bar{s}A)^{0.5}}, \tag{3.31}$$

where $\bar{s}$ denotes the separation between the sample and the mesoscopic part of the tip at closest approach, R is the radius of the tip, V_{bias} the sample bias voltage, V_{cpd} the contact potential difference between tip and sample, and H the Hamaker constant of the system. These formulas indicate a $A^{3/2}$ dependence of the frequency shift. Therefore, Giessibl introduced a reduced frequency shift γ_{freq},

$$\gamma_{freq} = \frac{\Delta f}{f}kA^{3/2}, \tag{3.32}$$

which is useful for comparing results recorded with different experimental parameters. The physical unit of γ is the geometrical mean of the units of force and potential, and indeed the frequency shift can be described as a geometrical mean of force and potential, as demonstrated by Ke et al. [342] and formally derived by Giessibl et al. [208].

Dürig has shown that the force versus distance curve $F(z)$ can be reconstructed from frequency shift versus distance curves without any assumption for the force law. In an iterative procedure, the calculated force curve and the measured frequency shift curve are brought into consistency [154]. The result of the procedure for the short-range forces between a silicon tip and a silicon surface is presented in Fig. 3.26. It has been shown that the force curve can also be recovered by a simpler matrix inversion

[206]. Several groups have solved the equation of motion numerically [12,399], even including the control circuit for a full representation of the experimental situation [224]. An accurate inversion formula has been introduced by Sader and Jarvis [565], which is relatively easy to be implemented and is found to be valid for large and small amplitudes compared to the interaction lengths. It is now often referred to as the Sader–Jarvis method and is the standard method to convert frequency shifts into interaction forces $F(z)$ or energies $U(z)$:

$$F(z) = 2k \int_z^\infty \left(1 + \frac{A^{1/2}}{8\sqrt{\pi(t-z)}} \right) \Omega(t) - \frac{A^{3/2}}{\sqrt{2(t-z)}} \frac{d\Omega(t)}{dt} dt, \qquad (3.33)$$

$$U(z) = 2k \int_z^\infty \Omega(t) \left((t-z) + \frac{A^{1/2}}{4} \sqrt{\frac{t-z}{\pi}} + \frac{A^{3/2}}{\sqrt{2(t-z)}} \right) dt, \qquad (3.34)$$

where $\Omega(z) = \Delta\omega(z)/\omega_{res}$. These formulas are not limited to the large amplitude case, but can be used for any amplitude A.

3.5.2 High-Resolution Imaging

The most important outcome of DFM is its ability for true atomic resolution. Such resolution is proven when point defects at surfaces and step and kink sites are imaged. True atomic resolution was first demonstrated by Giessibl for the Si(111)7×7 reconstructed surface [203]. This semiconductor surface with its open structure has evolved into a test system that allows one to understand the imaging mechanisms of DFM [241,384,410]. In a low-temperature experiment, even the rest atoms of the 7×7 reconstruction have been imaged (Fig. 3.27a) [384]. Morita and co-workers have studied the differences in contrast formation for the Ag- and Sb-covered Si(111)7×7 surface and interpreted the result in terms of the chemical force arising from the overlap of a dangling bond at the tip with the different types of bond found at these surfaces. Contrast formation due to chemical forces at the Si(111)7×7 surface has been discussed by several other authors [384,513]. A study of monatomic steps on the Si(111)7×7 surface revealed the role of long-range forces in high-resolution imaging. The increased long-range interaction on the lower terrace close to a step causes a lifting of the tip, while the reduced long-range force on the upper terrace makes the tip approach the surface in order to maintain the constant frequency shift (Fig. 3.27a). Therefore, measured step heights are distorted and may even be inverted [240].

Other semiconductor surfaces have been successfully studied, including GaAs, InP, TiO_2, InAs (Fig. 3.27c). The last study brought out the role of tip-induced charge rearrangement in atomic-scale contrast formation on the basis of differences in the contrast found for different tips [589].

While all of the above-mentioned semiconductor surfaces have been studied by STM before, surfaces of insulators can be imaged only by force microscopy. The first

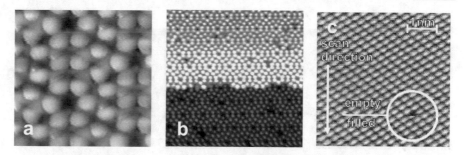

Fig. 3.27 High-resolution DFM images in the constant amplitude mode. **a** Si(111)7×7 surface imaged at low temperature. **b** Tunneling current recorded while imaging a step on a Si(111)7×7 surface at constant frequency shift. The increased current on the upper terrace reflects the reduced long-range forces, which cause a decreased tip–sample distance at constant frequency. **c** Single vacancy on a InAs surface. The vacancy is filled in the course of imaging, as it disappears after a few scanlines over the site. **a** Reprinted from [384], https://link.aps.org/doi/10.1103/PhysRevLett.84.2642, with permission from AIP Publishing. **b** Reprinted from [240], https://doi.org/10.1016/S0039-6028(00)00592-6, with permission from Elsevier. **c** Reprinted from [589], https://link.aps.org/doi/10.1103/PhysRevB.61.2837, with permission from AIP Publishing

image of the surface structure of an insulator including point defects was presented by Bammerlin et al. with the NaCl(100) surface [42,69]. Reichling and Barth presented images of the CaF$_2$(111) surface, where point defects could be resolved after dosing the surface with oxygen [549]. Rectangular monatomic pits at the KBr(100) surface created by electron irradiation have been resolved including step and kink sites (Fig. 3.28a) [66]. For all alkali halide surfaces it was found that only one type of ion is imaged by DFM, and that the atomic corrugation height depends on the relation between short-range and long-range forces [42,68]. These findings have been explained by the displacement of surface ions in the field of a charge at the tip end [399]. Atomic displacement is also assumed to explain the large corrugation of a frozen Xe layer, as found in a low-temperature DFM experiment [28,208]. Surface sites with lower coordination such as step and kink atoms are particularly subject to displacement (Fig. 3.28b). They thus appear as protrusions in DFM images and can cause instabilities in the cantilever oscillation [64,66]. For the important class of insulating oxides, atomically flat surfaces are difficult to prepare. Examples are the cleavage face of NiO and of Al$_2$O$_3$, which have been imaged with atomic resolution [53,286].

The role of forces has also been discussed for the atomic contrast in STM experiments on metal surfaces. Therefore, force microscopy with atomic resolution on metallic surfaces could contribute to their understanding. Measurements on thick aluminum films by Orisaka et al. [502] and on Cu(100) and Cu(111) surfaces by Loppacher et al. [404] yielded atomic corrugation comparable to what is found in STM experiments. The corrugation height is strongly dependent on the frequency shift used and thus on the force between tip and surface atoms.

The high resolution of DFM has attracted research groups working in the field of organic molecules. Although the functionality of organic molecules is changed in a vacuum environment, structural information can be obtained from DFM images.

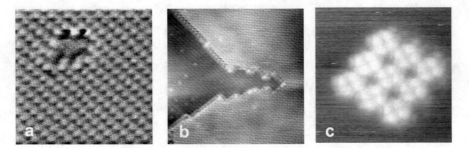

Fig. 3.28 High-resolution DFM images in the constant amplitude mode. **a** Initial stages of desorption on a KBr(001) surface after irradiation with low-energy electrons. **b** One monolayer of NaCl on a Cu(111) substrate. The step and kink sites appear elevated due to increased forces at these sites of lower coordination and due to a displacement of the ions. **c** A group of nine porphyrin molecules on a Cu(100) substrate. **a** Reprinted from [66], https://doi.org/10.1016/S0039-6028(00)01053-0, with permission from Elsevier. **b** Reprinted from [64], https://link.aps.org/doi/10.1103/PhysRevB. 62.2074, with permission from AIP Publishing. **c** Reprinted from [405], https://link.aps.org/doi/ 10.1103/PhysRevB.62.16944, with permission from AIP Publishing

Imaging of DNA strands in ultrahigh vacuum did not show enhanced resolution compared to studies in aqueous solution [415]. Self-assembled monolayers of organic molecules, however, allow molecular resolution and even differentiation of end groups of molecules [656]. The inner structure of porphyrin molecules on a Cu(100) surface (Fig. 3.28c) has been resolved by Loppacher et al., who were also able to determine the force exerted by the tip when this elevated molecule is pinned to the surface [405].

A tremendous improvement of the lateral resolution of force microscopy has been achieved by Gross et al. [231, 233], who decorated the tip with a single CO molecule. It is known that the CO points with the oxygen side towards the sample. Subsequently, pentacene molecules were imaged with submolecular resolution as seen in Fig. 3.29. Comparative studies with metallic tips or other attached atoms, such as chlorine or noble gases, show that CO gives the best resolution and is less reactive than the metallic tips. Therefore, it is possible to enter the repulsive regime without forming strong bonds, which would lead to unwanted manipulation of the molecules. The current understanding is that submolecular resolution of these relatively flat molecules is based on the repulsive forces, whereas van der Waals and electrostatic forces give the long-range background. The used microscope is based on a tuning fork AFM (Q-plus setup) [209, 210], which combines small amplitudes ($A = 10-100$ pm) and large spring constant (typically k = 1800 N/m). An important limitation is the restriction to low temperatures (liquid He temperatures), because the CO molecule desorbs or diffuses away from the tip apex at higher temperatures. It is also advantageous to use relatively sharp probing tips (cf. Fig. 3.7), where the long-range forces are reduced and the front-end atom does not have to counter-balance too large forces. The main atomic contrast is due to Pauli repulsion forces, but the CO molecule often tilts sideways, which leads to distortions of the intramolecular contrast. Nonetheless, the method is so sensitive that the bond order can be determined [232], where higher bond order leads to larger Pauli repulsion and to a smaller apparent bond length due

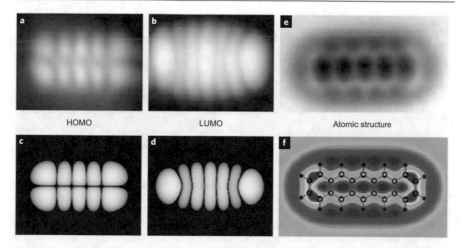

Fig. 3.29 a STM image of a pentacene molecule on a thin NaCl-film at a bias corresponding to the HOMO level. **b** STM image at a bias corresponding to the LUMO level. **c** and **d** Corresponding theoretical orbital images. **e** Submolecular resolution imaging of pentacene by a CO-decorated tip imaged at constant height. Frequency shift variations are plotted. **f** Chemical structure of the pentacene molecule. Reprinted from [233], https://doi.org/10.1038/nchem.1008, with permission from Springer Nature

to CO tilting. At intermediate distances (>500 pm) electrostatic forces have to be taken into account [163].

Force microscopy imaging with CO functionalization has become a valuable tool to characterize single molecules or molecular assemblies with submolecular resolution. Different stages of on-surface chemical reactions can be monitored in detail, where AFM is used to determine the structure, charge distribution and bond order. Examples are shown in Fig. 3.30, where molecular chains were assembled by Ullmann coupling and subsequently a graphene nanoribbon is formed by dehydrogenation. A second example shows two metal coordinated terpyridine molecules. The high resolution AFM images are often accompanied by STM imaging and DFT calculations to characterize the electronic structure of the molecules (see Fig. 3.29). Alternatives to CO functionalization are chlorine, xenon [339] or oxygen (CuO) [464]. So far, CO terminated tips yield the best resolution with least disturbance of the molecules under investigation. In contrast, metal terminated tips show instabilities when entering the Pauli repulsion regime. The flexibility of CO has the disadvantage that the appearance of the molecules is distorted, but has the advantage that the edges of the molecules are imaged in more detail compared to the more rigid tips. Deconvolution algorithms or machine learning algorithms can be used to determine the true structure of molecules or molecular assemblies [25,257,314].

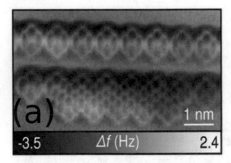 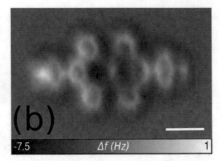

Fig. 3.30 Examples of submolecular resolution images of molecular assemblies with CO terminated tips. **a** Pyrenylene chains on Au(111) were assembled by on-surface chemistry (Ullmann coupling). The lower half shows a graphene nanoribbon formed by dehydrogenation reaction of two pyrenylene chains. **b** Two terpyridine molecules and a gold atom, forming a metal coordination complex. Such molecules, also called Swiss Nanodragster, were used for the first nanocar race held in 2017 at CEMES, Toulouse. Adapted from [510], distributed under the terms of the Creative Commons Attribution License (http://creativecommons.org/licenses/by/4.0, and [509] with permission from American Chemical Society

3.5.3 Spectroscopic Measurements

The most important spectroscopic measurements in DFM are frequency shift versus distance curves. The contribution of different types of force to the total force can be revealed by the differences in their distance dependence. When the electrostatic interaction is minimized by compensating the contact potential difference between tip and sample, the short-range contribution to the frequency shift can be separated by subtracting the van der Waals contribution, which is determined by a fit of the long-range part of the curve to (3.31) (see Fig. 3.31). The short-range part of the frequency shift can usually be described well by exponential force laws such as the Morse potential. The decay lengths found in room temperature experiments are larger than expected for a chemical force originating in the overlap of atomic wave functions [241,404,576]. The additional short-range force component can be ascribed both to contributions of electrostatic interactions, which may arise from the inhomogeneous contact potential, and to force contributions from the first few atomic layers of the tip. Alternatively, it has been proposed that the significant non-linearity of the distance dependence close to the surface can account for the large frequency shifts [19].

A different approach to recording frequency shift versus distance curves is to perform constant height scans at different heights, where even local variations can be measured. This method is restricted to low-temperature experiments due to the requirements on drift stability. In this way, Lantz et al. found a very small decay length for the force between single atoms at the Si(111)7×7 surface and a silicon tip [384].

Improvements in data acquisition and drift correction procedures, such as the atom tracking method [13,334], have made it possible that thousands of frequency shift versus distance curves, $\Delta f(z)$, curves can be acquired on a mesh of atomic dimensions. In this way, three-dimensional data sets $\Delta f(x, y, z)$ are acquired on time

Fig. 3.31 Analysis of a Δf versus distance curve. The long-range part of the recorded frequency shift (*open circles*) is fitted to (3.31). This contribution is subtracted from the whole curve (*open squares*). The remaining chemical force can be fitted by the frequency shift expected for a Morse potential

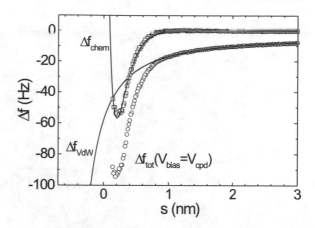

scales of several hours. Using integral equations, as described by the Sader–Jarvis method Sect. 3.5.1, one can derive an array of force values $F(x, y, z)$, and by further integration the potential energy $U(x, y, z)$. Atom-specific forces can be measured [21] and variations of the interactions at defect sites or element-specific contrast [629] can be investigated. Figure 3.32 represents of a 3D-AFM data set, where the atomic-scale variations of chemical forces above an oxidized copper surface are revealed. The strongest attraction is observed on the oxygen sites with local variations at defects due to lateral shifts of Cu atoms [54].

Lateral forces $F_{lat}(x, y, z) = -\frac{\partial U}{\partial x}$ can be calculated by differentiation of the potential energy. With this method it is even possible to measure the force to manipulate a single atom or molecule along a surface [639]. In this case, the forces are increased until a discontinuity is observed, at which the adatom starts to move.

3.5.4   Kelvin Probe Microscopy

The electrostatic force between tip and sample depends quadratically on the difference between the sample bias voltage V_{bias} and the contact potential difference V_{cpd} between tip and sample (3.30). The latter can be determined by recording the frequency shift Δf as a function of the sample bias V_{bias} (see Fig. 3.33). A more elegant method which allows a continuous mapping of V_{cpd} is the the so-called Kelvin probe force microscopy. In this technique, the bias voltage is modulated by a small AC voltage: $V_{bias} = V_{DC} + V_{AC} \sin \omega t$. If the modulation frequency ω is lower than the bandwidth of the frequency demodulator, the frequency shift Δf exhibits two harmonic components:

$$\Delta f_\omega \propto (V_{DC} - V_{cpd}) V_{AC} \sin \omega t , \tag{3.35}$$

$$\Delta f_{2\omega} \propto V_{AC}^2 \cos 2\omega t . \tag{3.36}$$

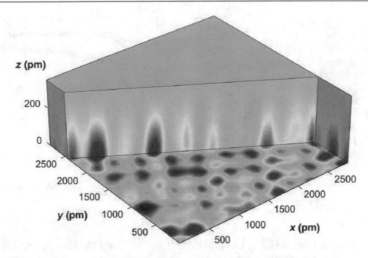

Fig. 3.32 3d-AFM data of a Cu(100)-O surface, where the color scale ranges from -12 to $12\,\mathrm{pN}$. The average force is subtracted. Variations due to chemical forces are shown. Reprinted from [54], https://link.aps.org/doi/10.1103/PhysRevB.87.155414, with permission from AIP Publishing

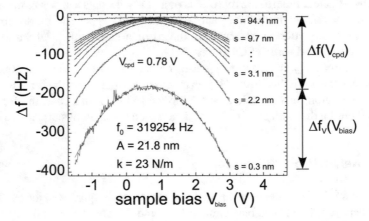

Fig. 3.33 Frequency shift versus bias voltage V_S for increasing distances s between tip and surface. The contact potential difference V_C does not show a significant distance dependence. Reprinted from [241], https://link.aps.org/doi/10.1103/PhysRevB.61.11151, with permission from AIP Publishing

Recording Δf_ω by means of a lock-in amplifier while scanning the surface yields local contrast of $V_{DC} - V_{cpd}$, i.e., local contrast of the sample contact potential. In a feedback circuit, V_{DC} can be adjusted to maintain $\Delta f_\omega = 0$. With this method, maps of V_{DC} represent the local contact potential difference with the tip independently of the actual measurement parameters [489,685]. In order to obtain absolute values for the local contact potential, well-defined tips have to be prepared and their contact potential calibrated. In ultrahigh vacuum, silicon tips cleaned by argon ion sputtering exhibit a reproducible work function of $4.7\,\mathrm{eV}$ [615].

The modulation frequency ω is usually chosen small compared to the bandwidth of the FM demodulator but large compared to the bandwidth of the height feedback.

Kelvin probe force microscopy has been used to profile a variety of surfaces such as metals, semiconductor devices, conducting polymers, Langmuir–Blodgett films, and self-assembled monolayers. When considering the lateral resolution of Kelvin probe force microscopy, one has to keep in mind the long-range character of the electrostatic interaction. Stray fields of the whole tip and of the cantilever can contribute to the total interaction [280]. On the other hand, the contact potential has been found to vary even on the atomic scale [355]. An alternative scheme of Kelvin probe force microscopy uses the eigenfrequency of the second mode of the cantilever oscillation as modulation frequency. The electrostatic interaction is detected by measuring the amplitude of this oscillation mode by means of a lock-in amplifier [348]. The high modulation frequency allows shorter integration times of the lock-in amplifier without loss of precision, and energy resolution as good as 5 meV has been obtained [615].

3.5.5   Dissipation Force Microscopy

Most of the forces acting between tip and surface are conservative forces. However, when part of the work done by these forces is converted into heat, the damping of the cantilever oscillation can be exploited to measure dissipation. The first experimental demonstration was reported by Denk and Pohl, who analyzed the resonance of an oscillating cantilever with a metallic tip in electrostatic interaction with a heterostructured semiconducting sample [146]. In this experiment, the cantilever oscillation was damped by Joule dissipation of charge carriers which were moved by the oscillating electric field produced by the tip vibration. The authors pointed out that the damping deduced from the resonance analysis could also be obtained from the excitation amplitude A_{exc} needed to maintain a constant oscillation amplitude. Lüthi et al. were the first to present data for atomic-scale variation of dissipation in a non-contact DFM on a Si(111)7×7 surface, where the strongest damping was found at the sites of the corner holes [412].

In all DFM measurements the power dissipation P_0 caused by internal friction in the freely oscillating cantilever is given by

$$P_0 = 2\pi f_0 \frac{kA^2/2}{Q} , \qquad (3.37)$$

with f_0 the eigenfrequency of the freely oscillating cantilever, k its spring constant, and Q the quality factor of the oscillation. This dissipation is independent of the sample and cannot be avoided. It produces a background signal in which variations in the dissipation have to be detected. Therefore, measurements with small oscillation amplitudes are highly desirable for dissipation force microscopy. The extra dissipation P_{ts} caused by tip–sample interaction can be calibrated by comparison with the intrinsic dissipation, once the Q-factor of the free oscillation is known.

There are different ways to determine the Q-factor of the cantilever far from the surface. One possibility is to record the amplitude spectrum of the thermal noise of

Fig. 3.34 Example of
a thermal spectrum, which
has been fitted by (3.38),
leading to the value
$Q = 142\,914$

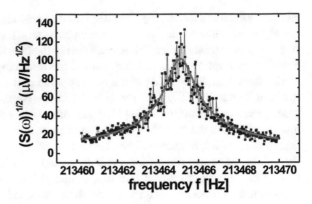

Fig. 3.35 Oscillation of the
cantilever as a function of
time. This is called the
ring-down method. The
curve has been fitted by
(3.39) and $Q = 142\,280$ has
been found, in good
agreement with the value
from the thermal spectrum

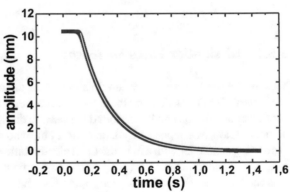

the cantilever. The form of this spectrum (see Fig. 3.34) is given by

$$S(\omega) = \frac{2k_{\mathrm{B}}T\omega_0^3}{Qk\left[(\omega^2 - \omega_0^2)^2 + \omega_0^2\omega^2/Q^2\right]} \,, \tag{3.38}$$

where $S(\omega)$ is the spectral amplitude density, $\omega = 2\pi f$ the radial frequency, and k
the spring constant.

An accurate method for cantilevers with high Q-factor is the ring-down method.
The amplitude A is measured as a function of time t after stopping the excitation.
An example is shown in Fig. 3.35. The curve is analyzed with the equation

$$A(t) = A(0)\mathrm{e}^{-\pi(f_0/Q)t} \,. \tag{3.39}$$

Finally, the Q-factor can also be determined in a phase variation experiment. The
frequency f and the excitation amplitude A_{exc} are recorded as a function of the phase
φ while the oscillation amplitude is kept constant. The advantage of this procedure is
that it can be applied with the tip in close proximity to the surface [403]. The relation
between frequency and phase can be derived from the equation of motion

$$\ddot{z} + \frac{\omega_0}{Q}\dot{z} + (\omega_0)^2(z - z_{\mathrm{exc}}) = 0 \,. \tag{3.40}$$

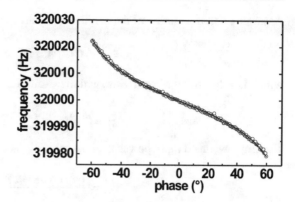

Fig. 3.36 Phase variation experiment, where (3.45) is used to determine the Q-factor

With the ansatz

$$z(t) = z_0 + A \cos(\omega t) , \tag{3.41}$$

$$z_{exc} = A_{exc} \cos(\omega t + \varphi) , \tag{3.42}$$

the excitation amplitude A_{exc} and the phase φ are related to the frequency by

$$A_{exc} = \frac{A_0}{f_0} \sqrt{(f^2 - f_0^2)^2 + \frac{f_0^2 f^2}{Q^2}} , \tag{3.43}$$

$$\varphi = \arctan\left(\frac{f^2 - f_0^2}{f_0 f} Q\right) , \tag{3.44}$$

which can be approximated for small frequency shifts ($\Delta f / f_0 \ll 1$) by

$$f = f_0 \left(1 - \frac{1}{2Q} \tan \varphi\right) . \tag{3.45}$$

Phase variation experiments, as shown in Fig. 3.36, are in good agreement with thermal noise spectra and ring-down experiments.

A simple way to interpret the local measurements of the excitation signal A_{exc} is to calculate the power dissipation. As suggested by Cleveland et al. and Gotsmann et al. [125, 224], the power P_{tip} dissipated by the interaction between tip and sample is given by the difference between the power which is delivered by the piezoactuator to the cantilever base P_{in} and the power which is used by the intrinsic damping of the cantilever (background dissipation) P_0:

$$P_{tip} = P_{in} - P_0 . \tag{3.46}$$

The power fed into the cantilever–tip system can be calculated from

$$P_{in} = k(z - z_{exc})\dot{z}_{exc} . \tag{3.47}$$

Assuming that the motion of the cantilever is harmonic,

$$z = A \cos(\omega t + \varphi) , \tag{3.48}$$

which is valid for DFM operation, and the excitation signal of the base is given by

$$z_{\text{exc}} = A_{\text{exc}} \cos \omega t . \tag{3.49}$$

The time-averaged input power is given by

$$\overline{P_{\text{in}}} = \frac{\omega k A A_{\text{exc}} \sin \varphi}{2} \tag{3.50}$$

and the internal damping of the cantilever is

$$\overline{P_0} = \gamma \dot{z}^2 = \frac{k\omega^2 A^2}{2\omega_0 Q_0} , \tag{3.51}$$

where $\gamma = k/(Q_0\omega_0)$ is the damping rate of the cantilever, ω is the resonance frequency at the separation of interest, ω_0 is the resonance frequency of the free cantilever, Q_0 is the Q-factor of the unperturbed cantilever and k is the spring constant. Finally, the power dissipated by the tip is given by

$$\overline{P_{\text{tip}}} = \frac{1}{2} \frac{k A^2 \omega}{Q_0} \left(\frac{Q_0 A_{\text{exc}} \sin \varphi}{A} - \frac{\omega}{\omega_0} \right) . \tag{3.52}$$

Since frequency shifts of the cantilever are relatively small, the term ω/ω_0 can be approximated by 1. Furthermore, the phase shift is given by $\varphi = 90°$ for dynamic force microscopy, which leads us to the equation

$$\overline{P_{\text{tip}}} = \frac{1}{2} \frac{k A^2 \omega}{Q_0} \left(\frac{Q_0 A_{\text{exc}}}{A} - 1 \right) = \frac{1}{2} \frac{k A^2 \omega}{Q_0} \left(\frac{A_{\text{exc}}}{A_{\text{exc},0}} - 1 \right) , \tag{3.53}$$

where $A_{\text{exc},0}$ denotes the excitation amplitude required to drive the oscillation with amplitude A far from the sample.

Typical power dissipation versus distance curves are presented in Fig. 3.37. These results demonstrate that the total dissipation is composed of a bias-dependent long-range contribution and a short-range contribution. The typical decay length of the short-range dissipation lies in-between the values for the tunneling current and for frequency shifts [402]. The maximum power loss $P_{\text{tip}} = 20 \times 10^{-15}$ W corresponds to a dissipation of 830 meV per oscillation cycle. Several contrast mechanisms can be observed in dissipation maps. In Fig. 3.38, there is a strong dissipation contrast between a Cu(111) substrate and surface areas covered by a NaCl monolayer, demonstrating that dissipation force microscopy can be a material-sensitive tool. Here, the contrast can probably be attributed to the variation in the contact potential. Sites with

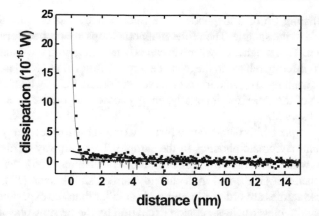

Fig. 3.37 Power dissipation versus distance curves for a silicon tip approaching a Cu(111) surface with two different bias voltages. *Squares* indicate a measurement with compensated contact potential, the in line for a potential difference of 0.9 V. The long-range part of the dissipation has been fitted with the function $\Delta P = c/(z - z_0)^2$, as indicated by the two *thick lines*

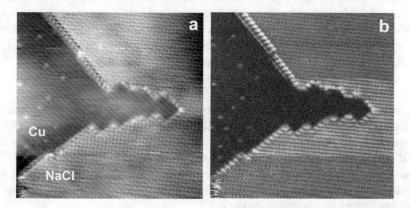

Fig. 3.38 Topography (**a**) and A_{exc} image (**b**) of a double layer of NaCl on Cu(111). The frame size is 18×18 nm^2. In the A_{exc} map, bright corresponds to an increased damping of the cantilever oscillation. Reprinted from [64], https://link.aps.org/doi/10.1103/PhysRevB.62.2074, with permission from AIP Publishing

low coordination like steps and kinks exhibit an enhanced dissipation. The dissipation map in Fig. 3.38b shows in the center a different contrast to the one found in the lower and upper parts of the frame. An atomic-scale reordering of the tip is suspected as being the cause of this shift in the dissipation contrast. Comparison with the topography image indicates the extreme sensitivity of the atomic-scale dissipation to the tip constitution.

Several mechanisms for the dissipation in DFM have been discussed. Localized charges induce mirror charges in the tip or in the sample, and these move during the oscillation cycle. Such moving charges are associated with dissipative currents via the resistance. A direct relationship between the dopant concentration and dissipation was found for semiconductor samples [146, 625].

The oscillating tip of a magnetic force microscope induces a local oscillating magnetic field at the sample. Therefore, magnetic loops are probed during approach and retraction. This magnetic hysteresis can be measured with the excitation signal A_{exc} which is directly related to the loss of energy. Grütter et al. [397] have observed that the main contrast in dissipation is observed at domain walls, which can be moved by the action of the tip. Experimentally, energy losses of about $\Delta E = 20$ meV per cycle were observed.

The sharp impact of the short-range forces acting at the lower turning point of the tip oscillation may create phonons in the sample. This pathway to dissipation has been studied within the framework of continuum mechanics [155], the fluctuation–dissipation theorem [198], and molecular dynamics simulations [12]. The values found in these studies are orders of magnitude smaller than the experimentally measured dissipation. Nevertheless, phonon excitation by the tip may play an important role for soft materials and force microscopy modes including repulsive contact formation.

The experimental results of dissipation force microscopy might also be influenced by non-linear effects [199]. A non-linear force law can cause a frequency spectrum with manifolds, where only one branch is accessible for the experiment. These asymmetric frequency spectra might mimic an increase in the Q-factor. The examination of frequency spectra could disentangle this effect from real dissipation effects. Experiments by Erlandsson [169] indicate the non-linear character of the resonance, while phase variation experiments by Loppacher et al. [403] are in accordance with the harmonic approximation.

Another possibility to explain the changes in the driving force at the resonance frequency is the excitation of higher oscillation modes by the short force pulse of the short-range forces. However, Pfeiffer et al. have shown that the higher bending modes are negligible with good accuracy during constant amplitude operation [518]. On the other hand, higher harmonics nf_0 may play a more important role. These harmonics are difficult to quantify due to their artificial appearance in the frequency spectrum caused by control electronics.

Tsukada et al. proposed that atomic-scale instabilities on the tip or on the sample may be important for understanding dissipation force microscopy [654]. In close analogy to the Tomlinson model in FFM, a second minimum of the potential is formed during the approach, where a tip or sample atom jumps out of its equilibrium position. During retraction the atom jumps back to the original position. This type of mechanism may explain the rather large energy losses without involving permanent changes to the tip–sample geometry. Kawai et al. [333] observed that atomic wires are formed between ionic crystal surfaces and the probing tip. The disruption of these atomic chains leads to hysteresis and energy loss and is intimately correlated with the average dissipation signal as measured by nc-AFM on ionic crystals.

Dissipation force microscopy is an operational mode with great relevance to atomic-scale studies of surfaces, particularly with respect to emerging nanomechanical technologies. Furthermore, the monotonic increase in dissipation with decreasing distance makes it a candidate for replacing the frequency shift as control parameter [198,311]. However, one has to keep in mind that frequency detection was invented for high Q-factor systems because of its fast response compared to amplitude detection schemes.

3.5.6   Non-contact Friction

To measure dissipative forces at relatively large separations (>1 nm) is of interest to understand the fundamentals of energy dissipation. Apart from the afore mentioned instabilities in the tip or sample, which lead to phononic dissipation, there are alternative ways to dissipate energy, e.g., electronic friction. The pendulum geometry is found to be more sensitive, because small spring constants can be used without risking jump-to-contacts. Up to a height of several hundreds of nm one can observe dissipation. Experiments on metals and insulators indicate damping coefficients of the order 10^{-13} to 10^{-9} kg/s [623]. A theoretical understanding was provided in the work of Volokitin and Persson [677], who derived a model based on the coupling between probing tip and sample via electromagnetic radiation, which excites phononic or electronic excitations. Charged surfaces or adsorbate-covered surfaces can enhance this mechanism and the model predicts the values in the same order of magnitude as the measured. Temperature dependent measurements on superconductors have revealed phononic and electronic contributions to non-contact friction Electronic contributions are reduced below the critical temperature, where electrons are bound as Cooper pairs [354]. An example of giant non-contact friction was found on the surface of $NbSe_2$, where charge density waves (CDW) are present. The local interaction of the tip with the CDW can create phase slips of the CDW, which then lead to hysteresis [380].

3.6   Tapping Mode Force Microscopy

3.6.1   Principles of Operation

As an alternative to frequency shift detection, the amplitude of the cantilever oscillation may serve as control parameter in force microscopy, as depicted in Fig. 3.40. A detailed description and analysis of this technique can be found in the review by Garcia and Perez [195]. The first realization of such a scheme was presented by Zhong et al. in 1993 [731]. A tip oscillation of 20–100 nm was excited close to the cantilever resonance frequency, and the root-mean-square value of the deflection was used as control parameter for the tip–sample distance. The amplitude is reduced due to an intermittent contact during each cycle and this scheme was therefore named tapping mode force microscopy. The advantage of the new mode is that lateral forces are greatly reduced compared to the contact mode, while the resolution is similarly limited only by the tip shape. Zhong et al. demonstrated these advantages by comparing contact and tapping mode measurements on an optical fiber surface containing residual polymer coating. Large oscillation amplitudes are a prerequisite for the tapping mode in air, in order to overcome attractive capillary forces by the restoring force of the cantilever spring.

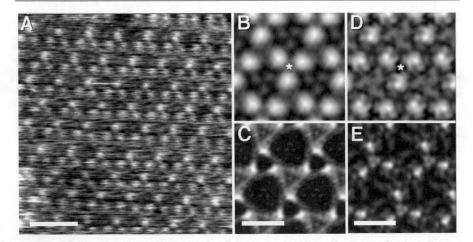

Fig. 3.39 Extracellular purple membrane surface imaged by tapping mode force microscopy. **a** Raw data. **b** Threefold symmetrized correlation average. **c** Standard deviation of the average. **d** and **e** Like **b** and **c** but for a contact mode measurement of the same surface. The correlation average allows one to average over all features showing the same regular structure. Tapping mode is demonstrated to be of comparable resolution to contact mode. *Scale bars* 10 nm (**a**) and 5 nm (**c**) and (**e**). Reprinted from [460], https://doi.org/10.1016/S0006-3495(99)76966-3, with permission from Elsevier

Shortly after its invention, Hansma et al. as well as Putman et al. reported tapping mode force microscopy of biological samples in liquids [253, 537]. Hansma et al. found a decreased frictional dragging of DNA molecules and decreased tip contamination. Putman et al. studied the complicated frequency response of the cantilever in liquid and measured amplitude versus distance curves. On the basis of their result, they discussed appropriate set points for the frequency and amplitude, and demonstrated that the tip does indeed form a contact with the surface at the lower turning point of the oscillation. It was also shown in similar experiments that for rather soft cantilevers and small vibration amplitudes, the tip does not tap the surface even in the amplitude feedback mode [138]. It should be mentioned that the frequency used for tapping mode force microscopy in liquids is not necessarily close to the eigenfrequency of the cantilever in air but may be chosen according to the acoustic modes formed in the liquid cell [537]. A practical test for discriminating cantilever oscillation modes from cell modes is to bring the tip into contact and to check the disappearance of the cantilever modes. Meanwhile, a multitude of biological systems have been successfully studied by tapping mode force microscopy. An example is shown in Fig. 3.39.

3.6.2 Phase Imaging

In the scheme presented in Fig. 3.40, the oscillation amplitude is detected with a lock-in amplifier. Therewith, additional information can be obtained from the phase shift between excitation and oscillation of the cantilever. In an early publication on such

Fig. 3.40 Instrumentation scheme for tapping mode force microscopy

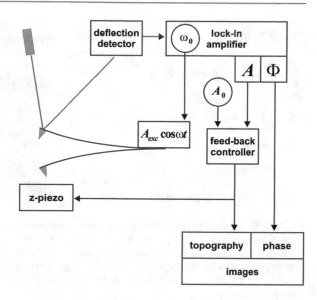

phase images, Magonov et al. derived a relation between the phase image and the stiffness of the surface, assuming that the interaction is dominated by a linearizable repulsive force [417]. Although this assumption can hardly be found in any experimental situation, the author proposed certain experimental parameters for stiffness imaging. According to their study of amplitude and phase versus frequency curves, large amplitudes and moderate set points would yield measurements of the stiffness, while very soft or very hard tapping would image adhesion or even result in contrast inversion.

Work done by Whangbo and Bar indicates that the use of the harmonic approximation to describe tapping mode experiments is limited to soft materials like elastomers and sharp tips. Amplitude versus frequency curves on hard materials like silicon and those recorded with dull tips exhibit truncated peaks in the frequency spectrum and hysteresis, which is characteristic for significant non-linearities [43,44,690]. The use of the harmonic approximation can be justified by verifying experimentally that the tip oscillation remains sinusoidal. In this case, the phase shift can be related to the tip–sample energy dissipation normalized to the internal cantilever dissipation [see (3.52)] as derived by Cleveland et al. [125] and experimentally confirmed by Bar et al. [46]. Generally, phase imaging can resolve material inhomogeneities at surfaces in great detail. However, quantification is often hampered by non-linearities, as discussed in the next section. Magonov et al. have reported a superior contrast in phase images compared to topographic images. However, the enhanced contrast of phase images may arise from the dependence of adhesion on the surface curvature [57].

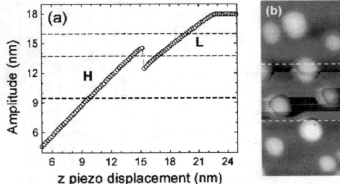
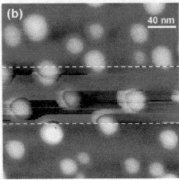

Fig. 3.41 Instabilities in tapping mode force microscopy due to non-linear tip–sample interactions. **a** The amplitude versus distance curve exhibits a jump between two stable oscillation states labeled L and H. **b** The imaging process becomes unstable for amplitude set points close to the value of the discontinuity in (**a**). Imaging with larger or smaller amplitude results in stable oscillations. Reprinted from [197], https://link.aps.org/doi/10.1103/PhysRevB.61.R13381, with permission from AIP Publishing

3.6.3 Non-linear Effects

The distance dependence of the tip–sample force is highly non-linear, particularly when the tip comes into repulsive contact with the sample. It is a basic result of mechanics that, for such non-linear systems, more than one state of oscillation can be stable. The amplitude detected in tapping mode force microscopy can show abrupt jumps for certain sets of experimental parameters. Consequently, artifacts like sudden height jumps or changes in the overall contrast appear in topography images. Such topography artifacts can be systematically correlated to a hysteretic behavior found in amplitude versus distance or amplitude versus frequency curves [45, 197, 372]. Numerical solutions of the equation of motion clearly reproduce the hysteresis in distance curves and frequency sweeps [110, 197, 431]. Even the instabilities during imaging can be modelled in simulations by assuming that experimental noise causes switches between different stable states of oscillation [431]. In Fig. 3.41, the correspondence between discontinuities in distance curves and instabilities in the imaging is demonstrated for tapping mode experiments on a structured semiconductor surface.

From an experimental standpoint, it appears desirable to control the occurrence of bistable oscillations. Experimental and theoretical results indicate that, in situations where either attractive or repulsive forces dominate, the imaging process exhibits a minimum risk of bistabilities [45, 110, 197]. For example, small oscillation amplitudes are more likely to result in a purely attractive interaction without repulsive contact, while for large amplitudes, the repulsion during contact dominates [490]. Determination of the phase shift can help to identify the respective interaction regime. Conditions that produce a phase shift above 90° are characteristic of an attractive regime, while phase shifts below 90° indicate a repulsive regime [196]. These findings allow one to formulate certain recipes for avoiding bistability conditions [110].

For a strong potential of the cantilever spring, accomplished by large amplitudes or the use of stiff cantilevers, the tip–sample interaction is a weak perturbation and the oscillation will be monostable. Furthermore, adjusting the driving frequency exactly to the resonance can help to minimize bistabilities, which arise from distortion of the resonance curve.

The best contrast in phase images is found for situations where attractive and repulsive interactions contribute similarly to the total interaction. Even slight changes in adhesion or stiffness will greatly affect the oscillation parameters. Unfortunately, this is exactly the regime of bistabilities, resulting in sudden changes in the phase contrast and a complicated quantification.

3.7   Further Modes of Force Microscopy

There have been numerous developments of other modes in force microscopy, which often meet the requirements of special sample characteristics like softness or strong adhesion. At the end of this chapter we would like to list some of these developments.

One of the oldest modes of force microscopy is scanning at constant height. Using this mode and static force detection, Ohnesorge and Binnig obtained the first force microscopy image with atomic resolution on calcite [499]. Tip and sample had been immersed in water in order to reduce forces and avoid a jump-to-contact. The experimental challenges of the constant height mode are thermal drift, piezocreep, and compensation for surface tilt. Its benefits are the redundancy of a controller and a direct comparison with most theoretical approaches. The negligible drift in low-temperature force microscopes has made it possible to image chemical forces with a height precision of one tenth of a nanometer [384]. Scanning at constant height plays an important role in magnetic force microscopy.

A very elegant method for studying the nature of tip–sample forces is the analysis of the thermal noise of the cantilever deflection. According to Boltzmann statistics, the logarithm of the probability for a certain cantilever bending should be proportional to its corresponding potential energy. With this mode of operation, solvation forces in water [124] and the force between the tip and soft polymer brushes [201] have been determined. The resonance analysis can also be done by exciting the cantilever oscillation, either with a sweeping frequency [167] or with white noise [363].

The nature of tip–sample forces is revealed in the distance dependence of the force. Therefore, the measurement of force versus distance curves is an important mode in force microscopy, as has been demonstrated for DFM in Sect. 3.5.3. The static detection of force curves in air or vacuum is hampered by the jump-to-contact instability. On the other hand, the pull-off force can be recorded to study the adhesion of the sample. The slope of the force versus distance curve during tip–sample contact is a measure of the sample compliance. These data can be collected even during scanning when using an electronic setup to record distance curves with high repetition rate (pulsed force mode) [370].

In the static mode, the recorded deflection versus distance curves have to be analyzed with respect to the cantilever spring constant and the difference between

Fig. 3.42 Schematic force spectroscopy setup for molecular recognition with a typical force versus retraction curve. Antibody molecules at the surface can link to antigen molecules attached to the tip. The strength of the molecular link can be determined from the rip-off force in the retraction curve and used to identify the binding partners. Courtesy of M. Hegner

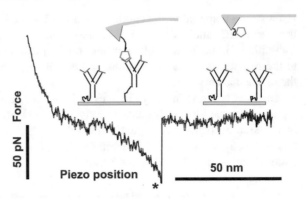

cantilever position and tip position. To overcome this problem and to avoid a jump-to-contact, instruments with force feedback have been designed, where an external adjustable force prevents the cantilever from bending. This can be done with a magnetic field acting on a small magnet on the cantilever [310] or using electrostatic forces [168].

In liquids, the tip–sample forces are small enough to prevent jump-to-contact by using stiff cantilevers. There is an emerging field of applications for force versus distance curves in biophysics. If one manages to attach biomolecules of interest to the sample surface and to the tip, interactions between single molecules can be determined directly [165]. A schematic setup for studying molecular binding is shown in Fig. 3.42. Furthermore, conformational changes of large molecular aggregates like DNA can be studied by recording the force while stretching the molecule [95]. It is important to note that the result of a force versus distance curve can depend on the velocity of tip retraction [170]. The average force for breaking a molecular bond appears lower in slow experiments due to the finite probability of a thermally activated process. This finding can be directly compared to the explanation of the velocity dependence of atomic friction described in Sect. 3.4.6.

3.8   Pulsed Force Mode

The pulsed force mode is a non-resonant intermittent contact mode, where a vertical distance modulation is applied to the sample [367,369]. The cantilever response is acquired with high speed data acquisition, where a full force distance curve is measured for every cycle. A snap shot of two cycles is shown in Fig. 3.43, where the snap into contact is observed at the beginning of the cycle. Then, the maximum force is related to the topographic height and the maximum adhesive force is observed on the retraction, followed by a ring-down of the free cantilever. The sample stiffness can be determined from the slope of the curve in the repulsive part. A number of physical parameters, such as adhesion, topography and sample rigidity can be determined from these force distance curves. In combination with a suitable elasticity model it is possible to get quantitative values of the sample elasticity and adhesion. The method is suitable for soft materials, such as biological materials or polymers.

Fig. 3.43 The pulsed force mode is an intermittent contact mode, where a sinusoidal modulation is applied to the sample. Fast data acquisition is used to measure the time response of the cantilever. A full force distance curve is acquired for every cycle. Reprinted from [367], https://doi.org/10. 1016/j.tsf.2003.11.254, with permission from Elsevier

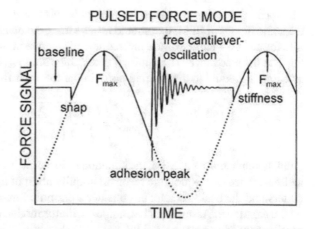

3.9   Force Resolution and Thermal Noise

A major source of noise in force microscopy is the thermal noise of the cantilever. In this section we suggest methods for estimating the force resolution of SFM at the thermal noise limit. The mean deflection of a cantilever without any force acting on the tip is zero. However, the thermal noise causes a non-zero mean square deflection $\langle z^2 \rangle$. With the help of the equipartition theorem, the thermal energy $k_B T$ can be related to the mean square deflection $\langle z^2 \rangle$ of the cantilever, given that the cantilever deflection has only one degree of freedom:

$$\frac{1}{2}k\langle z^2 \rangle = \frac{1}{2}k_B T , \tag{3.54}$$

where k is the spring constant of the cantilever. Hence, the mean deflection is given by

$$\sqrt{\langle z^2 \rangle} = \sqrt{\frac{k_B T}{k}} . \tag{3.55}$$

It is important to note that the thermal noise is not white noise with equal contributions at all frequencies, but that the frequency spectrum of the thermal noise is dominated by peaks at the mechanical resonances of the cantilever. The quality factor Q of the cantilever determines the amplitude of thermal noise at the resonance. Many experiments record the cantilever deflection with a certain bandwidth $\Delta\omega$. As a consequence of the mechanical resonance structure, the mean square deflection of the cantilever within a small bandwidth $\Delta\omega$ is larger at the resonance ω_0 than in off-resonance measurements [545]:

$$\langle z^2 \rangle_{\omega=\omega_0} = \frac{2k_B T Q \Delta\omega}{\pi \omega_0 k} , \tag{3.56}$$

$$\langle z^2 \rangle_{\omega\neq\omega_0} = \frac{2k_B T \Delta\omega}{\pi \omega_0 k Q} . \tag{3.57}$$

The mean deflection translates into thermal noise in the force via $F = k\sqrt{\langle z^2 \rangle}$. In the static modes of SFM, one should always limit the bandwidth of the measurement to a frequency well below the mechanical resonance in order to avoid the enhanced noise at the resonance. The minimal detectable force $F_{\min}$ in static modes is defined as being just equal to the thermal noise in the force far from the resonance:

$$F_{\min}^{\text{static}} = \sqrt{\frac{2k_{\text{B}}Tk\Delta\omega}{\pi\omega_0 Q}} . \tag{3.58}$$

In dynamic modes of SFM, the situation is different. Measurements at the mechanical resonance can exploit the resonant amplification of a periodic force. In order to understand the basic idea, let us consider a periodic force with amplitude F_ω acting on the cantilever. Assuming a harmonic oscillator model, the amplitude $z_{\omega=\omega_0}$ of the cantilever at its resonance will be

$$z_{\omega=\omega_0} = Q\frac{F_\omega}{k} .$$

The minimal periodic force $F_{\min}^{\text{dynamic}}$ that causes a detectable cantilever oscillation is defined by $z_{\omega=\omega_0} = \sqrt{\langle z^2 \rangle}_{\omega=\omega_0}$, finally resulting in

$$F_{\min}^{\text{dynamic}} = \sqrt{\frac{2k_{\text{B}}Tk\Delta\omega}{\pi\omega_0 Q}} . \tag{3.59}$$

The minimal detectable force is the same in static and dynamic modes. However, in the static case, measurements are made off-resonance, and thermal noise is reduced by a high Q-factor. In the dynamic case, the resonant amplification of the periodic force by the factor Q surpasses the enhanced noise at the resonance and on-resonance experiments are favorable. It is important to recall that these results have been derived for the ideal case where thermal noise limits the force resolution. The noise in the cantilever deflection signal can easily be dominated by other contributions than thermal excitation. In particular, electronic noise can be several orders of magnitude stronger than the contribution of thermal noise to the measured signal far from the resonance. Consequently, minimal detectable forces in the static mode have to be determined after measurement of the actual noise spectrum by the method presented above [207].

In the following, we present an estimate for the minimal detectable force gradient in dynamic force microscopy. In this method, described in Sect. 3.5, the tip–sample force is detected as a frequency shift in the mechanical resonance. The uncertainty in a frequency measurement of an oscillation $z(t) = A\sin(\omega t)$ is proportional to the uncertainty in the amplitude $\sqrt{\langle z^2 \rangle}$ according to

$$\frac{\sqrt{\langle \Delta\omega^2 \rangle}}{\omega_0} = \frac{1}{2\pi}\frac{\sqrt{\langle z^2 \rangle}}{A} . \tag{3.60}$$

This relation can be derived via the slope of $z(t)$ at $t = 0$. The best approximation we can obtain by simple means is to relate the force gradient to the frequency shift by

$$\frac{\partial F}{\partial z} = 2k \frac{\Delta \omega}{\omega_0} .$$

The minimal detectable force gradient is obtained by combination with (3.57):

$$\frac{\partial F}{\partial z} = \frac{k}{\pi A} \sqrt{\frac{2k_B T \Delta \omega}{\pi \omega_0 k Q}} = \sqrt{\frac{2k_B T k \Delta \omega}{\pi^3 \omega_0 A^2 Q}} . \tag{3.61}$$

This simple derivation is close to formulas found by similar approaches [22,434].

In this section, we have modelled the cantilever as a single harmonic oscillator, where the actual deflection is measured. A real cantilever, however, has an infinite set of resonance frequencies corresponding to higher bending modes [541]. Furthermore, in a beam-deflection instrument, the bending of the cantilever is measured rather than the deflection of the tip. Although the measured bending is proportional to the tip deflection, the proportionality factor varies for higher bending modes. Two questions arise here: firstly, do higher modes increase the thermal noise with respect to (3.55), and secondly, might the minimal detectable force be lower when using the resonances of higher modes?

The individual, time-averaged deflection of the higher modes follows [96]

$$\sqrt{\langle x_i^2 \rangle} = \sqrt{\frac{12 k_B T}{k \alpha_i^4}} , \tag{3.62}$$

where α_i are constants determined by

$$\cos \alpha_i \cosh \alpha_i = -1 . \tag{3.63}$$

The first modes are $\alpha_1 = 1.8751$, $\alpha_2 = 4.6941$, $\alpha_3 = 7.8548$ and approximately for higher modes $\alpha_i = (i - 1/2)\pi$. The contributions of higher modes to the thermal noise decrease rapidly, and Butt and Jaschke demonstrated that a summation of the thermal noise of all higher modes results in the original (3.55) [96]. Their work takes into account the experimental situation of either an interferometric or a beam-deflection scheme. The difference is important when spring constants are calibrated by means of the thermal noise spectrum. The relation given in (3.55) is correct only for interferometric measurements, while the deflection determined by means of a beam-deflection scheme has to be divided by a factor of $\sqrt{4/3}$.

The minimal detectable periodic force for higher modes can be described in analogy to (3.59) as [546]

$$F_{\text{min},i}^{\text{dynamic}} = \sqrt{\frac{2k_B T k_i \Delta \omega}{\pi \omega_{0,i} Q_i}} , \tag{3.64}$$

where the k_i is the spring constant and $\omega_{i,0}$ the mechanical resonance of the respective mode. It can be shown that $k_i \propto \alpha_i^4$ increases much faster for higher modes than $\omega_{i,0} \propto \alpha_i^2$ [545]. Furthermore, experimental results indicate that Q_i decreases for higher modes [518,546], so that the minimal detectable force in higher modes is actually clearly increased and no sensitivity is gained.

3.10 High-Speed AFM

In comparison to other microscopes, such as electron microscopy or optical microscopy, scanning probe microscopy and in particular AFM is a slow technique, because every line has to be rastered by mechanical motion of the probing tip. The limiting factor is the resonance frequency of the cantilever. One way to overcome this limitation is to use the parallel operation of AFM, which has been demonstrated by Quate and coworkers [459]. The other approach is to increase the resonance frequency of the cantilever and to optimize all the components for high speed AFM. During the last decades, tremendous progress was made. One essential ingredient was the development of so called nanocantilevers, which are significantly reduced in dimensions compared to conventional AFM. E.g., a length of 5–10 μm is combined with a width of a few μm and thickness of about 100 nm. The corresponding resonance frequency is 1–10 MHz (cf. Fig. 3.44). In high-speed AFM, the deflection of these small cantilevers can be measured by dedicated optics, which allows to optimize the focus and align on the nanocantilever. The mechanical relaxation time

$$\tau = \frac{Q}{\pi f_1} \tag{3.65}$$

is determined by the Q-factor of the cantilever and the first resonance frequency f_1. With $f_1 = 1$ MHz and $Q = 2$ (typical for operation in fluids) we get 0.6 μs. Relaxation times of the order of some fractions of μs correspond to a maximum bandwidth of the feedback loop of the order of 100–500 kHz. The typical benchmark in high-speed AFM is the number of frames per seconds (fps). Early examples by the Hansma group showed DNA imaging with 0.6 fps [667]. The Ando group demonstrated the Brownian motion of myosin at 12.5 fps [32]. The Miles group showed ultrafast imaging with 1300 fps of collagen fibers [521]. However, this work was performed in constant height mode without feedback control. The main focus of more recent developments was on the high-speed imaging of fragile objects, such as living cells or loosly bound proteins [31]. For this purpose, high speed AFM with full feedback control, tapping mode operation and implementation of sharp probing tips were essential. Key developments are the construction of high speed scanners, which have high resonance frequencies and dedicated electronics to reduce oscillations [175]. The data acquistion and novel ways to excite the oscillations of the cantilevers in liquids [376] were other essential ingredients. One of the hallmarks in this field is the imaging of a myosin protein running along a actin filament (cf. Fig. 3.44). This motion is driven by ATP hydrolysis. The study is outstanding,

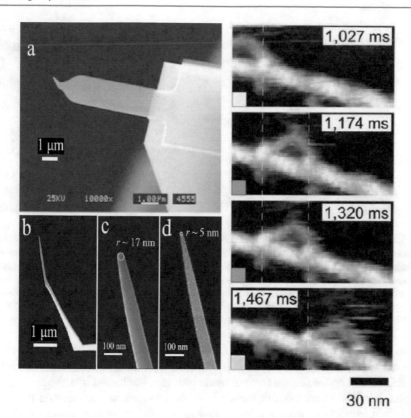

Fig. 3.44 Left side: **a** Microfabricated cantilever for high-speed AFM. The length is 6 μm, the width is 2 μm and the thickness is 90 nm, which gives a resonance frequency of 3.5 MHz in air and 1.7 MHz in water. **b** Electron beam deposited tips have a radius of 17 nm (**c**), but can be sharpened to radius of 5 nm by exposure to nitrogen or oxygen plasma (**d**). Right side: Sequence of high-speed AFM images: An individual myosin protein (M5-HMM) is walking along an actin filament as observed by high speed AFM. The motion is driven by ATP hydrolysis. Frame rate is 7 fps and scan area is 30×65 nm^2. Reprinted from [31], https://doi.org/10.1088/0957-4484/23/6/062001, with permission from IOP Publishing

because it directly visualizes a fundamental biological process, a motor protein in action [31]. The kinetics of antimicrobial activity was observed by high-speed AFM on live *Escherichia coli* bacteria, where disruption of the cells was observed after injection of the antimicrobial peptide CM15 [174].

3.11   Multifrequency AFM

So far, either the static deflection of the cantilever (contact AFM) or the amplitude/resonance frequency of the first flexural mode (tapping mode AFM/dynamic force microscopy or non-contact AFM) were described as input parameters for the feedback loop. In addition, it is possible to measure properties of other modes, such

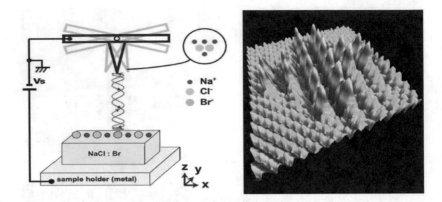

Fig. 3.45 Left side: Principle of bimodal AFM: The first flexural mode is excited with relatively large amplitude of about 1 nm, whereas the first torsional mode is excited with small amplitude. Right side: An image of the frequency shift of the torsional mode gives a clear contrast between the bromine ions and the chlorine ions. The bromine ions were manipulated by vertical manipulation beforehand. The experiments were performed at room temperature. The cross consists of 20 bromine ions and is found be stable on a time scale of hours to days. Reprinted from [336], https://doi.org/10.1038/ncomms5403, with permission from Springer Nature

as the second flexural mode or the first torsional mode. These techniques are called bimodal AFM. Often the z-feedback is operated with the frequency shift of the first flexural mode and the second flexural mode provides additional interaction. E.g., the amplitude of the second flexural mode is excited by voltage modulation for detection of electrostatic interactions. This signal can be fed into a second feed-back loop to compensate variations of the local contact potential and is called AM-KPFM. It is also possible to run the first mode with larger amplitude (typically 1 nm) and to excite the second mode with small amplitude. It was shown that the second mode is then proportional to the local force gradient [331,332], whereas the first mode is an average of the normal force, as described above by the Sader–Jarvis method. An alternative is to excite the first mode with relatively large amplitude (typically 1 nm) and excite the first torsional mode with small amplitude. The constant frequency shift images recorded with this bimodal AFM are not altered compared to conventional nc-AFM, but the information from the torsional signal is very valuable. An example is shown in Fig. 3.45, where single bromine ions in a matrix of NaCl(001) are observed with increased contrast [336]. The lateral resolution of the torsional signal is improved compared to conventional nc-AFM. To first approximation the torsional signal is related to the lateral force gradient [332].

There are a number of possible realizations for multifrequency AFM [194]:

Multiharmonic AFM is performed by the acquisition of the higher harmonics, while the topography is imaged in conventional dynamic AFM mode. The detection of higher modes in air or vacuum is challenging, because the signal levels of the higher harmonics are extremely low at moderate imaging forces. In liquid, the higher harmonics are easier to be detected, because the flexural eigenmodes have low Q-factors (Q of the order of one). The acquistion of several higher harmonics gives valuable information about the stiffness and viscoelastic damping of biological material. In vacuum, it was shown that the simultaneous acquisition of 7 modes provides higher resolution [273]. Kawai et al. [335] were able to reconstruct force versus distance curves by the the acquistion of higher harmonics. Compared to conventional force versus distance curves, this mode is faster and gives instantaneous information about the local interactions during imaging.

Intermodulation AFM is a technique, where two frequencies in the vicinity of a cantilever resonance are excited [526]. Tip surface interactions generate a new set of frequencies, the intermodulation products, which include information about the interaction forces. The advantage of this method is the real-time acquistion of non-linear interactions [301].

Torsional harmonic AFM is based on a specially designed cantilever with a small platform at the end of the cantilever. On one side the probing tip is attached. Conventional dynamic AFM mode is used to probe the surface. Since the tip is offset from the cantilever axis, forces result in high torques on the cantilever and the higher torsion harmonic modes are excited. The acquisition of the higher modes and subsequent Fourier synthesis gives time resolved information about the interactions [569].

Band excitation A digital signal, which spans a continuous band of frequencies, is synthesized and applied as excitation, while the response of the cantilever in this frequency range or in an even larger frequency band is detected [315]. High speed data acquistion and Fourier transformation provide information about amplitude and phase versus frequency curves. These data are collected at each location on the surface. Band excitation was used to probe elastic properties and energy dissipation of polymeric and biological systems and was also used to study ion diffusion in electrochemical cells. The large amount of data generated in this mode is a major obstacle for a wide use of band excitation.

Nanomechanical holography is a combination of ultrasonics and dynamic force microscopy, where both the sample and the cantilever are mechanically excited and local interactions are modified by the scattered ultrasonic waves [601]. The method gives information about subsurface structures in the spirit of sonar. It was possible to detect nanoparticles in the interiour of biological cells or buried electrical contacts of microelectronic devices.

For more details about multifrequency AFM the reader is referred to the excellent review from Garcia and Herruzo [194].

Magnetic Force Microscopy

4

Abstract

This chapter is related to the field of magnetic force microcopy (MFM), where the probing tip is covered with a magnetic layer and the magnetic domain structure of the sample can be characterized with high lateral resolution. In addition, other scanning probe methods mapping the magnetic field at a microscopic scale and techniques providing access to atomic scale magnetism are reviewed.

4.1 Principles of Magnetic Force Microscopy

4.1.1 Early Work

Magnetic Force Microscopy is a versatile technique used to map the stray field emanating from a sample surface with high spatial resolution and sensitivity. The magnetic force microscope (MFM) is an atomic (or scanning) force microscope [86] with a tip that is made sensitive to magnetic stray fields. First results by Martin and Wickramasinghe [433], Abraham et al. [15], and Mamin et al. [423] demonstrated a lateral resolution of about 100 nm on thin film magnetic recording heads and magnetization patterns in longitudinal recording media. Cantilevers were fabricated by etching iron and nickel wires and bending these into an L-shape. These cantilevers were oscillated at a fixed frequency near their fundamental mode resonance frequency. The tip-sample interaction arising from the sum of the van der Waals, electrostatic and magnetic force gradients then shifts the cantilever resonance frequency, and leads to a corresponding change of the cantilever oscillation amplitude or phase between the cantilever oscillation and drive signals. These signals were measured by a lock-in amplifier and the amplitude was used for the tip-sample distance feedback [Fig. 4.1a]. Depending on the reaction speed of the z-feedback loop, the oscillation amplitude signal can be kept constant locally or on average. The local force gradient acting on the cantilever is then reflected by the change of the z-position [out B in Fig. 4.1a] or the change of the cantilever oscillation amplitude [out A in Fig. 4.1a [423]]. Mamin

© Springer Nature Switzerland AG 2021
E. Meyer et al., *Scanning Probe Microscopy*, Graduate Texts in Physics,
https://doi.org/10.1007/978-3-030-37089-3_4

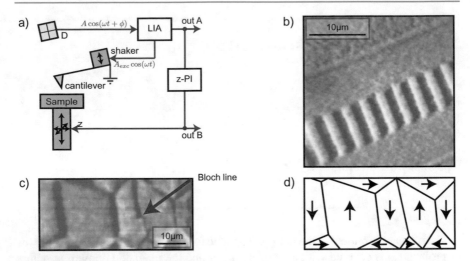

Fig. 4.1 a Block of the electronic circuitry of the MFM operated in the dynamic mode. The cantilever is driven near its resonance. **b** MFM image of an 8 μm wide track written in a CoPtCr media with in-plane magnetic anisotropy (adapted from [561], with permission from AIP Publishing). **c** MFM image of a permalloy thin film. The domain walls are visible either as dark or bright lines. The red arrow points a Bloch line (adapted from [424], https://doi.org/10.1063/1.101898, with permission from AIP Publishing)

et al. already pointed out several aspects that remained of high relevance also for current MFM work. First, accurate tip-sample distance control is required and a higher lateral resolution can be obtained if the tip is scanned at a small tip-sample distance. Second, the differentiation between the contrast arising from the topography of the sample and that from the magnetic tip-sample interaction can be challenging.

It is noteworthy that at a time where all atomic force microscopes used for topographical imaging were operated in static modes, dynamic scanning force microscopy operation modes were used in all of the early work of the IBM [423,424,433,435] and Philips [141] research groups. With the latter operation modes, reliable non-contact imaging was possible, and reproducible MFM results of stray fields of recording heads [433], written bits in recording materials [141,435,561] [Fig. 4.1b], permalloy elements [424] [Fig. 4.1c and d] and natural domains [141] in CoPt multilayers were obtained.

Other early work either scanned the tip in contact with the sample [235,568] or used tunneling between the tip and sample [221] to control the tip-sample distance. Under such operation conditions large non-magnetic forces act on the tip. The force constant relevant for the magnetic tip-sample interaction then becomes

$$c_{\text{eff}} = c_{\text{L}} - \frac{\partial F_{\text{nm}}}{\partial z}, \tag{4.1}$$

where c_{L} is the force constant of the cantilever and $\frac{\partial F_{\text{nm}}}{\partial z}$ is the z-derivative of the non-magnetic forces. Typically $|c_{\text{eff}}| \gg c_L$. Hence, the deflection of the cantilever [235,568] or change of its resonance frequency arising from the magnetic forces become small and may easily be dominated by variations of the non-magnetic tip-sample interaction forces F_{nm}. Consequently, most MFM results obtained with the tip in contact with the sample or while tunneling from the tip to the sample remain unconvincing, particularly because most work did not image clearly identifiable magnetic domain structures or bit patterns. Reliable imaging with a tunneling-stabilized magnetic force microscopes was demonstrated a few years later by Moreland and Rice [467,468], Gomez et al. [222], Katti et al. [329], and also by Wadas, Hug, and Güntherodt [682]. Although these studies demonstrated the feasibility of tunneling-stabilized MFM, the magnetic images suffered from substantial cross-talk from the topography, and stable tunneling under ambient conditions with compliant tunneling tips (cantilevers) scanned over large areas as typically required for magnetic imaging remained challenging.

4.1.2 Tip-Sample Distance Control

In scanning force microscopy the cantilever deflection, d,

$$d = \frac{F_{ts}}{c_L} , \tag{4.2}$$

or the change of resonance frequency of the cantilever, Δf, in the small oscillation amplitude approximation given by

$$\Delta f = -\frac{f_0}{2c_L} \cdot \frac{\partial F_{ts}}{\partial z} , \tag{4.3}$$

where F_{ts} is the tip-sample force, f_0 is the resonance frequency of the free cantilever, and $\frac{\partial F_{ts}}{\partial z}$ is the z-derivative of the tip-sample force. A scanning force microscope measures the sum of all forces or force derivatives acting on the tip. For the case of an MFM, i.e. if the magnetic tip-sample interaction is of interest, all other contributions must be kept small. For this, the tip must *not* be in contact with the sample otherwise the cantilever deflection caused by the magnetic forces (or the frequency shift) becomes minute, i.e.

$$d_{mag} = \frac{F_{mag}}{c_{eff}} , \tag{4.4}$$

where c_{eff} is the effective cantilever force constant given by (4.1) which depends on the derivative of the non-magnetic tip-sample forces. If the tip is in contact with the sample, the latter are dominated by the derivative of the Pauli repulsive force, which is orders of magnitude larger than the typical force constants of cantilevers used in MFM.

Figure 4.2a, and b show an MFM image and cross-section of a domain structure of a $BaFe_{12}O_{19}$ single crystal acquired in 1992 by Hug et al. [298]. The data shown in Fig. 4.2a were obtained by scanning the tip parallel to the average tilt of the sample surface, i.e. at $z(x, y) = z_0 + \alpha x + \beta y$ without an active tip-sample distance control.

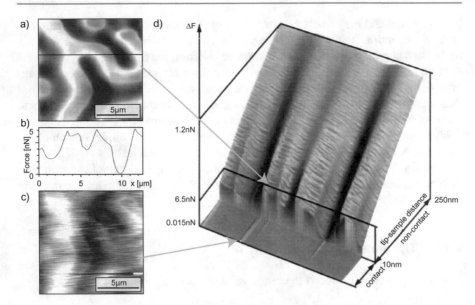

Fig. 4.2 **a** MFM image of $BaFe_{12}O_{19}$ single crystal measured in non-contact. **b** Cross-section of the force variation taken at the scan-line indicated by the black line in (**a**). **c** MFM image of $BaFe_{12}O_{19}$ single crystal measured with the tip in contact with the sample surface. **d** 3-dimensional representation of scan lines taken at the black line shown in (**a**) for different z-positions (adapted from [298])

The constants, α and β were determined from a previously recorded topography data. The magnetic signal size and the lateral resolution both improve at reduced tip-sample distances, but the cantilever deflection reduces by orders of magnitude, if the tip is brought in contact with the sample surface [see Fig. 4.2c]. Figure 4.2d shows the evolution of the contrast as a function of the tip-sample distance. Clearly the best contrast and highest spatial resolution are obtained if the tip is scanned at small tip-sample distances but the tip does not make contact with the sample.

Tip-sample distance control is thus crucial for a reliable MFM operation. Rugar et al. [561] pointed out that the tip-sample distance servo loop [Fig. 4.1a] is stable only as long the total force derivative remains attractive. More generally, the necessary condition is that the tip-sample interaction used for the z-feedback remains monotonous. In the presence of a magnetic tip-sample interaction that can generate repulsive or attractive force derivatives, this can be achieved by applying a sufficiently larger tip-sample bias, such that the negative electrostatic force derivative dominates the largest positive magnetic force derivative. However, when scanning at small tip-sample distances the topographical texture of the sample becomes visible in the MFM image.

A first experimental approach to disentangle topographic and magnetic forces was presented by Schönenberger and Alvarado [588] [Fig. 4.3a]. The tip-sample distance was modulated at a frequency f much smaller than the cantilever resonance frequency. Because the sample interacts with the tip, the cantilever then oscillates at the frequency f. Its oscillation amplitude then is a good measure of the tip-

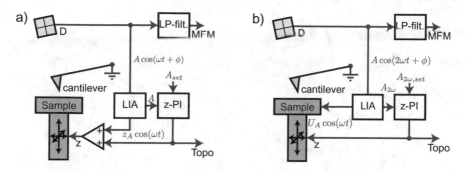

Fig. 4.3 a The tip-sample distance is sinusoidally oscillated at a frequency of a few kilohertz. Via the magnetic and non-magnetic tip-sample forces, this oscillation of the tip-sample distance leads to a corresponding oscillation of the cantilever deflection that can be measured with a lock-in amplifier. A feedback system adjusts the sample z-position to keep the cantilever oscillation amplitude constant. The latter is a proxy for the tip-sample distance. Hence the output signal of the feedback is a measure of the sample topography. The dc-deflection of the cantilever reflects the variations of the magnetic forces [588]. **b** In this alternative setup, the tip-sample potential is sinusoidally modulated

sample distance provided that the electrostatic and van der Waals forces dominate the magnetic tip-sample force, which is typically the case at small tip-sample distances or if a sufficiently large sample bias voltage is applied. The oscillation amplitude of the cantilever was measured by a lock-in amplifier and kept constant by the z-feedback. Its output then reflects the topography of the sample, while the low-pass filtered signal of the cantilever deflection is a measure of the magnetic force.

In their work, Schönenberger and Alvarado [588] further pointed that the cantilever oscillation occurring at 2ω from an applied oscillating bias, $V_T = V_0 + V_1 \sin(\omega t)$, is independent from local magnetic forces and even from the local contact potential. The 2ω can thus be used for tip-sample distance control. Data obtained in this mode were presented in their paper discussing MFM contrast formation [586].

The latter idea was taken up recently by Schwenk et al. [592,593], and Zhao et al. [727] (see Sects. 4.4.2.2 and 4.4.2.3), who developed tip-sample distance control modes suitable for MFM operation under vacuum conditions. The benefits of these more advanced MFM operation modes for the resolution and sensitivity will be discussed in Sect. 4.4.1.

A robust method to control the tip-sample distance for magnetic force microscopy performed under ambient conditions was demonstrated by Hosaka et al. [285] [Fig. 4.4a]: at each point (x, y) of the scanned image, the tip-sample distance was reduced until the oscillating tip made intermittent contact with the sample surface, and a preset oscillation amplitude smaller than that of the freely oscillating cantilever was obtained. The z-position required to obtain this condition then is a measure of the sample topography. Then, the tip is retracted from the sample by a preset distance Δz (the lift-height) to measure the magnetic tip-sample interaction. Images of the topography and magnetic structures of a magneto-optical disk were shown.

A different implementation of this technique was reported by Giles et al. [212] [Fig. 4.4b], where each scan line was scanned twice. To record the topography, a

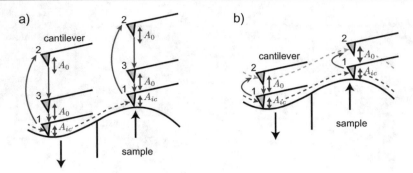

Fig. 4.4 a The topography of the sample is imaged with the intermittent contact mode (1). At each point of the scan-line the tip is retracted from the surface (2) and reapproched to a preset tip-sample distance (3) to measure the magnetic forces, before the tip-sample distance is reduced such that the oscillating tip again intermittently contacts the sample surface [285]. **b** Lift-mode operation: Each scanline is scanned twice. First the topography is recorded in the intermittent contact mode with an oscillation amplitude setpoint A_{ic} that is smaller than the free resonant amplitude A_0. The magnetic tip-sample interaction is then measured in the second scan which is performed at a preset lift height of the tip above the previously recorded topography scan line [212]

first scan is carried out with the oscillating tip intermittently touching the sample surface. The z-feedback adjusts the relative cantilever-sample z-position such that the oscillation amplitude remains constant at a level smaller than the free oscillation amplitude of the cantilever. A subsequent scan of the same line is then performed at a predefined lift height that is sufficiently large such that the tip does no longer touch the surface of the sample. Consequently, variations of the cantilever oscillation amplitude or phase then predominantly reflect the magnetic tip-sample interaction. Lift-mode operation became the most-widely used distance control mode in magnetic force microscopy and was key for the use of MFM for magnetic materials research, albeit the separation of the magnetic and topographic signal remains incomplete for various reasons that will be discussed in detail in Sect. 4.4.3.

Note that with modern MFM instrumentation, peak-force tapping [485, 486, 720], multiharmonic off-resonance intermittent contact modes [485, 512] instead of the classical intermittent contact mode can be used to record the topography before the tip is lifted above the surface to record the magnetic signal.

Recently, single-passage MFM distance control methods have been developed, where topographical and magnetic signals are recorded simultaneously. Among them are multifrequency methods [194] allowing single passage imaging of magnetic signals. Li et al. [391] mechanically excited the cantilever on its first and second flexural resonance mode simultaneously. The response of dynamic properties of the cantilever (its oscillation amplitude, phase or frequency) is proportional to the tip-sample force derivative signals covered within the oscillation path of the tip averaged with a semi-circular weighting function [211]. In the experiment of Li et al. [391], the first mode drive amplitude was selected much smaller than that of the second mode, making the first mode more sensitive to the shorter-range van der Waals forces (see also Sect. 4.4.2.1). Consequently, the first mode oscillation amplitude was used for the tip-sample distance feedback, whereas the phase-shift of the second mode became sensitive to the longer-ranged magnetic forces. Dietz et al. [149] have later used the same bi-modal operation mode but with a first mode oscillation amplitude selected larger than that of the second mode to map the topography of superparamagnetic particles on a mica surface. In their work, Dietz et al. [149] used the first-mode amplitude to control the tip-sample distance. The second mode phase shift was shown to depend on the magnetic tip-sample interaction when a field of 80 mT was applied to magnetized the superparamagnetic ferritin particles. An alternative bi-modal operation mode was used by Schwenk et al. [592] to perform MFM under vacuum conditions (Sect. 4.4.2.1). The latter and other multi-modal operation modes suitable for MFM operation under vacuum conditions [593, 727] are discussed in Sect. 4.4.2.

Recently, more complex AFM modes that record larger data sets have been introduced: Forchheimer et al. performed intermodulation AFM [525] to obtain a force-distance curve in every image point, which allows to disentangle magnetic from topographic signals [186]. Intermodulation AFM [525], peak-force [720] or off-resonance intermittent contact modes [512] rely on a smart experimental set-up to strongly reduce the measured data to a subset relevant to answer specific experimental questions. In contrast to these experimental methods, Kalinin et al. recently presented the so-called Google mode or G-mode [324] that records the cantilever position data at the full bandwidth of the 4-quadrant photo-detector which is several MHz together with the xyz position of the scan piezo and all other instrumentation data. Collins et al. [128] have later used big data analytics to extract magnetic and electrostatic tip-sample forces from a 4 Gbyte G-mode data set recorded on a 256×256 pixel^2 image taking about 18 min. to capture.

4.1.3 Magnetic Force Microscopy Tips and Cantilevers

In scanning tunneling microscopy (Sect. 4.5.2.1) or in atomic force microscopy experiments with true atomic resolution (Sect. 4.5.2.2), the strong exponential decay of the tip-sample forces limits the volume of the tip apex that is relevant for the tip-sample interaction force. Consequently, the macroscopic and mesoscopic geometry of the tip has only a minor effect on the achievable lateral resolution. This is different for longer-ranged forces. Then, an extended part of the tip apex contributes to the measured tip-sample interaction force (or force gradient). Although, magnetic fields of periodical magnetization distributions decay exponentially with increasing distance from the sample surface (Sect. 4.3.1), the spatial wavelengths of such magnetic patterns are generally much larger than those of atomic structures. As a result, the spatial resolution that can be obtained in a magnetic force microscopy experiment depends strongly on the geometrical extension of the magnetic moment distribution of the tip.

In early work, L-shaped wires with electrochemically etched tips made from ferromagnetic materials [561] or electrochemically coated by a ferromagnetic materials [141] were used for MFM [Fig. 4.5a]. Such tips have a high magnetic moment, which leads to an increased magnetic signal, but also generates a strong field at the sample, which may influence the micromagnetic structure of the latter. Moreover, such hand-

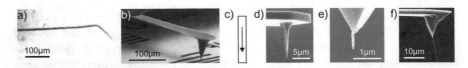

Fig. 4.5 **a** Bent micro-wire cantilever with electrochemically etched tip (adapted from [561], https://doi.org/10.1063/1.346713, with permission from AIP Publishing). **b** Microfabricrated single crystalline Si cantilever (for example as coated with a magnetic layer by [236]). **c** Schematics of ideal flat-ended extended dipole tip (see [530]). **d** High-aspect ratio tip coated on its back side with a 3 nm-thick Co layer deposited onto a 1 nm-thick adhesion layer, protected against oxidation with a 2 nm Ti layer. **e** Early non-magnetic electron-beam induced growth of a cylindrical nanotip on the top of a microfabircated tip. **f** Iron-filled carbon nanotube tip attached to the tip of a conventional microfabricated cantilever (adapted from [703], https://doi.org/10.1063/1.3459879, with permission from AIP Publishing)

crafted tips were challenging to fabricate and the reproducibility remained poor. Consequently, the coating of non-magnetic tips with a thin ferromagnetic layer was proposed [236, 681, 682]. The latter became the method of choice once microfabricated silicon, silicon-oxide, or silicon nitride cantilevers became available [236] [Fig. 4.5b]. Such cantilevers are further beneficial for obtaining high mechanical quality factors, particularly when operating under vacuum conditions (see Sect. 4.4.1).

Obtaining a high quality factor with soft microfabricated cantilevers that still have a reasonably high resonance frequency is key for most magnetic force microscopy experiments, because magnetic force typically are much smaller than typical interatomic forces relevant for atomic resolution imaging. Therefore, tuning-fork AFM methods [210] which had a strong impact for atomic resolution imaging work, are not suitable for magnetic force microscopy, although a few studies presented some MFM results [112, 584, 598].

As discussed in the previous section, MFM experiments are performed with the tip *not* in contact with the sample surface. Because the stray fields arising from geometrically smaller magnetic moment distributions of the sample decay more rapidly with the distance z from the sample surface, the lateral resolution of MFM images strongly degrades at increased tip-sample distance [Fig. 4.2c]. Even though this is the case, the decay of the stray field with tip-sample distance can be precisely described (see Sect. 4.3.2), and thus higher lateral resolution data can in-principle be recovered from MFM data measured at larger tip-sample distances [295]. This however requires a sufficiently high signal-to-noise ratio such that the signal of the high-spatial frequency components of the stray field remains above the noise level. The latter can be achieved by increasing the measured signal through MFM tips with a larger magnetic moment, or by decreasing the noise level of the measurement (Sect. 4.4.1). Because the stray field emanating from high magnetic moment tips can lead to noticeable reversible and even irreversible local perturbations of the micromagnetic structure of the sample, the second strategy is the better choice. The tip magnetic moment is thus preferably kept sufficiently small to keep the influence of the tip-stray field on the sample micromagnetic structure negligible. Once, the material of the magnetic tip coating and its thickness have been selected, the optimization of its spatial distribution remains the relevant task. Porthun et al. [530] have performed a transfer-function analysis (Sects. 4.3.2 and 4.4.4) and reported that a flat-ended extended dipole tips have the best performance [Fig. 4.5c].

A strategy for obtaining such MFM tips is to attach high-aspect ratio structures to standard microfabricated tips, for example by electron beam induced growth (EBID) [181, 295, 531, 564, 608]. These tips are subsequently coated with magnetic materials. More recently, EBID has been used to directly fabricate ferromagnetic tips [59, 659]. Alternatively, carbon nanotubes either coated with a magnetic material [145, 375] or filled with iron [703] have been attached to tips of microfabricated cantilevers. However, ultrathin nanotube tips often show a large lateral flexibility which may lead to image distortions.

Alternatively, focussed ion beam (FIB) milling can be used to obtain needle-like tip geometries from more pyramidal tips on standard cantilevers [520]. More recently, Neu et al. [483] have used FIB to cut needle-shaped pieces from a magnetically hard, $SmCo_5$-film that was epitaxially grown onto a MgO substrate. These tips allowed imaging in fields of up to 0.7 T applied opposite to the magnetization of the tip. Note, that tips with much lower coercivities are often used even in very high fields up to several Tesla, but with a tip magnetization aligned parallel to the field (see for example Figs. 4.43, 4.44, 4.46, and 4.51 in Sect. 4.4.6). However, high coercivity tips may be useful for MFM measurements with fields applied in the plane to avoid a field-dependent canting of the tip magnetization and resulting from it a more complex interpretation of the measured data. For the measurement of in-plane field components, MFM tips with a perforated magnetic coating have been proposed [184]. Note however that the direct measurement of the in-plane field components is not necessary because these can also be calculated from a measured map of the z-component of the stray field above the sample surface. This is because the magnetic stray field is a conservative field arising from a scalar magnetic potential [662] (see (4.26) in Sect. 4.3.1 for more details).

A possibility to obtain a magnetic tip is by attaching microscopic particles of hard magnetic materials or magnetic beads [341]. However, all these tips are handcrafted, which limits the reproducibility.

From the transfer-function approach (Sects. 4.3 and 4.4.4.2), it becomes apparent that tips generating a small dipole at the tip apex, instead of a magnetic monopole, would lead to an increased signal for smaller spatial wavelengths of the stray field, and should consequently achieve an improved spatial resolution. Such dipole tips have been proposed [503] and fabricated [506, 706, 714]. Amos et al. [30] have coated a pyramidal tip with two 10 nm-thick layers of an FePt (45/55) alloy separated by a 6 nm MgO layer. An $L1_0$ phase was obtained by annealing the tip at 650 °C and to achieve a demagnetized tip state, characterized by an antiferromagnetic alignment of the two ferromagnetic layers. The smallest period visible in their MFM image acquired with such a tip was about 40 nm, corresponding to a bit length of 20 nm [Fig. 4.6a], which is considerable better than what was achieved with the tip in a saturated state (parallel magnetization of both layers, see Fig. 4.6b. It was claimed that this antiferromagnetic balancing is incomplete at the tip apex, and that this small region is responsible for the superior resolution. The loss of lateral resolution observed when the tip was saturated by a strong external field supports this explanation. Although it is obvious that such

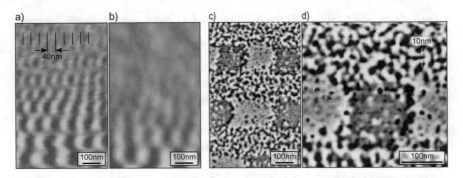

Fig. 4.6 500 nm × 800 nm MFM image of bit tracks in a hard disk sample recorded under ambient conditions using a tip consisting of two ferromagnetic layers in a demagnetized state (**a**) and after magnetization. **b** Adapted from [30], https://doi.org/10.1063/1.3036533, with permission from AIP Publishing. **c** 500 nm × 800 nm and **d** zoomed high-resolution MFM image of a prototype hard disk sample acquired under vacuum conditions with an ultra-high aspect ratio tip coated with a single ferromagnetic thin film on one side of the tip [Fig. 4.5d]. The yellow arrows in **d** highlight a down domain with a width of 10 nm

a concentration of unbalanced magnetic moment at the tip apex is beneficial for the lateral resolution, it also limits the tip-sample interaction force. In other words, in their work Amos et al. [30] achieved a maximized tip-sample interaction with the minimal active magnetic volume of the tip.

However, an excellent spatial resolution could be obtained already in 2005 by Moser et al. [472] who used an MFM operated under vacuum conditions. The superior signal-to-noise ratio (see Sect. 4.4.1) allowed the use of ultrasharp, high-aspect ratio tips [8], coated with a single, thin ferromagnetic layer on the backside of the tip [Fig. 4.5d]. Bits of a track recorded at 1016 kfci (kilo flux changes per inch), i.e. with a bit period length of 25 nm (12.5 nm bit length) were imaged with best signal-to-noise ratio. MFM images with a resolution of 10 nm were already obtained two years earlier, using the prototype of the MFM later used by Moser et al. Figure 4.6c and d shows MFM data of a prototypical perpendicular recording material acquired in vacuum. It seems apparent that tips with a similar radius and geometry but with two ferromagnetic layers that are antiferromagnetically coupled, could obtain an even higher lateral resolution.

MFM probe characterization is of fundamental importance for the understanding of the obtained MFM contrast and consequently for the assessment of the sample micromagnetic structure. Apart from an SEM analysis of the tip geometry, the magnetic characterization of the tip is of highest importance. High-resolution techniques such as Lorentz microscoscopy [441,594] or electron holography [441,506,607] provide information on the tip's microscopic magnetic stray field and micromagnetic structure, but the complexity of these techniques limits their usefulness. Calibrating the response of MFM tips with suitable calibration measurements thus is a more practical approach [662] (Sect. 4.4.5).

4.2 MFM Contrast Formation

The magnetic force microscope is one of the magnetic imaging techniques that map the stray field emanating from a sample surface. The measured MFM contrast then arises from the cross-correlation of the tip magnetization with the gradient of the stray field of the sample [561]. A detailed description of the contrast formation is given in Sects. 4.3, 4.4.4, and 4.4.5. In most cases, the stray field exists even without the presence of the tip [Fig. 4.7a]. The sources of the stray field of ferromagnetic samples are discussed in Sect. 4.3. The situation is different if a tip interacts with an high permeability sample [Fig. 4.7b] or with an ideally diamagnetic sample [Fig. 4.7c] like a (type 1) superconductor: the tip is attracted to in the first and repelled from the surface of the sample in the second case (see Sect. 4.2.2 for further details).

Further, as in any scanning force microscopy technique, the cantilever measures the sum of all forces acting on the tip. Apart from the magnetic field-mediated sample-tip interaction, van der Waals and electrostatic forces arising from both the applied tip-sample potential and the contact potential act on the tip. These different contributions must be disentangled and suitable methods are discussed in Sect. 4.4.3. To understand contrast formation in magnetic force microscopy, it is generally convenient to distinguish the cases discussed in the next three sub-sections.

4.2.1 Negligible Perturbation

The stray field of the tip or that of the sample will also perturb the magnetization of the sample or the tip, respectively. However, in many cases the observed MFM contrast can be modeled with great precision even under the assumption that such a perturbation does not occur. In this case, the perturbation is negligible. The MFM contrast can then be calculated from the cross-correlation of the tip magnetization with the gradient of the sample stray field. For this, the stray field arising from the magnetic moment distribution inside the sample must be understood. This is conveniently done by introducing magnetic volume and surface charges arising from

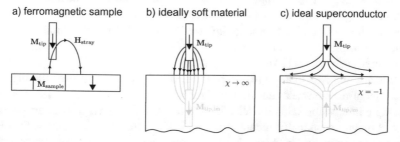

Fig. 4.7 a Stray field of ferromagnetic sample interacts with magnetic moments of tip. **b** Ideally soft sample with $\chi \to \infty$: all B-field-lines are perpendicular to the sample surface. An attractive force acts on the tip that can be modeled by the tip interacting with the field of an image-tip inside the sample. **c** Ideal superconductor with no flux penetration. The tip is repelled from the sample surface

the divergence of the magnetization inside the sample or at the surfaces of the sample (Sect. 4.3.1). Because the tip magnetization distribution is generally not known, models to describe the net magnetic moment distribution of the tip can be used. Then the cross-correlation of the magnetic moment distribution of the tip with the sample stray field can be performed. This is an easy task for simple tip magnetic moment distribution models, but can require extended calculations for more complex distributions. However, the magnetic moment distribution inside the tip is generally unknown. Then a calibration of the response of the MFM tip in a two-dimensional Fourier space and the use of a transfer-function approach becomes more convenient (Sects. 4.4.4.2 and 4.4.5).

Note that a description of the MFM contrast formation based on a transfer function theory relies on a linear response of the tip to the sample stray field. As a result, such an approach will fail if the perturbation of the sample or tip magnetization by the stray field of the tip or the sample, respectively, becomes irreversible. This is typically the case for bulk magnetic tips (etched magnetic wires) but can also when using thin film magnetic tips, when imaging magnetically soft samples.

4.2.2   Reversible Perturbation

A reversible modification of the sample magnetic moment distribution for example occurs if a magnetic tip interacts with a sample that has an infinite permeability [Fig. 4.7b]. If a tip is approached to such a sample, the field lines of the tip at the surface of the sample must remain perpendicular to the sample surface. For this situation, the interaction of the tip with the sample can be described by the interaction of the tip with an *image* tip having an opposite magnetization located inside the sample at the same separation from the surface as the real tip outside the sample. The tip will be attracted to the surface because the interaction between the tip and the *image* tip is obviously attractive. The opposite is true for a type-1 superconductor (or type-2 superconductor with no or pinned vortices or a graphite surface) that screens the tip field [Fig. 4.7c], because of its diamagnetic response.

However, even if the sample does not have an infinite permeability or is not ideally diamagnetic, the field of the tip can lead to a (rotary) response of the sample magnetization. The magnetic energy involved with such a rotation process depends on the tip-sample distance. Hence, a force different from that arising from the non-perturbative case occurs [14].

Abraham and McDonald [14] proposed a simple model that uses the magnetic anisotropy of the sample as an additional term in the energy equation describing the tip-sample interaction. This model was used to explain the MFM contrast contours above a magneto-optically recorded bit in a TbFe thin film imaged with a high magnetic moment Fe tip. Many MFM images taken on thin films samples with patterns of up/down domains of a size considerably larger than the film thickness show an up/down (domain) contrast, rather than the a domain wall contrast that one would expect from the stray field or its derivate that shows a strong signal only at and near the domain walls. Belliard et at. [58] used the approach of Abraham and McDonald [14] to explain their MFM results on domains in Co/Pt-multilayer thin films which showed an up/down domain contrast. Similar work was done by Hubert et al. [293]. They imaged the same area of a pattern of up/down domains in a permanent magnet material with two opposite magnetization states of the tip. The contrast of the difference and sum between the two images were then attributed to the magnetic charge pattern and the susceptibility response of the sample, respectively. However, as discussed in Sect. 4.4.3, spatial variations of the electrostatic and van der Waals force can also contribute to the contrast visible in the sum data.

The model developed by Abraham and McDonald [14] may serve as a first approach for more modeling a mutual response of the sample magnetization on the stray field of the tip. More complex situations, e.g. the interaction of the tip field with a domain wall in the sample, can however only be handled by micromagnetic calculations that take into account the sample and the tip, as well as the different tip-sample positions of the scanned image. Scheinfein et al. [575] for example modeled the interaction of an Fe (bulk) tip scanning 20 nm above a 200 nm thick permalloy film, and showed that the magnetic moment distribution of the Néel-capped Bloch wall in the permalloy film is strongly influenced by the presence of the tip. In such a case, the magnetization distribution may be reversible in the sense that there is one well-defined state of the tip and sample magnetization for a given position of the tip above the sample.

A good example of such a subtle, reversible perturbation was presented by Foss et al. [187], who studied domain walls in Fe_3O_4 single crystalline samples. Inside the sample, typically a few tens to hundreds of nanometers away from the surface, the magnetization of the domains is parallel to the sample surface, i.e. in-plane. The domain walls are Bloch walls with magnetic moments that rotate out of the sample plane and are thus perpendicular to the surface at the positions of the walls. This would lead to magnetic charges at the surfaces and, consequently, to a local demagnetization field that would act on the wall magnetization. The system can reduce the total magnetostatic energy by blending the Bloch wall existing inside the sample smoothly into a Néel wall at the surface. Such a magnetization structure is called a Néel cap. Such a complex wall structure will trivially adapt to the local field applied by the MFM tip.

Schematical drawings of the alterations of the different types of domain walls arising from the locally applied field of the MFM tip are shown in Fig. 4.8b–e. Experimental evidence supporting this general behavior is shown in Fig. 4.9. The two MFM response profiles shown in Fig. 4.9a and b were measured above the same 180° DW using the same MFM tip at a lift height of 30 nm: (a) was measured with the tip magnetization antiparallel to the bulk Bloch DW magnetization (repulsive case), while (b) was measured with the tip magnetization parallel to the bulk Bloch DW (attractive case). Both profiles are asymmetric, but their asymmetries are markedly different from each other. The influence of the tip field produced a more antisymmetric profile in the repulsive case and a more symmetric profile in the attractive case. A way to measure the alteration of the DW is demonstrated in Fig. 4.9c in which the repulsive profile Fig. 4.9a was inverted and superimposed on the attractive profile Fig. 4.9b [187]. Note that the unperturbed DW profile should lie between these

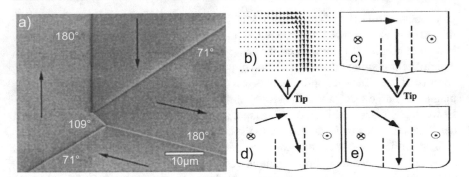

Fig. 4.8 a A liftmode MFM image of an area of an Fe_3O_4 single crystal containing three types of domain walls (DWs): $180°$, $109°$, and $71°$. The arrows indicate the direction of magnetization in the bulk of each domain. **b** Results of a 2D micromagnetic simulation of a $180°$ DW in Fe_3O_4. This cross section of the sample shows that the magnetization is out of the page on the right of the wall and into the page on the left. The vertical arrows in the sample interior indicate that the DW is a Bloch DW in this region, but near the surface, the spins in the DW gradually rotate into the surface plane to form a Néel-like DW portion referred to as a Néel cap. **c** A cartoon of the $180°$ DW structure in (**b**). The bulk Bloch DW is represented by a perpendicular magnetic dipole with the top pole about one Bloch DW width below the surface. The surface Néel cap is represented by an in-plane dipole perpendicular to the Bloch DW plane. **d** A schematic of the repulsive MFM measurement of this DW structure in which the tip is magnetized antiparallel to the bulk dipole moment. In this case, the tip field enhances the Néel cap. **e** A schematic of the attractive DW measurement for which the tip is magnetized parallel to the bulk dipole moment. The tip field reduces the Néel cap in this case. Adapted from [187], https://doi.org/10.1063/1.117281, with permission from AIP Publishing

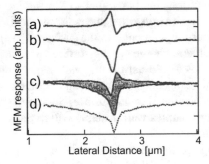

Fig. 4.9 Two MFM response profiles, **a** and **b**, measured above the same $180°$ DW using the same MFM tip at a lift height of $30\,nm$. **a** Tip magnetization antiparallel to the bulk Bloch DW magnetization, (repulsive case). **b** Tip magnetization parallel to the bulk Bloch DW (attractive case). In **c**, profile **a** was inverted and superimposed on profile (**b**). Their difference has been shaded. Profile **d** is the difference profile, which shows the additional, attractive MFM response due to the effect of the tip field on the DW structure. Adapted from [187], https://doi.org/10.1063/1.117281, with permission from AIP Publishing

two profiles. The difference, between these two profiles shaded in Fig. 4.9c has been plotted in Fig. 4.9d, giving the combined modifications of the DW from both profiles.

MFM results obtained on a c-axis-oriented $BaFe_{12}O_{19}$ single crystal recorded with a Co- and Ni-coated tip are displayed in Fig. 4.10a and b, respectively. The image

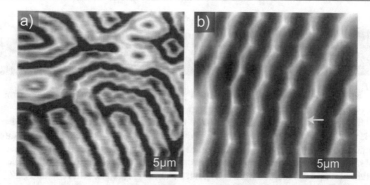

Fig. 4.10 MFM images of a c-axis-oriented $BaFe_{12}O_{19}$ single crystal recorded with a Co (**a**) and Ni (**b**) coated tip

recorded with the Co coated tip [Fig. 4.10a] shows a repulsive/attractive force contrast with the maximum contrast near the domain walls as expected for the stray field of the up/down domain pattern [see bottom panel of Fig. 4.20b]. As discussed by Wadas et al. [682], the domain walls in $BaFe_{12}O_{19}$ become wavy above a critical thickness of the crystal to reduce the overall magnetostatic energy. The image recorded with the magnetically softer Ni tip is however shows a very different contrast: all domains appear with a dark attractive force contrast, and the domain walls generate a repulsive force. Such a contrast arises if the magnetization of the tip follows the local direction of the stray field such that the tip-sample force becomes attractive for both the up and down domains. Moreover, the image of the domain wall consists of a sharp zig-zag line in the center of a wider white stripe. Such sharp lines are typically observed, if the magnetization of tip switches reproducibly between an up and a down state at locations above the sample where the stray field overcomes the coercive field of the tip. For an up/down domain pattern, this typically occurs at the domain walls, while the stray field is weaker in the center of the domains [see Sect. 4.3.2, Fig. 4.20b]. It is noteworthy that the appearance of such sharp structures in MFM images [yellow arrow in Fig. 4.10b] should not be misinterpreted as a true image of the stray field with a high spatial resolution. In fact, the stray field is not correctly imaged, and the sharpness of the line arises only from the (rapid) switching of the tip magnetization that occurs at a well-defined location above the sample, where the field surpasses the coercive field of the tip.

Similar to the switching of the tip magnetization, a domain wall can be dragged by the tip field, and then switch back to its equilibrium position defined by a specific local minimum of the magnetostatic energy of a given domain structure of the ferromagnetic sample in the local stray field of the tip. The distortion of the micromagnetic structure of the sample then typically depends on the scan direction and on the tip-sample distance. An example was presented already in the early work of Mamin et al. [424], where the dependence of the distortion of domain wall in a microfabricated permalloy square on the tip-sample distance was studied [Fig. 4.11]. The least distorted MFM image of the typically symmetric equilibrium domain structure expected for such a permalloy element is obtained at the largest tip-sample distance

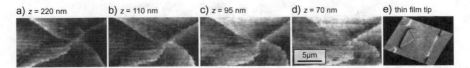

a) $z = 220$ nm   b) $z = 110$ nm   c) $z = 95$ nm   d) $z = 70$ nm   e) thin film tip

5µm

Fig. 4.11 MFM images of a $20\,\mu$m, 30 nm thick permalloy square, acquired with an iron wire tip (**a**) to (**d**), and a thin film tip (**e**). The latter was fabricated by sputter-deposition of a nominally 15 nm-thick $Co_{71}Pt_{12}Cr_{17}$ film onto the tip of a microfabricated cantilever. The MFM images **a** to **d** taken with the higher-magnetic moment iron wire tip show a noticeable perturbation of the micromagnetic pattern of the sample that becomes particularly apparent for the data taken at smaller tip-sample distances. In contrast, an image distortion is not noticeable in the MFM image taken with the thin film tip (**e**). Adapted from [424], https://doi.org/10.1063/1.101898 and [236], https://doi.org/10.1063/1.104030 with permission from AIP Publishing

$z = 220$ nm [Fig. 4.11a]. At smaller tip-sample distances [Fig. 4.11b–d] substantial distortions of the domain walls become apparent. Note that only one year later, the same patterns were image again, but with a microfabricated silicon cantilever, made sensitive to magnetic fields by a thin magnetic coating deposited onto the tip [236]. Using such a thin film tip, with a lower magnetic moment and consequently a smaller tip stray field, the distortion of the domain wall remains negligible.

4.2.3   Irreversible Perturbation

An irreversible perturbation of the sample magnetization occurs if a domain disappears, nucleates or a domain wall irreversibly changes position during MFM data acquisition. In most MFM experiments, the goal is to avoid such irreversible changes of the micromagnetic state of the sample (or the tip). However, by appropriately adjusting external parameters such as the tip-sample distance, sample temperature or external magnetic field, the field of the tip can also be used to modify the sample micromagnetic state in a controlled way. Kong et al. [362] imaged the poles of microfabricated Ni-bars, and flipped their magnetization by a reduction of the tip-sample distance below that used for imaging with the tip position above the pole giving a repulsive magnetic force contrast. The writing of magnetically-harder materials was demonstrated by Manalis et al. [425], who used a homogenous background field in addition to that emanating from the tip to write magnetic bits with a diameter of about 180 nm on a perpendicular media and an amorphous TbGdFeCo magneto-optical recording media. Other examples are the depinning or writing of vortex lines in superconductors [299, 471, 558, 559, 626]. Figure 4.12a displays $\Delta f(z)$-curves acquired on a 100 nm-thick Nb film at 7.9 K in zero external field. Because the Nb film is in the superconducting state, the magnetic flux arising from the MFM tip (here 30 nm of Co deposited is expelled from the sample (Meissner effect) at larger tip-sample distance. As a result, an increasingly positive shift of the resonance frequency [red arrow between the dashed red lines of curve (i)], if the tip is approached from 250 nm to about 140 nm. If the tip-sample distance is further reduced, sudden drops of the frequency shift towards more negative frequencies were observed [black arrows in curves (i) and (ii)], which could be attributed to the formation of vortices.

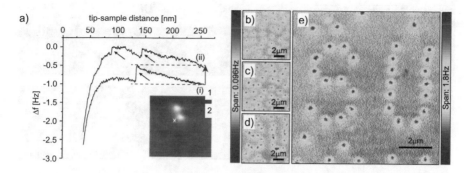

Fig. 4.12 **a** Δf versus tip-sample distance curves obtained by approaching an MFM tip to a super-conducting Nb thin film at 7.9 K. The increasing frequency shift at larger tip-sample distances is caused by the repulsive Meissner force arising from the expulsion of the magnetic flux of the MFM tip from the sample (red arrow between the red dashed lines). The discontinuities highlighted by the black arrows occur when vortices are nucleated. The inset shows the two vortices nucleated after recording the the curve (**i**). Note that the image is displayed with a white color corresponding to a more negative frequency shift. Figure adapted from [559] with permission from AIP Publishing. **b** to **e** Lateral manipulation of vortices. Figure adapted from [626] with permission from AIP Publishing. **b** Vortex state imaged after cooling to 7 K in an external field of a few Gauss imaged with a tip-sample distance $z = 300$ nm. **c** to **d** Intermediate configurations after vortex manipulation. e) final configuration imaged at $z = 120$ nm and at $T = 5.5$ K for improved resolution and vortex pinning

In the case of curve (i), two vortices are nucleated simultaneously resulting in a -0.4 Hz step, whereas in case (ii) two vortices appear successively with decreasing tip-sample distance.

A controlled lateral manipulation of vortices in a 300 nm-thick Nb film was performed by Straver et al. [626] [Fig. 4.12b–e]. To image the vortices, the tip-sample distance was set to 300 nm and the temperature was kept between 7 and 7.2 K. For their manipulation, the tip was approached to $z = 10$ nm, then moved parallel to the surface to drag along the vortex and finally retracted again to $z = 300$ nm to re-pin the vortex. To image the final state after the manipulation of many vortices to form the letters "SU", the sample was cooled to $T = 5.5$ K to improve the vortex pinning and the tip sample distance was reduced to $z = 120$ nm to obtain a higher lateral resolution [Fig. 4.12e].

More recent MFM work addressed the manipulation of skyrmions in metallic multilayer samples with interfacial Dzyaloshinskii–Moriya interaction (DMI) [725]. The ground state of such samples is governed by a helical magnetization state consisting of a pattern of small up/down magnetic domains with Néel walls with either counterclockwise or clockwise chirality depending on the sign of the DMI. Figure 4.13a shows such a ground state in a Ta(4.7 nm)/[Pt(4 nm/Co(1.3 nm)/ Ta(1.9 nm)]$_{20}$ multilayer sample. The image was acquired at room temperature, under ambient conditions using a Nanosensor PPP-MFMR large magnetic moment tip in tapping/lift mode operation. The stray field of the tip at its apex was estimated to be about 135 mT. A lift height of 100 nm was used to obtain the data in Fig. 4.13a and in the successive scans from Fig. 4.13b to e. Because of the strong stray field of the tip, skyrmions

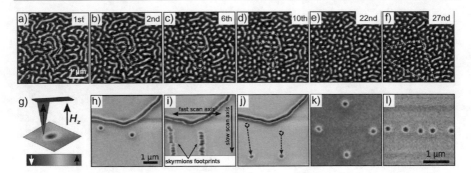

Fig. 4.13 **a** to **f** MFM data obtained on Ta(4.7 nm)/[Pt(4 nm/Co(1.3 nm)/ Ta(1.9 nm)]$_{20}$ multilayer sample. **a** Data acquired in zero field showing the helical ground state with the labyrinth up/down domains. **b** to **f** Successive scans: the tip field increases the number of skyrmions and reduces the average length of the worm-like domains. Figure adapted from [725] with permission from AIP Publishing. **g** Schematics of the tip-up tip magnetization state in an up external field over a skyrmion with a down magnetization. **h** to **l** MFM data obtained on a Pt/CoFeB/MgO multilayer sample. **h** MFM data acquired while keeping the tip at constant lift height. Two skyrmions and a single stripe domain are visible. **i** MFM image mapped in the double passage mode, where each line is scanned twice, first in tapping mode to measure the topography and a second time keeping the tip at the set lift height. **j** Image acquired like **h** showing the two initial skyrmions at a lower position in the image. **k** and **l** Controlled manipulation of selected skyrmions performed by rotating the fast scan axis to move the skyrmions perpendicular to the fast scan axis direction. Figure adapted from Casiraghi et al. [101], distributed under the terms of the Creative Commons Attribution License (http://creativecommons.org/licenses/by/4.0)

are gradually generated in successive scans. After a first scan [Fig. 4.13b], a mixed state of worm-like domains and skyrmions appears. By a repetition of the scans, more and more skyrmions are generated and the average length of the work-domains shortens [Fig. 4.13b–e]. To obtain a higher resolution MFM image, the lift-height was reduced to 50 nm for the final image [Fig. 4.13f] displaying an almost perfect hexagonal skyrmion lattice at least in a part of the scanned area. It is noteworthy that in zero field the skyrmions represent a metastable state, whereas the initial state displayed in Fig. 4.13a represents the systems ground state. Because of the repulsive skyrmion–skyrmion interaction, the system remains locked in the metastable skymion state.

The manipulation of individual skyrmions was reported by Casiraghi et al. [101]. Again, tapping/lift mode MFM operated under ambient conditions was used for imaging and manipulation. Figure 4.13h shows an MFM image obtained at 71 mT on a Pt/CoFeB/MgO-multilayer sample in a single passage mode, where the tip was kept at a set lift height. For the recording of the partial image displayed in Fig. 4.13i the tapping/lift mode operation was used, scanning the sample from the top to the bottom and stopping before the scanning of the selected image area was completed. The image shows the scanlines obtained while the tip was lifted above the sample surface. The field of the tip has moved both skyrmions towards the bottom of the image, as becomes apparent in Fig. 4.13j that is acquired with the tip remaining at constant lift height. Using different slow axis scan directions individual skyrmions could be moved to arbitrary positions [Fig. 4.13k and l].

While the stray field of the MFM tip in combination with external fields or changes in temperature or other external parameters influencing the sample magnetic state can be used to modify the sample micromagnetic state, in most cases the goal remains to avoid irreversible changes of the micromagnetic state of the sample or tip. This can be achieved by reducing the tip-sample interaction by using tips with sufficiently small magnetic moments but sufficient coercivity to obtain a stable tip magnetization. A smaller tip-sample interaction however requires an improved sensitivity which is in the best case limited by the thermodynamical properties of the cantilever. These can be improved by an appropriate cantilever design and by MFM operation in vacuum preferably at low temperatures (see Sect. 4.4.1). In an ideal case, the state of the MFM tip would be switchable to choose for example, between a low moment state for imaging and a high moment state for manipulation. Such tips with a switchable magnetic moment have been fabricated for example by Panchal et al. [506].

4.2.4   Dissipation Contrast

4.2.4.1 Energy Dissipation Mechanisms

For a cantilever oscillating on one of its resonance modes with frequency f_0, the internal friction in the cantilever and its acoustic radiation into its support leads to an intrinsic dissipation of energy that can be calculated from $E_0(A) = \pi c_L A^2 / Q_0$, where c_L is the force constant, Q_0 is the quality factor of the free cantilever (intrinsic quality factor), and A is the oscillation amplitude of the cantilever [279]. If the tip interacts with the sample, the conservative part of the tip-sample interaction leads to a shift of the resonance frequency of the cantilever. If the cantilever is driven by a PLL (Fig. 4.14), the frequency of the excitation always tracks the actual resonance frequency of the cantilever. A dissipative tip-sample interaction affects the cantilever quality factor and causes a change of the amplitude of the cantilever excitation (Fig. 4.14). If the phase feedback keeps the phase error sufficiently small, the cantilever excitation signal tracks the actual cantilever resonance with a high precision. Under this condition, the energy loss per oscillation cycle arising from the tip-sample interaction becomes

$$E_{ts} = E_0 \cdot \left[\frac{A_{exc}}{A_{exc,0}} - \frac{f}{f_0} \right] \approx E_0 \cdot \left[\frac{A_{exc} - A_{exc,0}}{A_{exc,0}} \right], \qquad (4.5)$$

where $A_{exc,0}$ and A_{exc} are the drive amplitudes required to keep the cantilever oscillation amplitude A constant for the free cantilever and for the case of a dissipative tip-sample interaction, respectively. The approximation is possible because in most MFM experiments $\frac{f}{f_0} \approx 1$. Note that if the PLL does not track the actual cantilever resonance frequency, A_{exc} will also contain contributions of the frequency shift. Such an apparent dissipation signal [396] occurs because the cantilever needs to be excited with a higher amplitude if the driving oscillator frequency does not match the cantilever resonance frequency (Fig. 4.14).

Another process possibly leading to an apparent dissipation can occur in scanning force microscopy setups that use a deflection sensor with a non-linear response, for

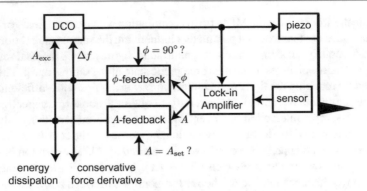

Fig. 4.14 Schematics of a phase-locked loop (PLL) as used for scanning force microscopy. The phase-feedback adjusts the frequency of the digitally-controlled oscillator (DCO) such that phase shift between the cantilever oscillation and actuation remains at 90°. Consequently, the DCO tracks the actual resonance frequency of the cantilever. The amplitude feedback adjusts the excitation amplitude A_{exc} to keep the oscillation amplitude of the cantilever at $A = Q \cdot A_{exc}$. The required excitation amplitude A_{exc} is a measure for the energy dissipated via the the non-conservative tip-sample forces

example a fiber-optical interferometer [396]. In such a setup, the fiber-to-cantilever distance and thus the operation point of the interferometer is typically controlled by an additional feedback (w-feedback). The operation point and with it the deflection sensitivity of the interferometer will change if this w-feedback is not sufficiently fast to correct for tip-sample force-induced changes of the cantilever deflection. This leads to an *apparent* change of the cantilever oscillation amplitude and to a corresponding change of the cantilever excitation amplitude (and thus to an apparent dissipation). The non-harmonicity of the cantilever potential arising from the tip-sample interaction may also lead to a change of the (harmonic) cantilever oscillation amplitude and hence to an apparent dissipation [19,200,279].

A *true* dissipation occurs if the area enclosed by the tip-sample force curve traced within the oscillatory motion of the tip does not vanish, i.e. if

$$E_{ts}(z_{ltp}, A) = \int_0^{2\pi/\omega} F_{ts}(z_{ltp} + A \cdot (1 + \cos(\omega t))) \, dt > 0. \qquad (4.6)$$

It is convenient to classify dissipative processes into hysteresis-related and velocity dependent mechanisms [155,279].

- Hysteresis-related dissipative processes.
  A loss of energy from the oscillating cantilever into the tip-sample system occurs when the tip-sample interaction force probed by the oscillating cantilever becomes hysteretic. Typically, such a hysteresis must occur in each or at least most oscillation cycles to become detectable. Hence, a tip stray field induced lateral hopping of a piece of a domain wall will only lead to a detectable signal if the hopping occurs repeatedly with the up/down oscillation of the tip. Such a domain wall hopping makes the tip-sample force curve that is traced within the oscillatory

path of the tip hysteretic and energy is lost from the oscillating tip into magnons or Eddy currents generated by the rapid hopping of the wall in the sample and possibly also in the tip. The latter occurs if the sudden change of the tip-sample interaction also generates a correspondingly rapid change of the orientation of the tip magnetization field. It is noteworthy that a (repeatedly occurring) domain wall hopping arising from the lateral scanning motion of the tip, also leads to a dissipation of energy. This energy loss is however compensated by additional power from the high-voltage amplifier driving the lateral motion of the scan-piezo. A loss of energy from the oscillating cantilever will presumably also occur because the tip-sample force will suddenly change. However, the hysteresis of the tip-sample force will occur only within *one* oscillation of the tip, such that the small loss of energy will presumably remain undetected.

- Velocity-dependent dissipative processes.
  Apart from energy dissipation arising from hysterestic processes, various mechanisms depending on the velocity of the tip oscillation exist. Velocity-dependent dissipation can be phenomenologically described by an equation of motion

$$m_{\text{eff}} \ddot{z} + \gamma \dot{z} + c_{\text{L}}(z - z_0) = F(t) \tag{4.7}$$

that contains a dissipation force $\gamma \dot{z}$ proportional to the velocity of the oscillating tip, $m_{\text{eff}} = c_{\text{L}}/\omega_0^2$ is the effective mass of the cantilever, and $F(t) = F(z_0 + A\cos(\omega t))$ is the conservative tip-sample interaction force. The dissipation force is the sum of an intrinsic dissipation term which is a property of the free cantilever and a tip-sample interaction-dependent dissipation term. Hence $\gamma = \gamma_0 + \gamma_{\text{ts}}$, where $\gamma_0 = \pi c_{\text{L}}/2\omega_0 Q_0 = m_{\text{eff}}\omega/Q_0$ is related to the intrinsic quality factor Q_0.

The dissipation term arising from the tip-sample interaction relates to the corresponding quality factor as

$$\frac{1}{Q_{\text{ts}}} = \frac{2\omega}{\pi c_{\text{L}}} \cdot \int_0^{\pi} \gamma_{\text{ts}}(z_0 + A\cos\phi)\sin^2\phi \, d\phi. \tag{4.8}$$

For small oscillation amplitudes, the standard relation $\gamma_{\text{ts}} = \pi c_{\text{L}}/2\omega_0 Q_{\text{ts}}$ holds. Typically, the cantilever is in thermal equilibrium, and the mean-square amplitude of the cantilever agrees with the equipartition theorem. Then, the fluctuation-dissipation theorem holds and the flat spectral density of the fluctuations of the tip sample interaction force S_{Fts} are related to the damping coefficient γ_{ts} as

$$S_{\text{Fts}} = 4\gamma_{\text{ts}} k_B T. \tag{4.9}$$

Hence, fluctuations of the tip-sample interaction force generate additional velocity-dependent dissipation described by (4.8) with the additional dissipation coefficient γ_{ts}. This additional dissipation then leads to a detectable change of the oscillation amplitude that can be obtained from (4.5).

The minimally measurable dissipation is given by

$$\gamma_{\min} = \frac{F_{th}}{\omega_0 A_{rms}} , \qquad (4.10)$$

where F_{th} is the minimally measurable force, given by (4.50), $\omega_0 = 2\pi f_0$, where f_0 is the free resonance frequency of the cantilever, and A_{rms} is the rms-oscillation amplitude of the cantilever. From (4.50) discussed in Sect. 4.4.1, it is apparent that the best sensitivity for the measurement of energy dissipation is obtained for a cantilever that has a low stiffness c_L, a high resonance frequency f_0, and a high intrinsic quality factor Q. For a high quality factor operation in vacuum is beneficial (see Sect. 4.4.1).

4.2.4.2 Magnetic Dissipation Force Microscopy

In MFM, the magnetic field emanating from the sample interacts with the magnetic moments of the tip, or conversely, the magnetic field emanating from the tip interacts with the magnetic moments of the sample. As discussed in Sect. 4.2.1, in many cased the observed MFM contrast can be modeled under the assumption that the magnetic moment distributions of the tip and sample are not modified by the presence of the stray fields of the sample and tip, respectively (negligible perturbation case). In Sect. 4.2.2, reversible modifications of the magnetic moment distributions of the sample or tip were discussed. In this case, imaging remains reproducible such that consecutively acquired images are identical. The observed contrast can however not be explained with models neglecting a field-induced change of the micromagnetic structure of the sample or the tip. Such changes of the micromagnetic structure arising from the interaction of the magnetic tip with the sample can lead to a measurable energy dissipation. The physics of the dissipation process can then either be of hysteretic- of velocity-dependent type.

A hysteretic-type dissipation can, for example, occur if the oscillatory magnetic field of the tip at the surface of the sample leads to lateral motion of a domain wall segment away and back to a domain wall pinning site. Note that the latter is required to obtain a hysteretic motion of the wall segment, such that the wall segment becomes depinned for a critical minimal tip-sample distance z_{c1}, but jumps back to this pinning center at a second critical tip-sample distance $z_{c2} > z_{c1}$ under the condition that both critical tip-sample distances are contained within the oscillation path of the tip. If the latter condition is met, the hysteretic domain wall motion occurs in every cycle of the tip oscillation, which then leads to a measurable dissipation. Figure 4.15a shows frequency shift data obtained on a domain in a Cu/Ni(8.5 nm)/Cu(001) sample. The solid and dashed white circles highlight domain wall segments above which a dissipation signal occurs [see white spots in Fig. 4.15b]. After a reduction of the tip-sample distance from 64.4 nm for the images in Fig. 4.15a and b to 61.1 nm for the images in Fig. 4.15c and d, the domain wall segment highlighted by the solid white circle moved and became pinned at another location. The pinning at this new location is sufficiently strong, such that the tip-field can no longer depin the wall segment. Consequently, the dissipation of this wall segment not longer occurs [see solid white circle in Fig. 4.15d], while the wall segment highlighted by the dashed

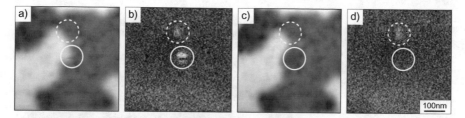

Fig. 4.15 MFM data obtained on a Cu/Ni(8.5 nm)/Cu(001) sample. **a** Δf frequency shift image, and **b** simultaneously acquired A_{exc} dissipation image with a tip-sample distance of 64.4 nm. The solid and dashed white circles highlight segments of the domain wall that lead to increased energy dissipation visible as white spots in the dissipation image (**b**). **c** and **d** Corresponding Δf and A_{exc} data acquired with a tip-sample distance reduced to 61.1 nm. The domain wall segment highlighted by the solid white circle has retracted and becomes pinned sufficiently that the oscillatory field of the tip does no longer depin this wall segment. Consequently, no dissipation is observed, while dissipation still occurs at the wall segment highlighted by the dashed circle

circle still gives rise to a weak dissipation signal. Note that the dissipation images show pronounced spots of higher dissipation [Fig. 4.15b and d] indicating locations where the tip field is sufficiently strong to depin a wall segment.

Often, strong dissipation is observed to occur along a (deformed) circular pattern. Such a dissipation pattern then indicates tip-positions that generate a sufficient field to depin a wall segment located close to the center of the pattern. Alternatively, the field of the sample can also induce instabilities of the tip magnetization. Igelsias-Freire et al. [305] have shown that the field emanating from an in-plane magnetized permalloy dot leads to a reproducibly switching of the magnetization in a small volume close to the tip apex. Figure 4.16a displays the magnetization distribution of the permalloy dot obtained from micromagnetic calculations, while Fig. 4.16b displays the 3-dimensional contours map, where $B_z = (-14.0 \pm 0.3)$ mT. The contours observed in the frequency shift and dissipation images at different tip-sample distances ranging from 250 nm to 100 nm correspond to the contours of constant $B_z = (-14.0 \pm 0.3)$ mT at the corresponding scan-heights [Fig. 4.16c–f]. Note that in some cases, contours of increased dissipation occur, while the corresponding contours are not observed in the frequency shift images.

A velocity-dependent type of magnetic dissipation was described by Grütter et al. [239]. Figure 4.17a shows a constant force gradient image of a 30 nm-thick, 20 μm-sized permalloy square, while Fig. 4.17b shows the simultaneously recorded dissipation image. The full scale variation in the latter image corresponds to a dissipation of 9 pN s/m or a power dissipation of $6 \cdot 10^{-18}$ W. The observed maxima showed a strong dependence on the tip-sample separation, becoming weaker at larger tip-sample distances. The observed energy dissipation was attributed to tip-field induced changes of the domain wall width that then lead to corresponding oscillations of the local strain via magnetoelastic coupling, and thus to the generation of phonons [396, 397]. Dissipation then depends quadratically on the magnetostriction coefficient. Indeed, a larger dissipation is observed on a track of bits written in a 40 nm-thick recording CoCrPt recording media [Fig. 4.17c and d] using the same MFM tip.

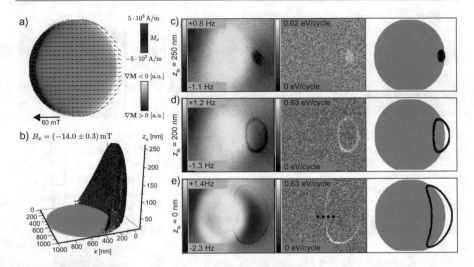

Fig. 4.16 **a** Simulated magnetization distribution of a permalloy dot under an external field of 60 mT along the horizontal direction. **b** The black point represent a field $B_z = (-14.0 \pm 0.3)$ mT emerging from the magnetization distribution shown in (**a**). The orange region represents the permalloy dot. **c**, **d**, and **e** Frequency shift (left column) and dissipation data (central column) compared to the simulations [xy cross-sections from (**b**)], at a tip-sample distance of 250 nm, 200 nm, and 100 nm, respectively. Figure adapted from Igelsias-Freire et al. [305] with permission from AIP Publishing

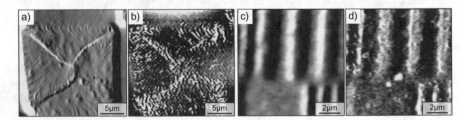

Fig. 4.17 **a** Constant force gradient image of a 30 nm-thick, 20 μm-sized permalloy square element. To enhance details, the image was differentiated along the fast scan-direction. **b** Dissipation image acquired simultaneously with (**a**). White corresponds to a larger cantilever damping, full scale variation in this image corresponds to 9 pN s/m ($\Delta Q/Q = 150/8000$). **c** Constant force gradient image of tracks written in a CoPtCr longitudinal magnetic recording medium. **d** Simultaneously acquired dissipation data Full scale variation corresponds to 45 pN s/m ($\Delta Q/Q = 30/8000$). Figure adapted from Grütter et al. [239] with permission from AIP Publishing

Even larger dissipation was observed on thin films of Terfenol-D, a material with two orders of magnitude larger magnetostriction than permalloy.

4.3   Magnetic Stray Fields

4.3.1   General Concepts

It is convenient to introduce the concept of magnetic charges to calculate the magnetic stray field **H** emanating from a magnetization pattern **M** inside a sample and to understand which information on the latter can be deduced from measurements of the former. Magnetic charges can be defined from one of Maxwell's equations as

$$\nabla \mathbf{B} = \mu_0 \nabla (\mathbf{H} + \mathbf{M}) = 0 \Rightarrow \nabla \mathbf{H} = -\nabla \mathbf{M} , \qquad (4.11)$$

where $\rho_m := -\nabla \mathbf{M}$ is the magnetic volume charge arising from a divergence of **M** inside a magnetic sample. In addition, magnetic surface charges $\sigma_m := \mathbf{n} \cdot \mathbf{M}$ occur at the surfaces of a magnetic sample if there is a magnetization component parallel to the normal vector **n** of the sample [Fig. 4.18]. The magnetic stray field emanating from the surface of a sample can then be written analogous to the electric field arising from electric charges [308] as

$$\mathbf{H} = \int_V \nabla \mathbf{M}(\mathbf{x}') \frac{\mathbf{x} - \mathbf{x}'}{|\mathbf{x} - \mathbf{x}'|^3} \, \mathrm{d}^3\mathbf{x}' + \int_A \mathbf{n} \cdot \mathbf{M} \frac{\mathbf{x} - \mathbf{x}'}{|\mathbf{x} - \mathbf{x}'|^3} \, \mathrm{d}^2\mathbf{x}' , \qquad (4.12)$$

where the first and second integrals arise from the magnetic volume, ρ_m, and magnetic surface charges, σ_m, respectively. Because the magnetization (as any vector field bounded in the domain $V \subseteq \mathbf{R}^3$ which is twice continuously differentiable) can be decomposed as $\mathbf{M} = \mathbf{M}_{irr} + \mathbf{M}_{div}$, where $\nabla \times \mathbf{M}_{irr} = 0$ is a curl-free and $\nabla \cdot \mathbf{M}_{div} = 0$ is a source-free component. It is obvious that the latter component does not contribute to the sample stray field and is thus not accessibe by MFM. Moreover, for all spatial distributions of the magnetic volume charge ρ_m inside a magnetic sample and magnetic surface charge σ on its boundaries, an effective magnetic surface charge $\sigma_{m,eff}$ generating the same magnetic stray field exists (see (4.44) in

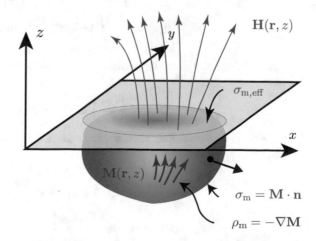

Fig. 4.18 The stray field, **H**(**r**, z) arises from the magnetic volume charge density, $\rho_m(\mathbf{r}, z) = -\nabla \mathbf{M}(\mathbf{r}, z)$ and from the magnetic surface charge density, $\sigma_m = \mathbf{M}(\mathbf{r}, z) \cdot \mathbf{n}(\mathbf{r}, z)$

Sect. 4.3.2 for further details). In addition, even in the absence of magnetic volume charges, different magnetization patterns $\mathbf{M}(x, y, z)$ inside a sample can generate the same magnetic surface charge pattern. Hence, without additional information it is not possible to uniquely derive the magnetization pattern from the magnetic stray field independent of the experimental method used to measure the latter. Quantitative MFM methods as discussed in Sects. 4.4.4 and 4.4.5 can thus be used to obtain an equivalent magnetic surface charge pattern.

Outside a magnetic sample, we can further assume that $\nabla \times \mathbf{B} = \mu_0 \mathbf{j} = 0$, i.e. no macroscopic and microscopic currents exist. Then the magnetic induction $\mathbf{B}$ and trivially the magnetic field $\mathbf{H}$ are conservative fields and can be obtained from a scalar potential ϕ_m as

$$\mathbf{H} = -\nabla \phi_m . \tag{4.13}$$

From the definition of the magnetic surface and volume charge distribution we find that the scalar magnetic potential ϕ obeys

$$\Delta \phi = 0 \tag{4.14}$$

$$\Delta \phi = \rho_m , \tag{4.15}$$

where (4.14) and (4.15) hold outside and inside the sample, respectively. In addition the magnetic potential must fulfill the boundary condition

$$\left. \frac{\partial \phi}{\partial n} \right|_{z \to 0^+} - \left. \frac{\partial \phi}{\partial n} \right|_{z \to 0^-} = \sigma_m . \tag{4.16}$$

In order to derive some general properties of magnetostatic fields, it is convenient to use a 2D-Fourier space, where the coordinates $(\mathbf{r}, z) := (x, y, z)$ are transformed to $(\mathbf{k}, z) := (k_x, k_y, z)$. The Fourier transform and the inverse Fourier transform are defined by

$$\mathbf{G}(\mathbf{k}) = \int \mathbf{g}(\mathbf{r}) e^{-i \mathbf{k} \cdot \mathbf{r}} \, dx \, dy \tag{4.17}$$

$$\mathbf{g}(\mathbf{r}) = \frac{1}{4\pi} \int \mathbf{G}(\mathbf{k}) e^{i \mathbf{k} \cdot \mathbf{r}} \, dk_x \, dk_y . \tag{4.18}$$

In this 2D-Fourier space the nabla operator takes the form, $\nabla_k = (ik_x, ik_y, \frac{\partial}{\partial z})$ because

$$\mathscr{F}\left(\frac{\partial^n f(x)}{\partial x^n} \right) = (ik_x)^n F(k_x) . \tag{4.19}$$

Outside a magnetic sample, the Laplace equation (4.14) holds. In the 2D-Fourier space

$$\Delta_k \phi = -(k_x^2 + k_y^2)\phi + \frac{\partial^2 \phi}{\partial z^2} = 0 \tag{4.20}$$

$$\phi(\mathbf{k}, z) = \frac{1}{k^2} \frac{\partial^2 \phi(\mathbf{k}, z)}{\partial z^2} , \tag{4.21}$$

is obtained, where $k = \sqrt{k_x^2 + k_y^2}$. The general solution for ϕ outside the sample becomes

$$\phi(\mathbf{k}, z) = \phi(\mathbf{k}, 0)e^{-kz} , \tag{4.22}$$

where z is the distance above an arbitrary xy-plane above (not intersecting) the sample. Hence the scalar magnetic potential $\phi(\mathbf{k}, z)$ and also the stray field $\mathbf{H}(\mathbf{k}, z)$

$$\mathbf{H}(\mathbf{k}, z) = - \begin{pmatrix} ik_x \\ ik_y \\ \frac{\partial}{\partial z} \end{pmatrix} \phi(\mathbf{k}, z) = \begin{pmatrix} -ik_x \\ -ik_y \\ k \end{pmatrix} \phi(\mathbf{k}, z) , \tag{4.23}$$

decay exponentially with increasing distance z from the sample surface. Outside the sample, the nabla operator can be written as

$$\nabla_k = (ik_x, ik_y, -k) . \tag{4.24}$$

Hence the x- and y-components of the stray field can be obtained from its z-component as

$$H_z(\mathbf{k}, z) = k\phi(\mathbf{k}, z) \tag{4.25}$$

$$H_{x,y}(\mathbf{k}, z) = -\frac{ik_{x,y}}{k} H_z(\mathbf{k}, z) . \tag{4.26}$$

It is therefore sufficient to measure the z-component of the stray field in an xy-plane above the sample. The other field components can then be obtained from it.

Inside a magnetic sample, the Poisson equation (4.15) holds. In the 2D-Fourier space

$$\Delta_k \phi = -(k_x^2 + k_y^2)\phi + \frac{\partial^2 \phi}{\partial z^2} = -\rho_m \tag{4.27}$$

$$\phi(\mathbf{k}, z) = \frac{1}{k^2} \left[\frac{\partial^2 \phi(\mathbf{k}, z)}{\partial z^2} + \rho_m(\mathbf{k}, z) \right] \tag{4.28}$$

is obtained. The general solution for ϕ inside the sample becomes

$$\phi(\mathbf{k}, z) = \phi(\mathbf{k}, 0)e^{\mp kz} + \frac{\rho_m(\mathbf{k}, z)}{k^2}, \tag{4.29}$$

where the minus and plus sign hold for $z > 0$ and $z < 0$, respectively.

4.3.2   Field of Thin Film Sample

For simplicity, we first calculate the field above a thin film sample with thickness t, a top surface located at $z = 0$, a pattern of magnetic surface charges $\sigma_{m,\text{top}}(\mathbf{r}) = -\sigma_{m,\text{bottom}}(\mathbf{r}) = \sigma_m(\mathbf{r})$, and a magnetic volume charge distribution $\rho(x, y, z)$ that is assumed to be independent of the z-position inside the sample [Fig. 4.19a]. Outside, i.e. above ($z > 0$) and below ($z < -t$) the thin film sample, the Laplace equation (4.20) holds, and the general solution (4.22) can be written as

$$\phi_{oa}(\mathbf{k}, z) = A_{oa}(\mathbf{k})e^{-kz} \quad \text{for} \quad z > 0 \tag{4.30}$$

$$\phi_{ob}(\mathbf{k}, z) = A_{ob}(\mathbf{k})e^{kz} \quad \text{for} \quad z < -t . \tag{4.31}$$

Inside the sample, the Poisson equation (4.27) holds, and the and the general solution (4.29) can be written as

$$\phi_i(\mathbf{k}, z) = A_{i1}(\mathbf{k})e^{kz} + A_{i2}(\mathbf{k})e^{-kz} + \frac{\rho_m(\mathbf{k})}{k^2} \quad \text{for} \quad -t < z < 0 . \tag{4.32}$$

Using the boundary conditions at the surface located at $z = 0$ and at $z = -t$

$$\phi_{oa,ob}(\mathbf{k}, z) = \phi_i(\mathbf{k}, z) \tag{4.33}$$

$$\frac{\partial \phi_{oa,ob}(\mathbf{k}, z)}{\partial z} - \frac{\partial \phi_i(\mathbf{k}, z)}{\partial z} = \sigma_m , \tag{4.34}$$

the coefficients $A_{oa,ob,i1,i2}(\mathbf{k})$ become

$$A_{oa}(\mathbf{k}) = -\frac{1 - e^{-kd}}{2}\left(\frac{\sigma_m(\mathbf{k})}{k} + \frac{\rho_m(\mathbf{k})}{k^2}\right) \tag{4.35}$$

$$A_{ob}(\mathbf{k}) = -\frac{1 - e^{-kd}}{2}\left(\frac{\sigma_m(\mathbf{k})}{k} - \frac{\rho_m(\mathbf{k})}{k^2}\right) \tag{4.36}$$

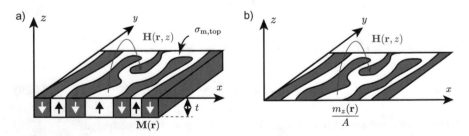

Fig. 4.19 a Stray field, $\mathbf{H}(\mathbf{r}, z)$ above a thin film sample of thickness t with a pattern of an up/down domains generating magnetic surface charge densities $\sigma_{m,\text{top}}(\mathbf{r}) = -\sigma_{m,\text{bottom}}(\mathbf{r}) = \sigma(\mathbf{r})$ and a magnetic volume charge density $\rho_m = -\nabla\mathbf{M}(\mathbf{r})$ that is independent of the z-position inside the sample. **b** Stray field, $\mathbf{H}(\mathbf{r}, z)$ above a pattern of magnetic moments with a magnetic moment per unity are density of $\frac{m_z(\mathbf{r})}{A}$

$$A_{i1}(\mathbf{k}) = \frac{1}{2}\left(\frac{\sigma_m(\mathbf{k})}{k} - \frac{\rho_m(\mathbf{k})}{k^2}\right) \tag{4.37}$$

$$A_{i2}(\mathbf{k}) = -\frac{e^{-kd}}{2}\left(\frac{\sigma_m(\mathbf{k})}{k} + \frac{\rho_m(\mathbf{k})}{k^2}\right) . \tag{4.38}$$

Inserting these coefficients into the general solutions, given by (4.30), (4.31), and (4.32), the magnetic potentials become

$$\phi_{oa}(\mathbf{k}, z) = \frac{1}{2}\left(\frac{\rho_m(\mathbf{k})}{k^2} + \frac{\sigma_m(\mathbf{k})}{k}\right) \cdot e^{-kz} \cdot [1 - e^{-kt}] \tag{4.39}$$

$$\phi_{ob}(\mathbf{k}, z) = \frac{1}{2}\left(\frac{\rho_m(\mathbf{k})}{k^2} - \frac{\sigma_m(\mathbf{k})}{k}\right) \cdot e^{kz} \cdot [1 - e^{-kt}] \tag{4.40}$$

$$\phi_{i}(\mathbf{k}, z) = \frac{\rho_m(\mathbf{k})}{k^2}\left(1 - e^{-kt/2}\cosh\left(k(z + t/2)\right)\right)$$

$$+ \frac{\sigma_m(\mathbf{k})}{k}e^{-kt/2}\sinh\left(k(z + t/2)\right) , \tag{4.41}$$

where the factors $e^{\pm kz}$ and $[1 - e^{-kt}]$ are called the distance and thickness loss factors, respectively. From these expressions the stray field in 2D Fourier space can easily be obtained from $\mathbf{H}(\mathbf{k}, z) = -\nabla_k \phi_{oa,ob,i}(\mathbf{k}, z)$.

For a sample with an up/down domain pattern, with a magnetization uniform through the thickness, and Bloch walls, i.e. $\rho_m := -\nabla\mathbf{M} = 0$ and $\sigma_m = M_z \cdot n_z$, the z-component of the stray field above the sample becomes

$$H_z(\mathbf{k}, z) = \frac{M_z(\mathbf{k})}{2} \cdot e^{-kz} \cdot [1 - e^{-kt}] , \tag{4.42}$$

where the other stray field components can be obtained from $H_z(\mathbf{k}, z)$ using (4.26).

Note that for the case of Néel walls, the magnetic volume charge, ρ_m, no longer vanishes and must be considered [643]. Equation (4.42) then evolves into

$$H_z(\mathbf{k}, z) = \left(\frac{M_z(\mathbf{k})}{2} + \frac{\nabla_k\mathbf{M}(\mathbf{k})}{2k}\right) \cdot e^{-kz} \cdot [1 - e^{-kt}] . \tag{4.43}$$

Hence, an effective magnetic surface charge density can be defined as

$$\sigma_{m,eff}(\mathbf{k}) := \sigma_m(\mathbf{k}) + \frac{\nabla_k\mathbf{M}}{k} , \tag{4.44}$$

where the second term reflects the propagation of the magnetic volume charge density ρ_m to the top surface.

Figure 4.20 highlights some important aspects of the stray field generated from a sample containing a periodical pattern of up/down domains. If the sample is of infinite thickness [Fig. 4.20a], the stray field close to the surface is reminiscent of the up/down magnetization pattern with sharp domain walls. With increasing distance

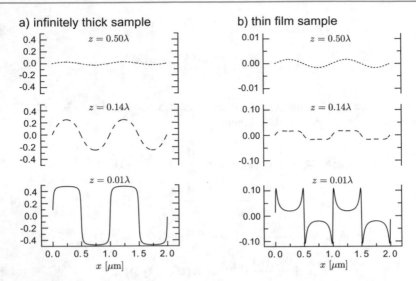

a) infinitely thick sample   b) thin film sample

Fig. 4.20 Stray field of a thin film sample with infinite thickness (**a**), and a thickness of 10 nm (**b**) containing a periodic pattern of up/down domains (domain size is 500 nm). For both samples, the stray field decays rapidly with increasing distance z from the sample surface. At larger distances from the surface, smaller spatial features of the stray field, such as the field above the sharp domain walls gradually thus appear more diffuse, or with a smaller lateral resolution. For the thin film sample in (**b**), the field above the center of the domain at small distances from the sample surface becomes smaller than that near the domain walls. This is because the stray field arising from the magnetic surface charge density at the sample top surface is almost compensated by that from the magnetic surface charge density at the sample bottom surface

z from the surface, the sharp domain walls gradually thus appear more diffuse, or with a smaller lateral resolution, until only the largest spatial frequency is visible, and the stray field appears sinusoidal. For the sample with a finite thickness of 10 nm [Fig. 4.20b], the stray field above the center of the domain at small distances from the sample surface becomes smaller than that near the domain walls. This is because the stray field arising from the magnetic surface charge density at the sample top surface is almost compensated by that from the magnetic surface charge density at the sample bottom surface.

If the thickness t of a sample becomes sufficiently small, the thickness loss term in (4.42) can be Taylor expanded. Then

$$H_z(\mathbf{k}, z) = \frac{m_z(\mathbf{k})}{2A} \cdot k \cdot e^{-kz} , \qquad (4.45)$$

where the first factor is the areal density of the z-component of the magnetic moment. Equation (4.45) is convenient, for example, when the field arising from a distribution of pinned uncompensated spins at the ferromagnet/antiferromagnet interface of an exchange biased sample needs to be described [327, 328, 579] [Fig. 4.19b].

4.3.3   Effects of the Domain Wall on the Field

Section 4.3.2 addresses the transfer function theory for arbitrary patterns of domains in ferromagnetic thin films with a magnetization that is perpendicular to the surface of the film and homogeneous through the film thickness. Figure 4.19a shows a typical pattern of up/down domains. The magnetization changes from the up to the down direction from one pixel to the next. For the case of Bloch domain walls, the angle between the vertical direction and the local magnetization in a Bloch domain wall running along the y-direction is given by

$$\theta(y) = \arctan\left[\sinh\left(\frac{\pi y}{\delta_{dw}}\right)\right] + \frac{\pi}{2} , \qquad (4.46)$$

where δ_{dw} is the domain wall width [and is displayed in Fig. 4.21a]. The z-component of the normalized magnetization is given by

$$m_z(y) = \tanh\left(\frac{\pi y}{\delta_{dw}}\right) \qquad (4.47)$$

$$m_z(k_y) = \frac{i\delta_{dw}}{\sinh\left(\frac{k_y \delta_{dw}}{2}\right)} , \qquad (4.48)$$

where (4.47) describes m_z in direct [see Fig. 4.21b] and (4.48) in Fourier space.

Calculations of the stray field arising from magnetization or equivalent magnetic surface charge patterns are advantageously performed in a 2D-Fourier space (Sect. 4.3.1). Typically, a pattern of up/down domains as depicted in Fig. 4.19a is available. Because of the rotation of the magnetization inside the domain wall, the component M_z, and with it the magnetic surface charges are reduced at the location of the walls. It is thus necessary to introduce the wall with its diameter, δ_{dw} into the

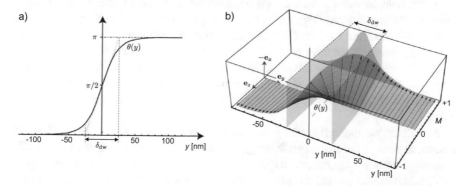

Fig. 4.21 Simulation of a Bloch wall for a material with a uniaxial magnetic anisotropy and a domain wall diameter $\delta_{dw} = 47$ nm. **a** The angle between the domain magnetization and the magnetization at distance y inside the domain wall $\theta(y)$ is given by (4.46). The definition of δ_{dw} is represented in the graph. **b** Simulation of the magnetization rotation through a Bloch wall. The definition of δ_{dw} is represented by the blue planes

model magnetization pattern. This is conveniently performed in 2D-Fourier space, where the magnetization pattern with the Bloch walls, $M_{z,\text{Bloch}}(\mathbf{k})$, can be obtained from the magnetization pattern with sharp, step-like walls, $M_{z,\text{step}}(\mathbf{k})$ by

$$M_{z,\text{Bloch}}(k_y) = \frac{\delta_{\text{dw}}}{2\sinh\left(\frac{k_y \delta_{\text{dw}}}{2}\right)} \cdot M_{z,\text{step}}(\mathbf{k}) \,, \tag{4.49}$$

as can be derived by dividing the 2D-Fourier transform of $\mathscr{F}(2\text{Hea}(y) - 1) = \frac{2}{ik_y}$, where $\text{Hea}(y)$ is the Heavyside step function, by the normalized magnetization in Fourier space given in (4.48). The expression in (4.49) allows the transformation of the step-like domain walls in a model magnetization pattern as displayed in direct space into a corresponding pattern that includes Bloch domain walls with the correct Bloch wall shape and diameter δ_{dw} as described by (4.47). The Bloch walls lead to a reduction of the surface charge at the wall locations. Note, however, that for the case of Néel walls, which for example occur in systems with perpendicular magnetic anisotropy with a sufficiently strong interfacial Dzyaloshinskii–Moriya interaction [38], the magnetization pattern in the wall has a non-vanishing divergence. Consequently, magnetic volume charges appear at the wall locations, $\rho_{\text{m}} = -\nabla \mathbf{M} \neq 0$. Depending on the direction of rotation of the magnetization in the Néel wall, the stray field arising from the volume magnetic charges at the walls, will amplify or attenuate the stray field generated from the surface charges [430,446,643] (Sect. 4.4.6.6). To calculate the field above the surface of a sample, an effective magnetic surface charge density as given in (4.44) can then be used.

4.3.4 Fields from Roughness and Thickness Variations

A extended magnetic thin film that is homogeneously magnetized does *not* generate a magnetic field outside the film. This is because the field-generated magnetic surface charge at the top surface compensates that arising from the opposite magnetic surface charge at the bottom surface. Note that the reduced field above the center of large domains [Fig. 4.20b, bottom panel] is caused by the same mechanism. Mathematically, this follows from the thickness loss factor $[1 - e^{-kt}]$ which vanishes for large wavelengths λ or small $k = \frac{2\pi}{\lambda}$. Hence, a magnetic stray field and with it an MFM frequency shift (or phase) contrast arises only when a pattern of up/down domains exist.

A realistic magnetic thin film is however neither completely flat nor uniform in thickness, even in the case of epitaxial growth on a crystalline substrate [446]. While a homogeneously magnetized flat film of constant thickness does not generate a stray field, this is no longer the case for a film with a roughness but constant thickness [Fig. 4.22a], or a film with a flat topography but a varying thickness [Fig. 4.22b, or both (not shown)]. The existence of a stray field for these cases can be understood from the graphical representations displayed in the middle and bottom panels of Fig. 4.22a and b. The schematical film structures depicted in these figures are

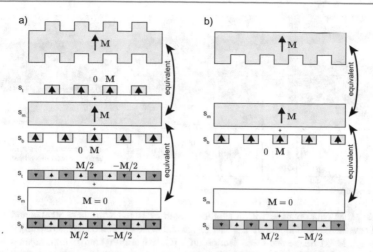

Fig. 4.22 **a** Sample with periodical roughness and decomposition into systems of layers generating the same stray field. **b** Sample with periodical thickness variation and decomposition into systems of layers generating the same stray field

considered to be composed of several thin film strata. For the case of a rough film [Fig. 4.22a] these are: stratum s_m representing the middle part of the film, strata s_{top} and s_{bottom} representing the up/down topography at the film top and bottom surfaces of the film, respectively. Similarly, the film depicted in Fig. 4.22b can be decomposed into the strata s_m and s_b. While the stratum s_m with constant thickness (and homogeneous magnetization) does not generate a stray field, the strata s_t and s_b displayed give rise to a stray field, because these consist of blocks of up magnetization. The stray field of these strata is equivalent to that originating from a $\pm M/2$ up/down magnetization pattern. Note that apart from the stray fields arising from the roughness and thickness variation of a magnetic thin film, local stray fields will also occur in case of a grain-to-grain variation of the z-component of the magnetization that can, for example, arise from a corresponding variation of the anisotropy axes of the grains. Figure 4.23b shows a domain in a Pt(2 nm)/Co(0.6 nm)/Pt(10 nm) trilayer deposited onto a thermally oxidized Si-wafer. Because the domain size is several micrometers in such thin films, the MFM contrast arising from the domains is localized to the vicinity of the wall. Under such conditions, the stray fields arising from the film thickness variations, film roughness, or grain-to-grain variations of the magnetization component M_z, become apparent in the MFM image provided the sensitivity of the instrument is sufficient to separate these contrast variations from the measurement noise (Sect. 4.4.1). The zoomed image displayed in Fig. 4.23c clearly shows that the granular contrast apparent in Fig. 4.23b and the cross-section below does not arise from measurement noise but from the morphology of the polycrystalline magnetic film. Note that the data shown in Fig. 4.23b and c was recorded in vacuum with a cantilever having a quality factor of about 10^6 to maximize the measurement sensitivity (Sect. 4.4.1). The contrast contributions arising from the film morphology can be disentangled from those arising from the domain pattern and non-magnetic contributions, see Sect. 4.4.3.

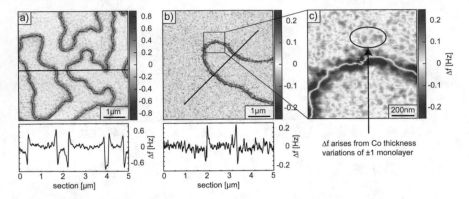

Fig. 4.23 MFM data obtained with a cantilever with $c_L \approx 0.3\,\text{N/m}$, $f_0 \approx 50\,\text{kHz}$ and a quality factor of several hundred thousand. **a** MFM image and cross-section on a $[\text{Co}(0.6\,\text{nm})/\text{Pt}(1\,\text{nm})]_9$-multilayer. **b** MFM image and cross-section on a $[\text{Co}(0.6\,\text{nm})/\text{Pt}(1\,\text{nm})]_1$-multilayer. **c** Shows a zoomed image of the area highlighted by the black square. The red/blue Δf-contrast arises from a ± 1 atomic monolayer thickness variation of the Co-film

4.4   Quantitative Magnetic Force Microscopy

MFM measures either a phase shift $\Delta\phi(\mathbf{r}, z)$ or a frequency shift $\Delta f(\mathbf{r}, z)$, which arise from the summed derivatives of all forces acting on the tip. Experimental methods for the separation of these force are discussed in Sect. 4.4.3. Three different cases for the magnetic field-mediated contrast have been described in Sect. 4.2. Here only the case where the influence of the stray field of the sample on the tip magnetization and that of the tip on the sample magnetization remains negligible (Sect. 4.2.1) is further analyzed.

For the magnetism community it is of particular importance that MFM can be performed with optimized lateral resolution and sensitivity (Sect. 4.4.1). Further, it should be possible to reproducably map the micromagnetic state of a sample in different external magnetic fields and temperatures (Sect. 4.1.2) and to disentangle the different contributions to the measured MFM contrast (Sect. 4.4.3). Furthermore, a calibration of the MFM tip is required to convert or relate the measured phase shift $\Delta\phi(\mathbf{r}, z)$ or a frequency shift $\Delta f(\mathbf{r}, z)$, arising from the magnetic field-mediated tip-sample interaction to a physical property more relevant for the magnetism community, such as stray field (Sects. 4.4.4 and 4.4.5). As discussed in Sect. 4.3, the stray field $\mathbf{H}(\mathbf{r}, z)$ emanating from a lower surface of a sample or from volume magnetic charges inside the sample, can be propagated to the surface of the sample, $z = 0$. As a result, an effective magnetic surface charge distribution $\sigma_{m,\text{eff}}$ (4.44) that generates the stray field above the sample surface can be obtained. The magnetization or the magnetic moment distribution inside the sample generating a specific $\sigma_{m,\text{eff}}$ is however not unique. It is thus generally not possible to obtain the magnetization inside the sample from the stray field above the sample. Relevant information on the micromagnetic state of the sample can however still be obtained, for example by fitting stray field or effective surface charge distributions arising from model magnetization patterns to the corresponding distributions obtained from the deconvolved MFM data.

The MFM transfer function theory presented in Sect. 4.4.4 is based on early work describing MFM contrast formation [104,586] but was then further developed by Hug [295] and van Schendel [662] two decades ago to obtain a precise quantitive description of the measured MFM contrast by including the calibration of the response of the tip to the relevant spatial wavelengths of the stray field (see Sect. 4.4.5). More recently various groups showed a renewed interest in the quantitative interpretation of MFM data [38,60,290,327,328,341,429,430,446,506,578,579,671–673,729]. Substantial efforts have addressed the limits of the tip-transfer function obtained from a calibration and best methods for the deconvolution process [290,481]. Unfortunately, all data for presented in these studies have been acquired with standard MFM equipment operated under ambient conditions, leading to a signal-to-noise factor that remained strongly limited by the low cantilever quality factor.

Although the MFM work by van Schendel [662] was still performed under ambient conditions, using static mode MFM operation, the full potential of quantitative MFM could only be explored by operation under vacuum, which is beneficial for the force derivative sensitivity of the instrument. Unfortunately, easy-to-use scanning force microscopes operating under vacuum conditions are only available from a small number of companies [1,3–5], and the resources invested into such products remained limited. The lack of an early availability of well-designed, robust, versatile and easy-to-use instruments presumably is the reason for the delayed and still lacking uptake of quantitative MFM techniques particularly by the community of researchers interested in magnetic thin film research. Possibly, recent interest of several groups specialized in MFM will change the situation and may lead to a revival of magnetic force microscopy for the study of magnetic samples.

4.4.1 In-Vacuum Operation for Improved Sensitivity

The common goal of most magnetic force microscopy experiments is to obtain a high spatial resolution. Hence, stray field components with small spatial wavelengths, λ or large spatial wave-vectors, $k = \frac{2\pi}{\lambda}$, need to be detected. Because of the exponential decay of the magnetic field with increasing wave-vector (see for example (4.42)), the magnetic force microscope needs to be able to detect these fields or the small forces or force derivatives arising from these fields. Trivially, the magnetic force can be increased by using MFM tips carrying a larger magnetic moment. This however implies that the tip stray field and with it the mutual influence of the tip on the micromagnetic state of the sample are also increased. Further, a larger magnetic moment of the tip is typically achieved with a larger thickness of the ferromagnetic coating, which would again limit the achievable spatial resolution. Hence, the thickness of the magnetic coating on the MFM tip and with it the magnetic moment of the tip need to be sufficiently small not to limit the spatial resolution but especially not to influence the micromagnetic state of the sample. Consequently, the force or force

Table 4.1 The minimum measurable force derivative (last line) at $T = 300\,\text{K}$ for an oscillation amplitude $A_{\text{rms}} = 5/\sqrt{2}\,\text{nm}$ is calculated for different cantilevers and a tuning fork, as typically used for MFM

Parameters	Cantilever in air	Tuning fork in air	Cantilever in vacuum	High-Q cantilever in vacuum
c_L (N/m)	3	1800	0.3	0.3
f_0 (Hz)	50'000	32500	50'000	50'000
Q	100	1870	30'000	1'000'000
$\left.\frac{\partial F_z}{\partial z}\right\|_{\text{min}}$ (μN/m)	20	140	0.36	0.063

gradient sensitivity of the MFM is crucial for obtaining a high spatial resolution particularly on samples with a small magnetic coercivity.

The thermal noise of the cantilever is given by [23, 356, 563]

$$F_{\text{th}} = \sqrt{\frac{4\pi k_B T c_L B_w}{\omega_0 Q}}\,, \tag{4.50}$$

$$\left(\frac{\partial F_z}{\partial z}\right)_{\text{th}} = \frac{1}{A_{\text{rms}}} \cdot \sqrt{\frac{4\pi k_B T c_L B_w}{\omega_0 Q}}\,, \tag{4.51}$$

where F_{th} and $\left(\frac{\partial F_z}{\partial z}\right)_{\text{th}}$ are the minimally measurable rms force and z-derivative of the z-component of the force, respectively, $A_{\text{rms}} = A_0/\sqrt{2}$ is the rms oscillation amplitude of the cantilever, $k_B = 1.38 \cdot 10^{-23}\,\text{JK}^{-1}$ is the Boltzmann constant, T is the temperature, c_L is the stiffness of the cantilever, B_w is the measurement bandwidth, $\omega_0 = 2\pi f_0$ with f_0 being the free resonance frequency of the cantilever, and Q is the mechanical quality factor of the cantilever. Using (4.3), the force derivative noise given in equation (4.51) can easily be converted into the minimally detectable tip-sample interaction force derivative.

Other noise sources are the detector and the oscillator noise which are given by [356]

$$\left(\frac{\partial F_z}{\partial z}\right)_{\text{det}} = \frac{2 c_L n_q B_w^{\frac{3}{2}}}{\sqrt{3} A_{\text{rms}} f_0}\,, \tag{4.52}$$

$$\left(\frac{\partial F_z}{\partial z}\right)_{\text{osc}} = \frac{c_L n_q \sqrt{B_w}}{A_{\text{rms}} Q}\,, \tag{4.53}$$

where n_q is the deflection noise sensitivity that is typically $\ll 100\,\text{fm}/\sqrt{\text{Hz}}$ for typical beam deflection sensors and can reach $1\,\text{fm}/\sqrt{\text{Hz}}$ for advanced interferometric sensors [284]. In any case, for the relatively soft cantilevers used in MFM with force constants $\ll 10\,\text{N/m}$, resonance frequencies around $50\,\text{kHz}$ and oscillation amplitudes $A \geq 5\,\text{nm}$, the thermal noise of the cantilever is the dominating source of noise (see also Table 4.1).

For an optimization of the thermal noise, it is useful to consider the stiffness and resonance frequency of a rectangular cantilever in its first flexural oscillation mode [541]

$$c_L = \frac{\alpha_1^4}{48} \cdot \frac{Etw}{L^3} \tag{4.54}$$

$$\omega_0 = \frac{\alpha_1^2}{\sqrt{12}} \cdot \sqrt{\frac{E}{\rho}} \cdot \frac{w}{L^2}, \tag{4.55}$$

where $\alpha_1 = 1.875$, $E = 1.69 \cdot 10^{11}\,\text{N/m}^2$ is the elastic modulus of silicon, $\rho = 2330\,\text{kg/m}^3$ is the density of silicon, t, w, and L are the thickness, width and length of the cantilever, respectively. It has been pointed out by Rugar et al. [563] that small width and long cantilevers and hence soft cantilevers are beneficial for a high force sensitivity because according to (4.54) and (4.55)

$$\frac{c_L}{\omega_0} = \frac{\sqrt{3}\alpha_1^2}{24} \cdot \sqrt{\frac{\rho}{E}} \cdot \frac{w}{L^2}. \tag{4.56}$$

Indeed, for cantilevers having stiffnesses in the low micro Newton per meter range, attonewton force sensitivities haven been obtained at low temperatures [563]. Such soft cantilevers are however not convient for MFM. Instead, cantilevers with stiffness of $0.1-10\,\text{N/m}$ and with resonance frequencies between 10 and $200\,\text{kHz}$ are typically used to obtain reasonable imaging speeds, and avoid a snap-to-contact if the tip of such a cantilever is approached to the surface of a sample. The most important factor limiting the sensitivity is the cantilever quality factor Q. Different factors that reduce the quality factor need to be considered. According to Lübbe et al. [408], the inverse of the quality factor depends on

$$\frac{1}{Q} = \frac{1}{Q_{\text{vol}}} + \frac{1}{Q_{\text{sup}}} + \frac{1}{Q_{\text{TED}}} + \frac{1}{Q_{\text{surf}}} + \frac{1}{Q_{\text{mount}}} + \frac{1}{Q_{\text{mol}}}, \tag{4.57}$$

which are the volume, support, thermoelastic, surface, mount and molecular friction losses. The latter occurs from the friction that the oscillating cantilever experiences in air and typically is the dominating factor limiting the cantilever sensitivity for magnetic force microscope instruments operated under ambient conditions. For ease-of-use, such instruments typically perform MFM measurements using a double passage operation mode, where each scan-line is recorded twice (see Fig. 4.4b in Sect. 4.1.2). First, the topography is measured using the intermittent contact mode, which requires cantilevers with reasonable high force constants (typically a few N/m) for stable operation. Then, the same line is scanned again but with the tip lifted by a preset height above the previously recorded topography. Because of the low quality factor $Q \approx 100$ and the slightly unfavorable stiffness-to-frequency ratio (with $c_L = 3\,\text{N/m}$ and $f_0 = 50\,\text{kHz}$), a sensitivity of about $20\,\mu\text{N/m}/\sqrt{\text{Hz}}$ (see Table 4.1) is obtained. A sensitivity that is a factor of seven worse, but at least in the same order of magnitude can be obtained with a tuning fork sensor that has also been used for MFM [157,598,648] because of the quality factor that remains reasonably high even under ambient conditions ($Q = 1870$ [584]). Note that tuning fork sensors are widely applied for atomic resolution work [210], but because of their high stiffness to frequency ratio are not well suited to map small magnetic forces. This becomes apparent, when comparing their sensitivity with that of a cantilever sensor optimally chosen for MFM operation under vacuum conditions (Table 4.1) that has an improved sensitivity by more than three orders of magnitude.

Unfortunately, most MFMs are still operated under ambient conditions [536], although it was pointed out already in 1991 by Albrecht et al. [23], that operation under vacuum conditions typically increases the quality factor of the cantilever, Q by by more than two orders of magnitude from about 100 to 30'000. With this, a sensitivity of about $0.36\,\mu N/m/\sqrt{Hz}$, about a factor of 50 better than that obtained in air, is reached (Table 4.1). In our work, we further reduced the surface losses and obtained cantilever quality factors of 10^6 and with it a sensitivity of $0.063\,\mu N/m/\sqrt{Hz}$ at room temperature. Figure 4.23b shows MFM data obtained on a Pt(3 nm)/Co(0.6 nm)/Pt(10 nm) sample. Micrometer sized domains are typical for such a sample with a low magnetic film thickness and perpendicular magnetic anisotropy. Consequently, only a part of an up and down domain are visible in the $5\,\mu m \times 5\,\mu m$ image shown in Fig. 4.23b. As derived in Sect. 4.3.2, the magnetic field away from the domain walls will then become extremely small. Hence, only the domain wall and not the domains generate a noticeable MFM contrast. The noticeable granular up/down red/blue contrast visible inside the domains is however not noise, as it becomes apparent from a reduced scan-size MFM image [Fig. 4.23c]. The granular contrast does clearly not arise from thermal noise but is caused areal variations of the magnetic moment density arising from local film roughness and variations of the Co film thickness (see Fig. 4.22 in Sect. 4.3.4). This clearly reveals that MFM performed with high-quality factor cantilevers is sufficiently sensitive to image the stray fields arising from a few monolayers of magnetic materials or even from a submonolayer of pinned uncompensated spins at the interface of a ferromagnet/antiferromagnet multilayer sample showing a magnetic exchange bias effect [579].

MFM operation in vacuum is clearly beneficial in terms of sensitivity, but comes at the cost of a more complex instrumentation, particularly if operated under ultra-high vacuum (UHV) conditions. UHV instruments are notoriously challenging to operate, and a sample or cantilever exchange can take several hours making such instrument not practical for the routine imaging of magnetic samples, but are used only if special conditions such as low temperatures and high fields are required [446,684,729]. However, to obtain a high quality factor is is sufficient to operated the MFM under moderate vacuum conditions, for example in the range of $10^{-7} - 10^{-6}$ mbar [408] which allows a sealing of the vacuum chambers with o-rings, a rapid pump down in a few tens of minutes and no bake-out. Nonetheless, apart from instrumentation, the electronics to drive the cantilever oscillation and to measure the cantilever's dynamic response on the local force gradients also becomes more complex and more challenging to operate. This is because the increased Q-factor reduces the bandwidth for the measurement of the oscillation amplitude correspondingly, making it inconvenient for distance control. Albrecht et al. [23] proposed to replace the lock-in amplifier by a phase-locked loop system that tracks the instantaneous resonance frequency of the cantilever (Fig. 4.14). Such a system contains two additional feedback loops. A first feedback (phase feedback) adjusts the frequency of the cantilever excitation to keep the phase between the cantilever excitation and oscillation at 90°, while a second feedback (amplitude feedback) keeps the cantilever oscillation amplitude at the selected setpoint by appropriately adjusting the amplitude of the cantilever excitation.

Moreover, imaging in vacuum typically requires a phase-locked loop system which requires expertise to be properly adjusted. In addition, topography imaging using the intermittent contact mode technique fails under vacuum conditions because of the high quality factor, such that true nc-AFM operation modes have to be used. For these, three interdependent feedbacks systems (two for the PLL) and one for the tip-sample distance) must be tuned, which requires expert knowledge. Recently, three new operation modes were developed that are suitable for distance control for an MFM operated under vacuum conditions [592,593]. Possibly, Q-control could be used to lower the high quality factor of a cantilever operated in vacuum such that intermittent contact mode would become applicable as an operation mode to map the topography of the sample. However, the feedback of the Q-control will introduce additional noise, such that the magnetic imaging should still be performed with the PLL system driving the high-Q-cantilever [133,630]. To date, such an operation scheme has however not yet been implemented.

4.4.2 Tip-Sample Distance Control Suitable for Vacuum

MFM operation in vacuum benefits from quality factors that are considerably higher than those obtained in air (Sect. 4.4.1). As mentioned before, most magnetic force microscopes use a two-passage method, where each line is scanned in a intermittent contact mode to measure the topography of the sample and subsequently with the tip lifted off the surface to record the magnetic signal [212,285] (Sect. 4.1.2). This measurement mode is robust and well applicable to samples with arbitrary topography, e.g., rough samples or patterned media. However, the intermittent contact mode is difficult to apply in vacuum. Because of the high quality factor of the cantilever, only a very small excitation amplitudes are required to obtain a reasonable oscillation amplitudes of a few nanometer, i.e. $A = Q \cdot A_{exc}$ when the cantilever is driven on its resonance. The energy loss per cycle, $\Delta E = \frac{1}{2Q} k A^2$ is thus small compared to the energy stored in the cantilever oscillation. If the tip of such a cantilever is brought in contact with the sample, a much larger energy loss occurs which cannot be compensated by the electronic control system driving the cantilever with an extremely small excitation amplitude A_{exc}. The implementation of any operation mode that requires an intermittent tip-sample contact thus becomes challenging.

For scanning force microscopy with atomic resolution, non-contact operation modes have been developed. Typically, the tip is brought in close proximity to the sample until a noticeable shift of the cantilever resonance frequency is detected. The z-feedback can then adjust the tip-sample distance such that the cantilever resonance frequency is kept at a selected setpoint. As already pointed out by Rugar et al. a necessary condition for a stable feedback operation is that the derivative of the measured interaction does not change sign [561], while scanning the sample surface. The net force derivative consists of two components

$$F' = F'_{\text{mag}} + F'_{\text{servo}} \, , \qquad (4.58)$$

where the derivative of the servo force, $F'_{\text{servo}} = F'_{\text{vdW}} + F'_{\text{E}}$ is the sum of the derivatives of the van der Waals and electrostatic forces acting on the tip. The latter can be tuned by the applied bias such that the total force derivative does not change sign at all xy-locations on the sample. If the z-feedback is operated with a sufficiently high gain, the actual resonance frequency of the cantilever and thus the interaction force derivative will be kept constant as the surface of the sample is scanned. A spatial variation of the magnetic tip-sample interaction will then lead to a corresponding variation of the tip-sample distance. To first order the variation of the tip-sample distance, Δz is proportional to F'_{mag} and related to

$$\frac{F'_{\text{mag}}}{\partial F'_{\text{mag}}/\partial z \Big|_{z=z_0}} \, , \qquad (4.59)$$

where z_0 is the average tip-sample distance. The actual tip-sample distance then depends on the expression given in (4.59) as well as on the speed of the z-feedback. This makes an interpretation of the measured data more challenging. Moreover, the obtained lateral resolution will vary considerably with the tip-sample distance. For this reason, the speed of the z-feedback is usually kept small, such that the feedback corrects drifts of the tip-sample distance while changes of the tip-sample distance (arising from a variation of the derivative of the magnetic forces) remain small. Nevertheless, if a domain wall runs parallel to the scan direction, the latter force will still lead to a noticeable change of the tip-sample distance even for a slow z-feedback. However, in a large number of earlier works, MFM was performed without a tip-sample distance feedback by scanning the tip parallel to an average tip-sample plane [89].

Clearly, an ideal control mode for the tip-sample distance for magnetic force microscopy should be independent of the magnetic interaction. Li et al. [391] reported a single passage method that makes use of bi-modal cantilever excitation introduced by Rodriguez and Garcia [553]. The cantilever was driven on its first mode at an amplitude of several tens of nanometers and simultaneously on its second mode at a much smaller amplitude. The first mode amplitude was kept constant by the feedback that controls the tip-sample distance, i.e. reflects the measured topography of the sample. The second mode amplitude and phase were used to record the magnetic signal. Note that for a scan of this kind to be carried out in a reasonable amount of time, the oscillation amplitude must stabilize quickly at every point. Hence, very high cantilever quality factors are impractical. Therefore, work was performed in

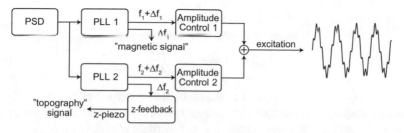

Fig. 4.24 Schematics of the dual-PLL system required for bi-modal oscillation of high quality factor cantilevers. PLL 1 tracks the first flexural oscillation mode resonance frequency f_1 of the cantilever. The output amplitude is adjusted by a PI feedback (Amplitude Control 1) to keep the (first mode) oscillation amplitude constant, e.g. at 12 nm. The shift of the first mode resonance frequency, $\Delta f_1^{(b)}$ predominantly reflects the longer-ranged magnetic part of the tip-sample interaction. PLL 2 tracks the second flexural oscillation mode resonance frequency f_2 of the cantilever. The PI feedback Amplitude Control 2 keeps the second mode amplitude constant, e.g. at 0.27 nm. At such a small amplitude, the second mode resonance frequency shift, $\Delta f_2^{(b)}$, predominantly arises from the van der Waals part of the tip-sample interaction, which depends on the tip-sample distance but not on the micromagnetic details of the sample. $\Delta f_2^{(b)}$ and thus the local tip-sample distance is kept constant by the z-feedback. Its output then reflects the topography of the sample. Figure adapted from Schwenk et al. [592] with permission from AIP Publishing

air, resulting in low quality factors of the cantilever, i.e. 120 and 500 on the first and second mode, respectively.

4.4.2.1 Bi-Modal Distance Control

Bi-modal cantilever excitation suitable for vacuum operation was first presented by Schwenk et al. [592]. A double phase-locked loop system is used to drive a high-quality factor cantilever on its first mode with an oscillation amplitude of several nanometer and simultaneously with sub-nanometer oscillation amplitude on its second mode (Fig. 4.24).

The force constant k_n and the resonance frequency ω_n of the nth mode relate to those of the first mode as

$$k_n = k_1 \cdot \left[\frac{\alpha_n}{\alpha_1} \right]^4 \quad \text{and} \quad \omega_n = \omega_1 \cdot \left[\frac{\alpha_n}{\alpha_1} \right]^2 , \tag{4.60}$$

with $\alpha_i = \{1.8750, 4.6941, 7.8548, 10.9955, \ldots\}$[96]. According to (4.53) and (4.60), the sensitivity of a higher mode n is lower than for the first mode by $\alpha_n/\alpha_1 \cdot \sqrt{Q_1/Q_n}$.

The first mode is hence better suited for the measurement of the small magnetic forces. Because the second mode is operated with a sub-nanometer amplitude and is *not* a harmonic overtone of the first mode, its simultaneous operation with the first mode does not alter the measurement of the tip-sample interaction performed with the first mode. As outlined by Giessibl et al. [206], the frequency shift Δf_i of a cantilever of stiffness k_i oscillating in a *single mode* i with an amplitude A_i in a force field derivative $\frac{\partial}{\partial z} F_z(z) = k_{ts}(z)$ becomes

$$\Delta f_i(z) = -f_{i,0} \frac{\langle k_{ts} \rangle}{2 k_i}, \tag{4.61}$$

where $z = z_{ltp}$ is the tip-sample distance at the lower turning point, $f_{i,0}$ is the free resonance frequency of the cantilever in the ith oscillation mode not interacting with the sample, and $\langle k_{ts} \rangle$, the averaged tip-sample stiffness is given by

$$\langle k_{ts} \rangle = \frac{2}{\pi A_i^2} \int_{-A_i}^{A_i} k_{ts}(z + A_i - q) \cdot \sqrt{A_i^2 - q^2} \, dq, \tag{4.62}$$

where A_i is the oscillation amplitude of the ith oscillation mode, and $k_{ts}(z)$ is the derivative of the tip-sample force at a tip-sample distance z. In Fig. 4.25a, the dependencies of repulsive and attractive magnetic forces, and of a van der Waals force given by

$$F_{mag} = F_0 \cdot e^{-\frac{2\pi}{\lambda}(z + z_p)} \cdot [1 - e^{-\frac{2\pi}{\lambda} t}] \tag{4.63}$$

$$F_{vdW} = -\frac{H R_{tip}}{z^2}, \tag{4.64}$$

are plotted. The magnetic force is assumed to arise from a sinusoidal variation of M_z with a wavelength of $\lambda = 20$ nm (10 nm feature size), in a thin film with a thickness of $t = 5$ nm and an oxidation protection layer thickness $z_p = 2$ nm. Because of the latter, the tip is 2 nm farther away from the magnetic layer, than from the sample surface. The magnitude of the magnetic force (determined by F_0) was adjusted such that the frequency shift obtained from it corresponds to that typically measured in an experiment. Similarly, a Hamacker constant $H = 10^{-24}$ J and a tip radius $R_{tip} = 5$ nm were chosen for the van der Waals force to obtain a reasonable agreement between the calculated and typically measured frequency shifts.

Figure 4.25b shows the corresponding z-derivatives of the forces. The shift of the resonance frequency of the fundamental cantilever oscillation mode arising from the repulsive/attractive magnetic forces and from their sum with the van der Waals force calculated from (4.61) is displayed in Fig. 4.26c. The cantilever stiffness is 0.3 N/m and its first mode free resonance frequency is 50 kHz. The van der Waals force generates a negative frequency shift background of -11.76 Hz at $z = 2$ nm. Repulsive/attractive magnetic force then lead to an additional positive/negative frequency shift, $\Delta f_1(z = 2$ nm$) = 4.404$ Hz highlighted by the arrow between the dashed lines in Fig. 4.25c.

The second mode resonance frequency shift, $\Delta f_2^{(u)}$ that would be obtained without a fundamental mode oscillation can also be derived from (4.61) [see wide gray line in Fig. 4.26d and the corresponding frequency shift scale on the right side of the graph]. The second mode resonance frequency shift obtained under bi-modal operation condition, $\Delta f_2^{(b)}$ is equal to the time-averaged $\Delta f_2^{(u)}(z)$-signal [see black line in Fig. 4.25d] and frequency shift scale on the left side of the graph. It has been

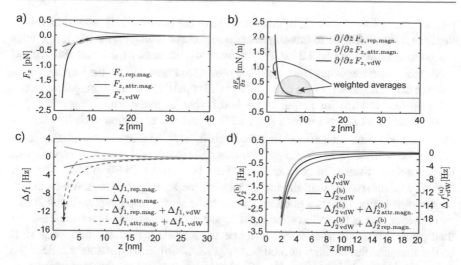

Fig. 4.25 a Van der Waals and repulsive/attractive magnetic forces. The equations describing the forces are given in the inset. The sample thickness is 5 nm. **b** Van der Waals and repulsive/attractive magnetic force z-derivatives. **c** Calculated shift of the resonance frequency of the fundamental cantilever oscillation mode arising from the repulsive/attractive magnetic forces and from their sum with the van der Waals force. The cantilever stiffness is 0.3 N/m and its first mode free resonance frequency is 50 kHz. The Δf_1-contrast arising from repulsive and attractive magnetic forces at a tip-sample distance of 2 nm is represented by the arrow between the dashed lines. **d** Time averaged frequency shift of the second mode under bi-modal operation conditions, δf_2^b, arising from the van der Waals force, and the sum of the van der Waals force with the repulsive/attractive magnetic forces

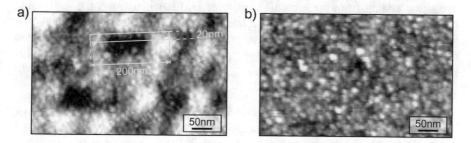

Fig. 4.26 a High resolution magnetic image of a modern hard disk recorded with $A_1 = 12$ nm. The 200 nm-wide and 20 nm-high yellow box serves as a guide for the eye to judge the lateral resolution of the MFM image. **b** Topography measured simultaneously with the magnetic signal with $A_2 = 0.27$ nm. The grain structure is nicely visible in both images. Figure adapted from Schwenk et al. [592] with permission from AIP Publishing

pointed out by Kawai et al. [330] that the shape of the bimodal frequency shift versus distance curve is similar to that of the unimodal curve [compare the black line with the wide gray line in Fig. 4.25d] The distance dependence of the bimodal second mode resonance frequency shift, $\Delta f_2^{(b)}$, is plotted in Fig. 4.25d for the van der Waals force (black line), and for sum of the latter with repulsive/attractive magnetic forces (blue line and red line for a repulsive and attractive magnetic force, respectively).

Assuming that the z-feedback (Fig. 4.24) adjust the z-position such that the bimodal second mode resonance frequency shift, $\Delta f_2^{(b)}$ is kept constant at $-2\,\text{Hz}$ during the scan of an MFM image, the magnetic force will lead to a variation of the tip-sample distance of only $\Delta z = 0.45\,\text{nm}$ [see arrows in Fig. 4.25d], which is typically much smaller than the topography of the sample. Hence, the z-feedback output signal (Fig. 4.24) is a good measure for the sample topography [Fig. 4.26a].

However, if the tip follows the local topography (as is also often the case for dual passage intermitted-contact-based operation mode MFM), the frequency shift of the fundamental oscillation mode, Δf_1 [Fig. 4.26b], does *not* solely reflect the magnetic tip-sample interaction but will also be affected by the sample topography. Hence, the separation of magnetic tip-sample forces from the sample topography remains incomplete. This is because the tip-sample distance averaged over an area relevant for the variations of the magnetic force is not constant, although the local tip-sample distance is kept constant (see section of separation of different contrast contributions).

Bimodal operation may be suited for MFM with highest lateral resolution because the tip-sample distance is kept constant at a value typically below 10 nm. This, together with the signal-to-noise ratio of the $\Delta f_2^{(b)}$-signal limited by the time-averaging arising from the bimodal cantilever oscillation narrows the achievable bandwidth of the z-feedback and with it the maximum scan speed. Furthermore, the interpretation of magnetic force microscopy data (see Sect. 4.4.4) is facilitated when the data were acquired with the average tip-sample distance kept constant. This is not possible with the bimodal operation mode.

4.4.2.2 Bimodal Capacitive Distance Control

The electrostatic tip-sample interaction arising from an applied tip-sample bias and the contact potential can be used advantageously for tip-sample distance control in magnetic force microscopy operation. A single-passage, bimodal magnetic force microscopy technique relying on capacitive tip-sample distance control has been developed by Schwenk et al. [593]. A double phase-locked loop (PLL) is used to drive the cantilever mechanically on its first flexural resonance f_1, typically with an oscillation amplitude $A_1 = 5\,\text{nm}$ (zero-to-peak), chosen to optimize the ratio between the measured magnetic force induced frequency shift and the frequency noise caused by thermal fluctuations [422,592]. The second cantilever oscillation mode at $f_2 = \alpha_1^2/\alpha_2^2 \cdot f_1 \approx 6.28 \cdot f_1$ is driven electrostatically by an oscillatory tip-sample bias at f_{ac} that is chosen to match half of the second mode resonance frequency, i.e. $f_{ac} = f_2/2$ (Fig. 4.27).

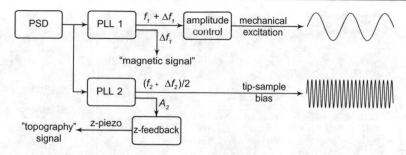

Fig. 4.27 Schematics of the dual-PLL system required for bimodal oscillation of high quality factor cantilevers. The first PLL mechanically drives the cantilever on its first mode, and tracks shifts of its resonance frequency. The second PLL excites the cantilever via an oscillatory electric field at half the resonance frequency of its second mode. The z-feedback then keeps the obtained second mode oscillation amplitude constant to map the sample topography. The required z-travel then reflects the topography of the sample. Figure adapted from Schwenk et al. [593] with permission from AIP Publishing

An oscillatory bias of the form $U(t) = U_{dc} + U_{ac} \cos(2\pi f_{ac} t)$ generates an electrostatic force given by

$$F_{E}(z, t) = \frac{1}{2} \frac{\partial C(z)}{\partial z} \cdot [\, U_{dc}^2 + 2U_{dc}U_{ac} \cos(2\pi f_{ac} t)$$
$$+ \, U_{ac}^2 \cos^2(2\pi f_{ac} t)\,], \qquad (4.65)$$

where $C(z)$ is the distance dependent tip-sample capacity, $U_{dc} = U_{dc}^{(K)} + U_{dc}^{(a)}$ is the sum of the contact and applied potential, and U_{ac} is the amplitude of the potential modulation. We see from (4.65) that F_E has components at frequency f_{ac} and $2f_{ac} = f_2$, the latter being:

$$F_{E,2f_{ac}}(z) = \frac{1}{4} \frac{\partial C(z)}{\partial z} \cdot U_{ac}^2 . \qquad (4.66)$$

This force oscillating at $2f_{ac} = f_2$ will thus lead to a second mode oscillation amplitude

$$A_2(z) \propto \frac{Q_2}{k_2} \cdot F_{E,2f_{ac}}(z) \qquad (4.67)$$

that depends on the second mode cantilever force constant $k_2 = \alpha_2^4 / \alpha_1^4 \cdot k_1$, the quality factor, Q_2, and on the tip-sample distance, provided that the frequency 2_{ac} tracks $f_2/2$ half the resonance frequency of the second oscillation mode. The latter is achieved with the second PLL. The second mode amplitude A_2 can then be kept constant by the tip-sample distance feeback. Note that the second mode amplitude is amplified by the second mode quality factor Q_2 and can thus be measured with an excellent signal-to-noise ratio. The bandwidth is however limited by $f_2/Q_2 \approx 300\,\text{kHz}/30'000 = 10\,\text{Hz}$. This limits the bandwidth of the z-feedback and with it the achievable scan speeds when operation at constant local tip-sample distance is required. However, for the quantitative interpretation of MFM data, it is convenient to scan at constant average tip-sample distance [38] (Fig. 4.28).

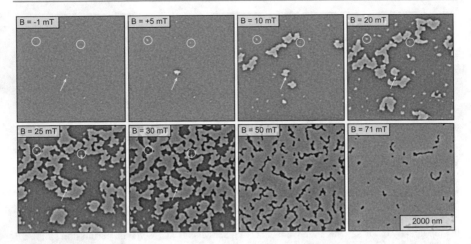

Fig. 4.28 MFM data of a $Si/SiO_2/Pt(5\,nm)/[Ir(1\,nm)Co(0.4\,nm)/Pt(1\,nm)]_5/Pt(2\,nm)$ multilayer sample in fields from $-1\,mT$ to $+71\,mT$ after saturation in negative fields. The data was acquired with the bimodal capacitive tip-sample distance control. At $-1\,mT$ isolated skyrmions appear. These expand into larger bubbles when a positive field parallel to the core magnetization of the skyrmions is applied. These domains expand in higher fields until at $+71\,mT$ the magnetization of the film is nearly saturated. Adapted from Ph.D. Thesis, Dr. Johannes Schwenk, https://edoc.unibas.ch/52283/1/Thesis_Schwenk_Online.pdf

The tip then scans parallel to an average sample surface plane but does not track local topographical height variations. These will then lead to a change of the local van der Waals force and consequently also to a shift of the first mode resonance frequency $\Delta f_{1,vdw}$ in addition to a frequency shift arising from the magnetic tip-sample interaction $\Delta f_{1,mag}$. This is a minor problem on sufficiently flat samples and when measuring at larger tip-sample distance beyond $10\,nm$. Experimental methods to obtain a complete separation between topography and magnetic tip-sample interaction induced frequency shifts are discussed in Sect. 4.4.3.

Note, that changes of the local Kelvin potential will also lead to a shift of f_1 but will not affect the distance control, because only U_{ac} not U_{dc} is responsible for the oscillatory tip-sample force at $2f_{ac}$ given by (4.66) and hence for the second mode amplitude $A_2(z)$. A spatial variation of the Kelvin potential could however still be compensated through an additional implementation of a Kelvin feedback controller. For operation under vacuum conditions an FM Kelvin method is particularly suitable. For this, an additional oscillatory tip sample bias could be applied at $f_{bias,K}$ leading to side bands at $f_1(x, y, z) \pm f_{bias,K}$. The tip-sample bias $U_K(x, y, z)$ required to null the side-band amplitudes then reflects the local Kelvin potential.

4.4.2.3 Frequency-Modulated Capacitive Distance Control

The previous section described the implementation of a tip-sample distance control mode that relied on the electrostatic actuation of the second mode flexural resonance of the cantilever. The obtained oscillation amplitude A_2, a measure for the tip-sample distance, can then be kept constant by the z-feedback. However, the second mode

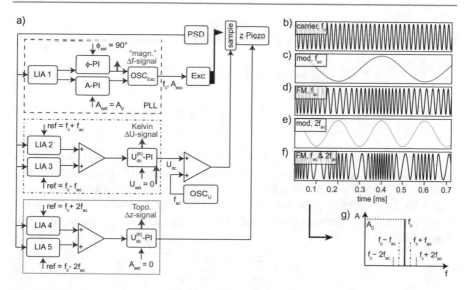

Fig. 4.29 a Schematics of the PLL and side bands detection systems required for frequency-modulated capacitive distance control in high resolution MFM. **b** Carrier at 50 kHz. **c** Modulation at 2.5 kHz. **d** FM modulated signal with 2.5 kHz (note: frequency-modulation greatly exaggerated). **e** Modulation at 5 kHz. **f** FM-modulated signal with 2.5 kHz and 5 kHz. **g** FM spectrum resulting from f). Higher order side band amplitudes are negligible. Figure adapted from Zhao et al. [729], distributed under the terms of the Creative Commons Attribution License (http://creativecommons. org/licenses/by/4.0.)

amplitude depends linearly on the second mode quality factor Q_2. The latter however often increases slowly over the course of several hours after a cantilever has been brought from air to vacuum without annealing under vacuum conditions. Furthermore, Q_2 changes with temperature and also if strong magnetic fields are applied. MFM operation at constant tip-sample distance over longer times therefore requires a repeated re-calibration of the actual second mode amplitude, A_2, to tip-sample distance relation. The latter can be obtained by measuring $A_2(z)$ until the tip makes contact with the surface. The precision of determination of the contact point in repeated measurements is typically about ± 1 nm. This limits the reliability of difference images taken from data measured, for example, with different magnetization directions of the tip, or data acquired in different external magnetic fields.

These limitations can be overcome with frequency-modulated capacitive tip-sample distance control. A schematic of the setup is presented in Fig. 4.29. The cantilever is driven mechanically on the first flexural oscillation mode with a PLL system (Fig. 4.14) that tracks changes of the cantilever resonance frequency Δf and also keeps the oscillation amplitude A_0 constant (dashed line box in Fig. 4.29). Changes of the first mode quality factor are thus compensated with an appropriate adjustment of the driving amplitude. Further, the tip-sample bias U_{ac} is oscillated at a frequency f_{ac} of a few hundred Hz, leading to a modulation of the cantilever resonance frequency.

A signal $A(t) = A_0 \cdot \cos(\omega_c t)$, which is frequency-modulated by $f(t) = A_m \cdot \cos(\omega_m t)$ can be written as

$$A_{\mathrm{FM}}(t) = A_0 \cdot \mathrm{Re} \left\{ e^{i\omega_c t} e^{i\beta \sin \omega_m t} \right\} \tag{4.68}$$

$$= A_0 \cdot \sum_{-\infty}^{\infty} J_n(\beta) \cos(\omega_c + n\omega_m)t, \tag{4.69}$$

where ω_c is the carrier frequency, J_n is the nth Bessel function [597], and β is the modulation index. The spectrum of the frequency-modulated (FM) signal (4.69) thus contains an infinite number of side-bands even for a single modulation frequency $\omega_m = 2\pi f_m$. Zhao et al. [728] however showed that all higher-order side bands can be neglected. This is because of the side band amplitudes are proportional to the Bessel functions

$$J_n(\beta) = \sum_{k=0}^{\infty} \frac{(-1)^k (\beta/2)^{n+2k}}{k!(n+k)!} , \tag{4.70}$$

where $\beta := \Delta f_{1,\mathrm{E}}/f_m \approx 5 \cdot 10^{-3}$ and $\Delta f_{1,\mathrm{E}}$ is the modulation of the first mode resonance frequency arising from the applied oscillatory bias with a frequency f_m. The spectrum of the frequency-modulated signal thus contains only first order side bands. However, because the electrostatic force (4.65) contains 2 ac-components, one at f_{ac} and one at $2f_{ac}$, the FM-spectrum [Fig. 4.29g] contains two pairs of side bands at $\pm f_{ac}$, and $\pm 2f_{ac}$ with amplitudes proportional to $U_{dc}^{(K)} + U_{dc}^{(a)}$, and $U_{ac} \cdot \frac{\partial C(z)}{\partial z}$, respectively. The amplitude of the first order sidebands can be nullified by compensating the Kelvin potential $U_{dc}^{(K)}$ by the applied potential $U_{dc}^{(a)}$ (although this is not required for the distance control). This is achieved for each image pixel scanned by the Kelvin feedback [shown in the dash-dotted box of Fig. 4.29a].

4.4.3   Separation of Forces

As in all scanning force microscopy experiments, the measured cantilever response arises from the sum of all forces acting on the tip. In magnetic force microscopy experiments, the dominant forces besides those arising from the magnetic field of the sample are the van der Waals and the electrostatic force, (apart from Pauli repulsion force when the tip is in contact or intermittent contact with the sample). The latter force typically dominates the much weaker magnetic tip-sample interaction (see Fig. 4.2) such that the acquisition of the magnetic force is typically performed without tip-sample contact (Sect. 4.1.2). Because of the rapid decay of the magnetic stray field with increasing tip-sample distance, the latter typically needs to be kept smaller that a few tens of nanometers. For this, various tip-sample distance control modes have been developed (Sects. 4.1.2 and 4.4.2).

For an MFM operated under ambient conditions, lift mode operation is typically used [212,285]. While the first scan performed with intermittent tip-sample contact

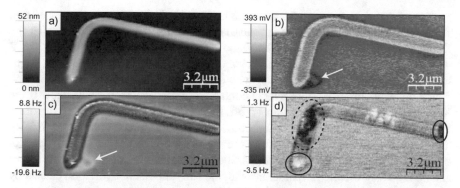

Fig. 4.30 **a** Topography of the 50 nm-thick L-shaped Co nanowire deposited onto a Si wafer measured in intermittent contact mode using an oscillation amplitude $A = 7.5$ nm. **b** Kelvin potential image acquired at a lift-height of 30 nm during the retrace scanlines. The white arrow highlights a pattern with a Kelvin potential even lower than that of the substrate (resumably arising from a contamination). **c** Frequency shift image recorded while the Kelvin controller was switched off, i.e. without compensation of the local Kelvin potential. No magnetic domains are are apparent, but the structure highlighted by the white arrow that is also visible in the Kelvin potential image is also detected. Clearly the visible contrast arises from local variations of the electrostatic force gradient caused by corresponding spatial variations of the Kelvin potential. **d** Frequency shift image with the Kelvin controller switched on. The local variations of the electrostatic force are compensated, such that much weaker frequency shift contrast arising from the domain structure inside the Co nanowire become visible. Adapted from Jaafar et al. [307], distributed under the terms of the Creative Commons Attribution License (http://creativecommons.org/licenses/by/4.0)

is a good measure for the topography, the phase shift recorded in the second scan passage does not necessarily reflect only the magnetic forces. Again, a contrast will be generated from electrostatic forces arising from a local variation of the contact potential, if the latter is not locally compensated. Moreover, a convolution of the topography with the magnetic contrast can also arise from the decay length of the magnetic forces, that is much longer than that of the Pauli repulsion forces that are used to image the sample topography (see Sect. 4.4.3.2) such that the contrast measured in lift-mode operation cannot be expected to be free from topographical artifacts.

4.4.3.1   Separation of Electrostatic Tip-Sample Forces

Electrostatic forces have been advantageously used to control the tip-sample distance in early MFM work [561, 586], and later also for tip-sample distance control under vacuum conditions [593, 728]. Jaafar et al. [307] pointed out the importance of separating electrostatic from magnetic contrast contributions when scanning microfabricated metal nanostructures on a silicon substrate.

Figure 4.30a shows a 50 nm-thick Co L-shaped nanostructure fabricated on a Si(111) wafer substrate by focussed electron beam induced deposition from a $Co_2(CO)_8$ precurser gas. The data was obtained in intermittent contact mode using a nanosensor PPP-FMR cantilever with a nominal stiffness of $k = 1.5$ N/m, and a nominal free resonance frequency of 75 kHz sputter-coated with a 25 nm thick Co/Cr film. During the retrace scan lines, the tip was lifted off from the surface by 30 nm. The spatial variation of the Kelvin potential depicted in Fig. 4.30b reveals that the Kelvin Potential difference is +320 mV above the Co nanowire, while it is −320 mV on the Si-substrate. At the right side of the lower edge of the Co nanowire, highlighted by the white arrow, a feature with an even lower Kelvin potential appears, which presumably arises from a local contamination of the Si wafer surface. If the Kelvin potential controller is switched off, the local variations of the Kelvin potential lead to a corresponding variation of the electrostatic force gradient dominating the frequency shift signal arising from the magnetic tip-sample interaction. Consequently, the frequency shift image [Fig. 4.30c] is dominated by the electrostatic forces and no domain structure is visible. Instead, the Kelvin potential variation caused by the contamination leads to a measurable frequency shift signal [see white arrow in Fig. 4.30c]. A magnetic contrast is visible in the data shown in Fig. 4.30d, that were acquired with an active Kelvin controller, i.e. with a compensation of the local contact potential differences. Measurements performed in applied in-plane magnetic fields confirmed that the white/black structures visible at the lower and right edge of the L-shaped nanowire arise from magnetic poles present at the nanowire edges (highlighted by black circles). Similarly, the dark pattern appearing below the 90° corner of the nanowire (inside the dashed black ellipse) are magnetic domains.

Another approach to disentangle electrostatic force contributions has been described by Angeloni et al. [33]. In their work, the electrostatic contrast contributions were removed by computing the difference between a phase images acquired with a magnetized and demagnetized tip. Alternatively, differences can be calculated between two data-sets acquired with opposite tip magnetizations (see Sect. 4.4.3.3 and Fig. 4.32).

It should be pointed out that Jaafar et al. [307] measured the oscillatory deflection of the cantilever arising from the applied oscillatory sample bias. The sensitivity for electrostatic signals can be improved when the frequency of the bias signal is matched with one of the resonances of the cantilever. Alternatively, the lateral resolution of the Kelvin image can be improved when implementing and FM-Kelvin technique, as for example implemented in the frequency-modulated capacitive distance control mode [728] (Sect. 4.4.2.3).

As any scanning force microscopy, MFM measures the sum of all forces acting on the tip. Neglecting the electrostatic contrast contributions can consequently lead to an incorrect interpretation of the observed contrast. Cervenka et al. [102] reported MFM images on HOPG and concluded that the observed contrast is of magnetic origin. Using in-vacuum MFM with a local compensation of the Kelvin potential, Martinez-Martin et al. [436] however showed that on HOPG local contact potential differences of up to 200 mV occur and did not observe any signal of magnetic origin. Moreover, Martinez-Martin et al. [436] pointed out that the data of Cervenka et al. [102] were misinterpreted. Clearly, a careful separation of magnetic and electrostatic forces is of highest importance when performing MFM on samples with local variations of the contact potential or containing other sources generating a local variation of the electric field. The latter can for example occur near step edges, even when the Kelvin potential remains constant.

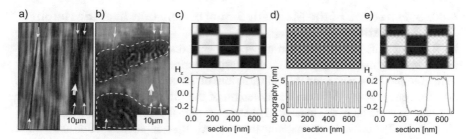

Fig. 4.31 a Topography, and **b** Phase signal attributed to the magnetic tip-sample interaction recorded by MFM lift-mode operation on duplex stainless steel (adapted from https://literature. cdn.keysight.com/litweb/pdf/5991-3185EN.pdf?id=2410217). Clearly the scratches in the topography image a) are also visible in the magnetic signal image (**b**). **c** z-component of the stray field of a pattern of up/down magnetic domains with magnetization ±1 A/m 10 nm above the sample surface. **d** 5 nm-high topography of the sample surface. **e** z-component of the stray field of a pattern of up/down magnetic domains with magnetization ±1 A/m calculated 10 nm above the local topography of the sample surface

4.4.3.2   Artifacts from the Topography in Lift-Mode Operation

In lift-mode MFM, the tip follows the previously recorded topography to measure the magnetic signal. The local tip-sample distance is thus kept constant.

Figure 4.31a and b show the topography and magnetic signal recorded by lift-mode MFM above a duplex stainless steel surface. The ferromagnetic nature of the ferrite phase becomes apparent from the domains visible in the enclosed dashed white lines in the "magnetic image" [Fig. 4.31b], while domains are absent in the paramagnetic austenite phase. The topography image shows narrow scratches (narrow white arrows) and a wider groove [wide white arrow in Fig. 4.31a]. Note that the narrow scratches become also visible in the "magnetic image" [Fig. 4.31b]. This documents that the separation of signals arising from the topography and magnetic stray field are not completely disentangled in lift-mode MFM operation, as can be understood from the data displayed in Fig. 4.31c–e. Figure 4.31c displays the calculated z-component of the stray field of a model checkerboard magnetization pattern 10 nm above a 50 nm thick magnetic film with a magnetization normalized to 1 A/m. Figure 4.31d displays a non-magnetic, checkerboard model sample topography with a height of 5 nm. Figure 4.31e, then shows the z-component of the stray field arising from the model magnetization pattern at a constant local tip-sample distance of 10 nm. The stray field image clearly contains a contrast variation arising from the topography, because the topography-induced change to the tip leads to a corresponding variation of the magnetic stray field. Note that such a change of the stray field would also occur when the particles at the sample surface are magnetic, because generally the decay lengths of the stray field is longer than that of the Pauli repulsion forces that determined the topography tracking during the intermittent contact operation. Clearly, lift-mode MFM operation does not allow a complete separation of the topography from the measured "magnetic signal", particularly, in the presence of smaller-scale topographical features.

4.4.3.3 Extraction of the Magnetic Signal in Constant Average Height Operation

Dual-passage lift-mode operation under ambient conditions is a robust method to simultaneously map the magnetic tip-sample interaction together with the topography of the sample. However, even if local variations of the contact potential (Sect. 4.4.3.1) are compensated, the different decay length of the different force acting on the tip can lead to artifacts in the measured phase shift signal arising from the topography (Sect. 4.4.3.2). Furthermore, mapping the magnetic tip-sample interaction at constant local tip-sample distance complicates the comparison of the measured data to the modeled MFM contrast, because the latter is conveniently performed in a plane above the sample surface, parallel to the average sample slope (Sects. 4.3, 4.4.4, 4.4.5, and 4.4.6).

Here it is assumed that the MFM is operated in vacuum to obtain a high Q-factor for improved sensitivity for small force gradients (see Sect. 4.4.1). The cantilever is driven on its actual resonance frequency using a PLL (Fig. 4.14) with an oscillation amplitude kept constant at a predefined setpoint value. The tip is scanned parallel to the average sample tilt at a preset average tip-sample distance, i.e. does not follow local topography variations. As described in Sect. 4.1.2, this can be done without feedback control, provided that the drift of the instrument is sufficiently small or is compensated by a preset z-drift rate. Alternatively, the bimodal capacitive distance control [593] (Sect. 4.4.2.2) or the frequency-modulated capacitive distance control [728] (Sect. 4.4.2.3) can be used to keep the average tip-sample distance constant. As described in Sect. 4.4.2.3, the tip-sample distance control is not affected by changes in the quality factor of the cantilever, if the latter mode is used. Independent of the tip-sample distance control mode, it is convenient to set the feedback speed sufficiently small that only the drift of the tip-sample distance, or a deflection of the cantilever arising from the application of an external field (but not tip-sample distance changes arising from local topography variations) are corrected. It is further convenient to compensate the average sample tilt by adding an appropriate fraction of the x- and y-scan signals to the z-feedback output. Then, the tip keeps scanning on the plane that is parallel to an average sample slope at a constant average tip-sample distance. Local variations of the topography then lead to variations of the van der Waals force and hence to a frequency shift signal related to the topography of the sample.

Figure 4.32a and b show MFM data obtained while scanning parallel to the average tip-sample tilt at an average tip-sample distance of 9.7 nm with a down and up tip-magnetization, respectively. The sample is a magnetic hard disk with an in-plane magnetic anisotropy. The location of the information track is highlighted by the black box, and the arrows show the direction of the in-plane magnetization. Images (a) and (b) show a white and black line, respectively, which is located at the transition of two bits that represents a source of the stray field. Furthermore, a granular contrast is apparent that arises from both grain-to-grain variations of the in-plane magnetization and from the sample topography.

While the magnetic contrast depends on the orientation of the tip magnetization, the contrast arising from topography-induced variations of the van der Waals force does not. Consequently, a separation of the magnetic from the van der Waals forces

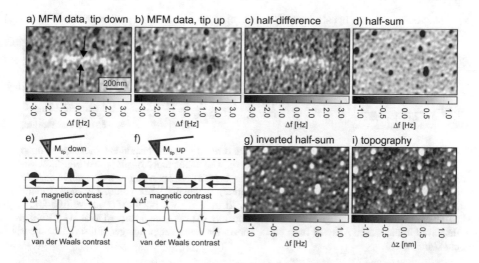

Fig. 4.32 **a** and **b** MFM images of a magnetic hard disk with in-plane magnetic anisotropy measured with a down and up tip magnetization, respectively. The black arrows give the direction of the magnetization along the track. **c** Half-difference of the data displayed in **a** and **b** showing the magnetic contrast. **d** Half-sum of **a** and **b** showing the van der Waals contrast arising from local topographical protrusions (particles). **e** and **f** Schematics of the contrast arising from the magnetic transitions (red and blue line color) and particles (black line color) with a tip with down and up magnetization, respectively. **g** Inverted half sum for comparison with the topography. **i** Topography of the sample acquired with the a frequency shift kept at −6.8 Hz. Adapted from Ph.D. Thesis, Dr. Peter Kappenberger, https://edoc.unibas.ch/474/

can be achieved by a calculation of the half-difference and half-sum of two MFM data sets acquired at the same sample location and with the same tip-sample distance but opposite directions of the tip magnetization. The magnetization of MFM tips consisting of a Co layer on one side of a high-aspect ratio silicon tip can be reproducibly reversed in fields of about ±50 mT. If the coercive field of the sample is larger than this, the tip magnetization reversal can be achieved without affecting the micromagnetic structure of the sample. This facilitates the acquisition of two successive images with opposite tip magnetization. These two images are then aligned with sub-pixel precision by an appropriate software procedure.

Figure 4.32c and d show the frequency shift data arising from magnetic and van der Waals force obtained from the half-difference and half-sum, respectively, of the data depicted in Fig. 4.32a and b. The latter data was acquired with an up and down magnetization of the tip. Figure 4.32e and f schematically display the dependence of the different sources of contrast for a down and up-orientation of the tip magnetization, respectively. A protrusion of the sample surface leads to an increased attractive van der Waals force (derivative) and thus to a more negative frequency shift (i.e. to a dark spot in the image). Figure 4.32g displays the inverted van der Waals contrast image for a better comparison with the topography image [Fig. 4.32i]. The latter was recorded at a much smaller tip-sample distance with the frequency shift kept constant at −6.8 Hz, corresponding to a tip-sample distance of 1.6 nm. At such a small

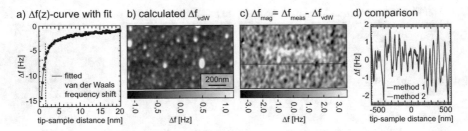

a) Δf(z)-curve with fit   b) calculated Δf$_{vdW}$ c) Δf$_{mag}$= Δf$_{meas}$- Δf$_{vdW}$ d) comparison

Fig. 4.33 a Average of $\Delta f(z)$-curves recorded at different sample locations (black square dots), and fitted van der Waals interaction $\Delta f(z)$-curve. The topography displayed in Fig. 4.32i was recorded with a frequency shift kept constant at -6.8 Hz corresponding to a tip-sample distance of about 1.6 nm (indicated by the dashed black lines). **b** Van der Waals contrast image calculated from the measured topography shown in Fig. 4.32i by using the fitted $\Delta f(z)$-curve [red line in (**a**)] **c** Magnetic contrast image obtained by subtracting (**b**) from the measured total $\Delta f(z)$-image displayed in Fig. 4.32a. Adapted from Ph.D. Thesis, Dr. Peter Kappenberger, https://edoc.unibas. ch/474/

tip-sample distance, the frequency shift is dominated by the van der Waals force derivative. Typical variations of the magnetic force then only lead to a change of the tip-sample distance that is smaller than about ± 0.2 nm. The recorded Δz-data thus is a good measure of the topography of the sample that is about ± 1.5 nm [Fig. 4.32i].

If the micromagnetic structure of the sample is affected in fields required to change the magnetization of the tip, the sample (or tip) must be removed from the MFM (or separated from each other) before the magnetization of the tip can be changed. This process requires a re-positioning of the sample (or tip) that is sufficiently precise to have an overlap of the scanned areas before and after the flipping of the tip magnetization. This considerably complicates the procedure to measure the same sample location twice, particularly when performing MFM at low temperatures. The removal of the sample is required to change the magnetization of the tip, followed by a reinsertion of the sample into the microscope, the re-positioning process, and the thermal equilibration can then take a considerable amount of time, making such a procedure inconvenient.

An alternative strategy to separate magnetic and van der Waals forces is based on the modeling of the van der Waals contrast from a previously measured image of the sample topography displayed in Fig. 4.32i. For this, $\Delta f(z)$-curves recorded at various positions, including those with attractive and repulsive magnetic forces, are averaged [black square dots in Fig. 4.33a]. The data is then fitted by a model van der Waals frequency shift versus distance curve (4.73). Here, the van der Waals interaction between a flat surface and a conical tip

$$F_{\text{cone}}^{\text{vdW}}(z) = C_F \left[\frac{1}{z} - \frac{1}{z+h} - \frac{h}{(z+h)^2} - \frac{h^2}{(z+h)^3} \right] \qquad (4.71)$$

$$C_F = -\frac{H_{\text{eff}} \tan^2(\alpha/2)}{6},$$

was used, where H_{eff} is the effective Hamacker constant of the tip-sample system, α is the full opening angle of the conical tip, h the tip-height, z the tip sample distance. The frequency shift versus distance curve calculated with (4.71) then becomes

$$\Delta f_{\text{cone}}^{\text{vdW}}(z) = C_{\Delta f}\left[I_1^A(z) - I_1^A(z+h) - hI_2^A(z+h) - h^2I_3^A(z+h)\right] \quad (4.72)$$

$$C_{\Delta f} = -C_F \cdot \frac{f_0}{2\pi Ak}$$

$$I_1^A(z) = \frac{2\pi}{a}\left(1 - \frac{z}{\sqrt{z^2 - a^2}}\right).$$

The fitted frequency shift versus distance curve displayed in Fig. 4.33a was obtained with the parameters $f_0 = 55'500\,\text{Hz}$, $k = 0.3\,\text{N/m}$, $A = 5\,\text{nm}$, $H_{\text{eff}} = 1.9 \cdot 10^{-20}\,\text{J}$, and $\alpha = 9.2^\circ$.

In order to maximize the lateral resolution and sensitivity, the smallest tip-sample distance to still pass the highest topographical feature must be selected. For a sample that generates only a weak magnetic stray field, the van der Waals contrast can easily dominate that arising from the magnetic stray field. Figure 4.34a shows MFM data of domains on a [Co(0.3 nm)/Pt(0.7 nm)]$_{\times 4}$/Co(0.3 nm)/Pt(2 nm) ferromagnetic (F) multilayer deposited on the top of a 1.5 nm-thick Co$_{0.8}$Cr$_{0.2}$ antiferromagnetic (AF) layer acquired at 8.3 K in a field of 190 mT. (Note that this F/AF bilayer was deposited on a Co/Pt-multilayer with 10 repetitions, which defined the small domain size in the thinner upper F layer at zero field, but was then driven into its saturated state at 190 mT [327]. When cooled from room temperature to 8.3 K, the domains in the F-layer imprint a corresponding pattern of pinned uncompensated spins into the AF layer [327]. The contrast of the MFM data displayed in Fig. 4.34a is however domi- nated by the strong stray field arising from the domains in the ferromagnetic layer. After saturating the ferromagnetic layer in a field of 500 mT [Fig. 4.34b], a granular contrast becomes apparent that is generated by topography-related van der Waals forces and magnetic fields arising from the roughness and thickness variations of the

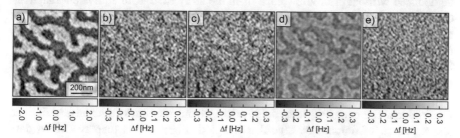

Fig. 4.34 MFM data obtained on an exchange-bias sample **a** MFM data dominated by the stray field of the ferromagnetic layer at 190 mT after zero-field cooling. **b** MFM data obtained on the saturated ferromagnetic layer at 500 mT and **c** at −500 mT. **d** Half-difference of (**b**) and (**c**). The contrast is caused by the stray field of the pinned uncompensated spins arising at the interface between the ferromagnetic and antiferromagnetic layer. **e** Half-sum of (**b**) and (**c**). The contrast is dominated by the topography induced van der Waals force and magnetic force arising from variations of the local magnetic moment density of the ferromagnetic layer. Adapted from Ph.D. Thesis, Dr. Sevil Özer, https://edoc.unibas.ch/19026/1/Sevil_thesis.pdf

ferromagnetic layer (see Sect. 4.3.4). The expected pattern of the imprinted pinned uncompensated spins is not visible. Figure 4.34c shows the MFM data obtained in a field of −500 mT that reverses the magnetization of the tip, and also that of the saturated F layer(s). Again, the pattern of pinned uncompensated spins is not visible. Figure 4.34d and e display the half-sum and half-difference of the MFM data shown in (b) and (c). The contrast of the half-difference image [Fig. 4.34d] solely arises from the stray field of the pinned uncompensated spins at the ferro/antiferromagnet interface. The van der Waals contrast as well as the magnetic contrast arising from roughness/thickness variations of the ferromagnet layer does not change sign when the magnetization of the tip and that of the ferromagnet layer is reversed, and does thus not appear in the half–difference image. The half-sum data [Fig. 4.34e] then shows the contrast arising from the topography and roughness/thickness variations of the ferromagnet layer. This demonstrates that such differential MFM techniques can be used to disentangle magnetic and topographical contributions to the measured contrast even when the latter dominates the contrast arising from the sample stray field.

4.4.3.4 Separation of Different Magnetic Contrast Contributions

The magnetic stray field arises from volume and surface magnetic charges as discussed in Sect. 4.3. In multilayer samples, the stray field may arise from domains in different magnetic layers. In many cases, a separation of the different contrast sources can be obtained if the coercivity of the layers is different. The MFM data can be acquired in fields that saturate some of the layers while domains still persist in other layers. Such techniques were, for example, applied to image the ferromagnetic domains and the corresponding patterns of pinned uncompensated spins in exchange-biased samples [327,579] (Fig. 4.34, and Sect. 4.4.6.2). Note that it is in principle not possible to obtain the 3-dimensional magnetic moment distribution from any kind of magnetic stray field measurement technique. The maximum information that can be deduced without prior knowledge from the stray field is an equivalent magnetic charge pattern at the surface of the sample (see Sect. 4.3.1). It is however possible, to match calculations of the MFM contrast of candidate model magnetization structures to the measured MFM data. This is conveniently performed with the MFM transfer-function method (Sect. 4.4.4) provided that the transfer-function of the MFM tip has been calibrated (Sect. 4.4.5).

Similarly, the magnetic stray fields arising from roughness/thickness variations and a pattern of domains can be disentangled. Figure 4.35a displays MFM data obtained at room temperature on a $[Co(0.6nm)/Pt(0.7nm)]_{\times 5}$-multilayer sample. Apart from the dominant contrast from the up/down domain pattern, a faint granular contrast is visible. After saturating the sample in a field of −50 mT, the MFM image [Fig. 4.35b] only reveals the granular contrast. The difference data shown in Fig. 4.35d does no longer reveal the granular contrast inside the dark down domains, but an amplified granular contrast in the white up domains. Consequently, the granular contrast does not arise from topography-induced variations of the van der Waals

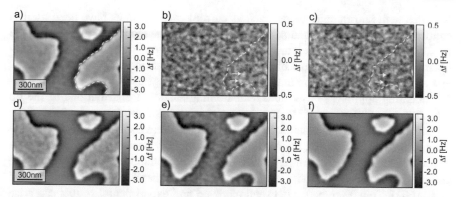

Fig. 4.35 a MFM data obtained at room temperature on a $[Co(0.6\,nm)/Pt(0.7\,nm)]_{\times 5}$-multilayer sample. The dashed white line shows the boundary of the white up-domain. The tip has a down magnetization. **b** MFM data obtained after saturation in a field of $-50\,mT$. **c** Data from (**b**) with the inverted contrast at the locations of the white domains. **d** Difference of (**a**) and (**b**): the contrast inside the dark down domains appear smooth, while that inside the bright up domains is granular. **e** sum of (**a**) and (**b**): the contrast inside the dark down domains appears smooth, while that inside the dark down domains is granular. **f** Difference between (**a**) and (**c**): the granular contrast arising from the thickness-induced local variation of the magnetic moment of the sample is subtracted. Adapted from Ph.D. Thesis, Dr. Johannes Schwenk, https://edoc.unibas.ch/52283/

force, but must be caused by variations of the local magnetic moment areal density. As discussed in Sect. 4.3.4, the latter arise from spatial variations of the thickness and roughness of the ferromagnetic layer. These stray field contributions can only be subtracted from the field arising from the domains (or domain walls) if the contrast of the data taken in saturation [Fig. 4.35b] is reversed at the location of the white up domains in Fig. 4.35a. The result of this local contrast inversion is depicted in Fig. 4.35c, where the white dashed line highlights the location of one of the up domains and the white arrows in Fig. 4.35b and c point to locations where the contrast inversion is particularly apparent. Figure 4.35f then shows the difference of the data depicted in Fig. 4.35a and c, and the granular contrast no longer appears.

4.4.4 MFM Transfer Function Theory

MFM measures either a phase shift, $\Delta\phi(\mathbf{r}, z)$, or a frequency shift, $\Delta f(\mathbf{r}, z)$, which arise from the summed derivatives of all forces acting on the tip. Experimental methods for the separation of these force are discussed in Sect. 4.4.3. Three different cases for the magnetic field-mediated contrast have been described in Sect. 4.2. Here, only the case where the influence of the stray field of the sample on the tip magnetization and that of the tip on the sample magnetization remains negligible (Sect. 4.2.1) is further analyzed. The MFM tip has often been modeled by simple combinations of point monopoles and point dipoles [360, 361, 400]. The force on the tip then becomes

$$\mathbf{F}_t = \mu_0 \left(\alpha q_t \mathbf{H}_s(x_t, y_t, z_t + \Delta z_q) + \beta \mathbf{m}_t \cdot \nabla \mathbf{H}_s(x_t, y_t, z_t + \Delta z_{\mathbf{m}}) \right). \quad (4.73)$$

Samples with known stray fields, such as current carrying lines have been used to calibrate the coefficients α and β, and the locations of the monopole Δz_q and dipole Δz_m [400]. Point-pole models are however often not a good description of the imaging properties of an MFM tip. Lohau et al. [400] found that the parameters α, β, Δz_q, and Δz_m depend strongly on the size of the current rings used for calibration. Better modeling approaches use more realistic magnetization distributions of the tip [561,586,680,682]. These rely on a known or simplified model-geometries of the tip and assume the magnetization to be uniform and along a specific direction inside the geometrical boundaries. Such an approach generally overestimates the response of the tip, and also the lateral resolution than can be obtained; see for example, the comparison between the model calculations and experiments given by van Schendel et al. [662]. This discrepancy between modeled and experimental results can arise from an inexact geometrical description of the magnetic layer boundaries, or from partial oxidation of the magnetic material on the tip. Moreover, the magnetization direction inside the magnetic layer often does not remain fixed because of the demagnetization fields that rearrange the magnetization distribution to reduced the surface magnetic charge density. Such problems that are inherent for a tip geometry-based description of the magnetic moment distribution of the tip, could in principle be overcome by micromagnetic modeling[575]. This however still requires an exact description of the layer geometry and grain structure which are not easily accessible experimentally. Calibrating the response of the tip in Fourier space does not require such microscopic knowledge. Schönenberger and Alvarado [586] were the first to describe the MFM contrast formation in Fourier space using an approximative description of the field of a thin film given by Mansuripur and Giles [428]. As pointed out by Schönenberger and Alvarado [586], the stray field above the sample can at least in principle be obtained from the MFM data once the response of the tip is known. In their work, the latter was however only obtained from model calculations assuming a given tip-shape and a single domain magnetization distributions inside the tip. Chang et al. [104] went a step further, and calibrated the transfer function of the tip from the measured MFM contrast of an array of microfabricated Ni strips, and the modeled stray field. It may seem surprising that this approach was not taken up by other researchers to calibrate the response of their MFM tips and to deconvolve the measured MFM contrast to obtain the stray field emanating from the sample surface. However, any procedure to deconvolve the response function of the tip requires a sufficiently high signal-to-noise ratio of the MFM data, reproducible imaging conditions during tip calibration and measurement, a linear tip-sample interaction, and a quantitative understanding of the dependence of the MFM instruments response on the used operation mode parameters. The latter is required if the true magnetic response of the tip is needed. These points were addressed by Hug et al. [295] by using static mode MFM imaging of domains in a Cu/Ni(10nm)/Cu(100) epitaxial system [89]. The experimental up/down domain contrast could be matched with an unprecedented precision using a model with a non-perturbed magnetization of the sample and tip. The tip was simply modeled by a magnetic point charge q_t retracted more than 100 nm from the lowest point of the tip, i.e. $\Delta z_q > 100$ nm. The excellent agreement between the experimental and modeled data confirms that an MFM image with up/down domain contrast can be explained in terms of a rigid magnetization or negligible perturbation model. It is thus not necessary to introduce any response of the sample magnetization pattern to the stray field of the tip to explain the observed up/down domain contrast, as suggested by Belliard et al. [58]. It is noteworthy that the MFM results of Hug et al. [295] were acquired with a static operation mode, where the magnetic force leads to a measurable deflection of the cantilever. Because of its simplicity, this mode facilitated the quantitative interpretation of the results. Although the static measurement mode is in principle less sensitive than the dynamic mode [237], even ultrathin Ni films with a thickness down to 2 nm could be measured with a good signal-to-noise ratio [89]. In order to model the MFM contrast for domains of very different size in films with different Ni thicknesses, a point charge model is insufficient. Instead, the response of the MFM tip for different spatial wavelengths of the stray field must be known. Van Schendel et al. [662] demonstrated that MFM data acquired on a suitable but still arbitrary pattern of up/down domains can be used to calibrate the tip. Further, model calculations using the calibrated response function of the MFM tip were compared to corresponding results obtained with a tip modeled by a flat magnetic layer with the geometry defined by one of the faces of the pyramidal tip. This reveals that geometry-based tip models overestimate the MFM contrast and spatial resolution of the domain walls. In addition, van Schendel et al. [662] described the effects of the canting of the cantilever for static and dynamic operation modes, but had not yet included a description of a finite oscillation amplitude of the cantilever (see (4.84), (4.85), and (4.90) to (4.95) in Sect. 4.4.4.2).

4.4.4.1 Force on MFM Tip Arising from Magnetic Field

The magnetostatic energy of a tip with a magnetization distribution $\mathbf{M}$ located at the position $(\mathbf{r}, z)$ in a stray field emanating from a sample $\mathbf{H}$, is given by

$$E_m(\mathbf{r}, z) = -\mu_0 \int \mathbf{M}(\mathbf{r}', z') \cdot \mathbf{H}(\mathbf{r} + \mathbf{r}', z + z') \, d\mathbf{r}' \, dz', \qquad (4.74)$$

where the integration is performed in a coordinate system attached to the tip [Fig. 4.36a]. The force acting on the tip is then given by

$$\mathbf{F}(\mathbf{r}, z) = \mu_0 \int \nabla \left[\mathbf{M}(\mathbf{r}', z') \cdot \mathbf{H}(\mathbf{r} + \mathbf{r}', z + z') \right] d\mathbf{r}' \, dz'. \qquad (4.75)$$

Using the vector identity

$$\nabla [\mathbf{M} \cdot \mathbf{H}] = [\mathbf{M} \cdot \nabla] \mathbf{H} + \underbrace{[\mathbf{H} \cdot \nabla] \mathbf{M}}_{= \text{①}} + \underbrace{\mathbf{M}(\nabla \times \mathbf{H})}_{= \text{②}} + \underbrace{\mathbf{H}(\nabla \times \mathbf{M})}_{= \text{③}}, \qquad (4.76)$$

where ① $= 0$ and ③ $= 0$ because the nabla operator, ∇, does not act on the dashed coordinates [Fig. 4.36a]; ② $= 0$ because, according to (4.13), the stray field $\mathbf{H}$ is conservative outside the sample.

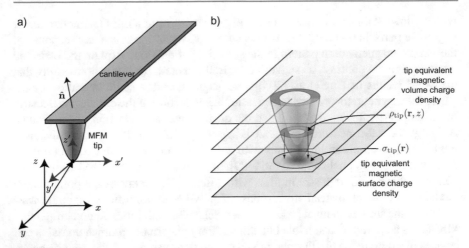

Fig. 4.36 a Coordinate systems and cantilever with normal vector $\hat{\mathbf{n}}$ canted towards the surface. **b** Magnetic charge distribution of the tip, ρ_{tip}, and its propagation to an xy-plane running through the tip apex with the tip equivalent magnetic surface charge distribution σ_{tip} that fully characterizes the magnetic imaging properties of the tip

The magnetostatic force acting on the tip given by (4.75) then becomes

$$\mathbf{F}(\mathbf{r}, z) = \mu_0 \int \left[\mathbf{M}(\mathbf{r}', z') \cdot \nabla\right] \mathbf{H}(\mathbf{r} + \mathbf{r}', z + z') \, \mathrm{d}\mathbf{r}' \, \mathrm{d}z', \qquad (4.77)$$

and in the two dimensional Fourier space already used to conveniently calculate stray fields (Sect. 4.3) the force is given by

$$\mathbf{F}(\mathbf{k}, z) = \mu_0 \int \left[\mathbf{M}(\mathbf{r}', z') \cdot \nabla\right] \left[\int \mathbf{H}(\mathbf{r} + \mathbf{r}', z + z') e^{-i\mathbf{k}\mathbf{r}} \, \mathrm{d}\mathbf{r}\right] \mathrm{d}\mathbf{r}' \, \mathrm{d}z'. \qquad (4.78)$$

Using the coordinate transformations $\tilde{\mathbf{r}} := \mathbf{r} + \mathbf{r}'$ and $\mathrm{d}\tilde{\mathbf{r}} = \mathrm{d}\mathbf{r}$, (4.78) can be re-written as

$$\mathbf{F}(\mathbf{k}, z) = \mu_0 \int \left[\mathbf{M}(\mathbf{r}', z') e^{i\mathbf{k}\mathbf{r}'} \cdot \nabla\right] \underbrace{\int \mathbf{H}(\tilde{\mathbf{r}}, z + z') e^{-i\mathbf{k}\tilde{\mathbf{r}}} \, \mathrm{d}\tilde{\mathbf{r}}}_{= \mathbf{H}(\mathbf{k}, z + z') = \mathbf{H}(\mathbf{k}, z) e^{-kz'}} \mathrm{d}\mathbf{r}' \, \mathrm{d}z'$$

$$= \mu_0 \cdot \mathbf{H}(\mathbf{k}, z) \int \underbrace{\int \mathbf{M}(\mathbf{r}', z') e^{i\mathbf{k}\mathbf{r}'} \, \mathrm{d}\mathbf{r}'}_{= \mathbf{M}^*(\mathbf{k}, z')} e^{-kz'} \, \mathrm{d}z' \cdot \begin{pmatrix} ik_x \\ ik_y \\ -k \end{pmatrix}$$

$$= \mu_0 \cdot \mathbf{H}(\mathbf{k}, z) \int \rho_{\text{tip}}^*(\mathbf{k}, z') e^{-kz'} \, \mathrm{d}z'$$

$$= \mu_0 \sigma_{\text{tip}}^*(\mathbf{k}) \cdot \mathbf{H}(\mathbf{k}, z), \qquad (4.79)$$

where we used that

$$\mathbf{M}^*(\mathbf{k}, z') \begin{pmatrix} ik_x \\ ik_y \\ -k \end{pmatrix} = \left[-\begin{pmatrix} ik_x \\ ik_y \\ -k \end{pmatrix} \mathbf{M}(\mathbf{k}, z') \right]^* = \rho^*_{tip}(\mathbf{k}, z'), \text{ and} \quad (4.80)$$

$$\sigma^*_{tip}(\mathbf{k}) := \int \rho^*_{tip}(\mathbf{k}, z') e^{-kz'} \, dz' , \quad (4.81)$$

where $\rho^*_{tip}(\mathbf{k}, z')$ is the the complex conjugate of the magnetic volume charge density at the height z'; $\sigma^*_{tip}(\mathbf{k})$ then is the complex conjugate of the tip equivalent magnetic surface charge density $\sigma_{tip}(\mathbf{r})$ in an xy-plane running through the apex of the tip which can be obtained from a propagation of the magnetic volume charge distribution ((4.81) and Fig. 4.36b).

In the two dimensional Fourier space, the force acting on the tip is therefore proportional to the product of the Fourier transform of the sample stray field and the complex conjugate of the Fourier transform of the tip-equivalent magnetic surface charge density $\sigma_{tip}(\mathbf{r})$. Once the latter is known, the stray field can be obtained from the measured force. Moreover, with the tip-equivalent surface charge density, the stray field of the tip is also known, i.e. the z-component of the stray field in a xy-plane running through the apex of the tip is given by

$$H_{z,tip}(\mathbf{r}) = \frac{1}{2}\sigma_{tip}(\mathbf{r}), \quad (4.82)$$

where the other field components and the propagation of the field to any distance Δz away from the tip can conveniently be calculated in Fourier space using (4.26) and the distance loss factor $e^{-k\Delta z}$.

4.4.4.2 Frequency Shift Arising from Magnetic Field

In a scanning force microscope the cantilever is typically tilted towards the surface such that the force is mapped along the direction of the normal vector $\hat{\mathbf{n}}$. In the limit of small oscillation amplitudes the frequency shift is given by

$$\Delta f_1 = -\frac{f_{1,0}}{2k_1}\frac{\partial F_n}{\partial n}, \quad (4.83)$$

where $f_{1,0}$ is the free resonance frequency and k_1 is the force constant of the fundamental oscillation mode of the cantilever. Using (4.79) and (4.26) the force acting on the cantilever along its normal direction defined by the vector $\hat{\mathbf{n}}$ arising from the Fourier transform of the z-component of the stray field $H_z(\mathbf{k}, z)$ becomes

$$F_n(\mathbf{k}, z) = \hat{\mathbf{n}} \cdot \mathbf{F}(\mathbf{k}, z)$$
$$= \mu_0 \sigma^*_{tip}(\mathbf{k}) \mathrm{LCF}_\eta(\mathbf{k}) H_z(\mathbf{k}, z), \quad (4.84)$$

where

$$\text{LCF}_\eta(\mathbf{k}) := -\frac{1}{k}\hat{\mathbf{n}} \cdot \nabla \qquad (4.85)$$

is the transfer function correcting for the canting η of the cantilever towards the surface. The derivative of the n-component of the force along the $\hat{\mathbf{n}}$-direction is then given by

$$\begin{aligned}
\frac{\partial F_n}{\partial n}(\mathbf{k}, z) &= \hat{\mathbf{n}} \cdot \nabla F_n(\mathbf{k}, z) \\
&= \hat{\mathbf{n}} \cdot \nabla \mu_0 \sigma_{\text{tip}}^*(\mathbf{k}) \text{LCF}_\eta(\mathbf{k}) H_z(\mathbf{k}, z) \\
&= -k\mu_0 \sigma_{\text{tip}}^*(\mathbf{k}) \text{LCF}_\eta^2(\mathbf{k}) H_z(\mathbf{k}, z) \\
&= \mu_0 \sigma_{\text{tip}}^*(\mathbf{k}) \text{LCF}_\eta^2(\mathbf{k}) \frac{\partial H_z(\mathbf{k}, z)}{\partial z}.
\end{aligned} \qquad (4.86)$$

Using (4.83) for the frequency shift in the case of small oscillation amplitudes, $A_1 \ll \frac{1}{k}$, the Fourier transform of the frequency shift becomes

$$\Delta f_1^{A_1 \to 0}(\mathbf{k}, z) = \text{TF}_{\text{tip}}(\mathbf{k}) \frac{\partial H_n(\mathbf{k}, z)}{\partial n}, \qquad (4.87)$$

where

$$\frac{\partial H_n(\mathbf{k}, z)}{\partial n} = \text{LCF}_\eta^2(\mathbf{k}) \frac{\partial H_z(\mathbf{k}, z)}{\partial z} \qquad (4.88)$$

is the derivative in the $\hat{\mathbf{n}}$-direction of the $\hat{\mathbf{n}}$-component of the stray field, and

$$\text{TF}_{\text{tip}}(\mathbf{k}) := -\frac{\mu_0 f_1}{2k_1}\sigma_{\text{tip}}^*(\mathbf{k}) \qquad (4.89)$$

is the tip transfer function describing the imaging properties of the cantilever in Fourier space.

For finite amplitudes A_1, the magnetic tip-sample force varies with the sinusoidal oscillation of the cantilever

$$\mathbf{q}(t) = \begin{pmatrix} 0 \\ q_y(t) \\ q_z(t) \end{pmatrix} = \hat{\mathbf{n}} \cdot q(t) = \hat{\mathbf{n}} \cdot A_1 \cos(2\pi f_1 t), \qquad (4.90)$$

where we assumed that the cantilever long axis runs along the y-direction. The frequency shift is obtained by an integration over one oscillation cycle (4.61). In Fourier space, the component $q_y(t)$ leads to a phase shift of $e^{ik_y q_y(t)}$, whereas the $q_z(t)$ component leads to a distance loss factor of $e^{kq_z(t)}$. According to (4.86), (4.62), and (4.61), the frequency shift of the first cantilever oscillation mode in Fourier space for a finite oscillation amplitude becomes

$$\Delta f_1^{A_1>0} = -\frac{f_1}{2k_1}\mu_0\sigma_{tip}^*(\mathbf{k})\mathrm{LCF}_\eta(\mathbf{k})H_z(\mathbf{k}, z)$$

$$\cdot \frac{2}{\pi A_1^2}\int_{-A_1}^{A_1} e^{[ik_y \sin(\eta)+k\cos(\eta)]q}\frac{q}{\sqrt{A_1^2-q^2}}\,dq$$

using: $ik_y\sin(\eta) + k\cos(\eta) = \nabla\cdot\mathbf{n}$

$$= -\frac{f_1}{2k_1}\mu_0\sigma_{tip}^*(\mathbf{k})\mathrm{LCF}_\eta(\mathbf{k})H_z(\mathbf{k}, z)\frac{2}{\pi A_1^2}\int_{-A_1}^{A_1}\frac{qe^{\nabla\cdot\mathbf{n}q}}{\sqrt{A_1^2-q^2}}\,dq$$

$$= \mathrm{TF}_{tip}(\mathbf{k})\cdot\mathrm{LCF}_\eta(\mathbf{k})H_z(\mathbf{k}, z)\frac{2}{\pi A_1^2}\int_{-A_1}^{A_1}\frac{qe^{\nabla\cdot\mathbf{n}q}}{\sqrt{A_1^2-q^2}}\,dq\ . \quad (4.91)$$

Note that $H_z(\mathbf{k}, z)$, the Fourier transform of the z-component of the stray field at the lower turning point, $z = z_{ltp}$ can be taken out of the integral of (4.61) describing the weighted averaging process, because the distance loss factor for obtaining the field along the positions of the tip oscillation remains within the integral. With the definition of the first order modified Bessel function of the first kind

$$I_1(\tilde{z}) = \frac{1}{\pi}\int_0^\pi e^{\tilde{z}\cos\psi}\cos(\psi)\,d\psi\ \ \text{with: } \tilde{z} = A_1\mathbf{n}\cdot\nabla\ \text{and}\ \cos\psi = \frac{q}{A_1} \quad (4.92)$$

$\mathbf{k}$ follows

$$\frac{1}{\pi A_1}\int_{-A_1}^{A_1}e^{\nabla\cdot\mathbf{n}q}\frac{q}{\sqrt{A_1^2-q^2}}\,dq = -I_1(\tilde{z}) \quad (4.93)$$

and (4.91) becomes

$$\Delta f_{1,mag}^{A_1>0}(\mathbf{k}, z) = \mathrm{TF}_{tip}(\mathbf{k})\cdot\frac{2}{A_1}\mathrm{LCF}(\mathbf{k})_\eta H_z(\mathbf{k}, z)I_1(\tilde{z})$$

$$= \mathrm{TF}_{tip}(\mathbf{k})\cdot\left.\frac{\partial H_n^{A_1}(\mathbf{k}, z)}{\partial n}\right|_{eff}, \quad (4.94)$$

where the expression

$$\left.\frac{\partial H_n^{A_1}(\mathbf{k}, z)I_1(\tilde{z})}{\partial n}\right|_{eff} := -\frac{2I_1(\tilde{z})}{kA_1}\cdot\mathrm{LCF}(\mathbf{k})_\eta\cdot\frac{\partial H_z(\mathbf{k}, z)}{\partial z} \quad (4.95)$$

is an effective stray field derivative along the $\hat{n}$-axis. With this, the frequency shift can be calculated from the Fourier transform of the z-derivative of the z-component of the stray field, using the small amplitude approximation denoted in (4.87).

Alternatively, the lever canting function (4.85), and the effective stray field derivative given in (4.95) to study the effects of the canting of the cantilever towards the surface of the sample on the measured contrast.

4.4.5   Calibration of the MFM Tip

The calibration of the tip transfer function [290, 481, 662] is performed as illustrated in Fig. 4.37a. First, MFM frequency shift data, $\Delta f(x, y)$, acquired on a suitable calibration sample, are disentangled to separate frequency shift contributions of topography induced variations of the van der Waals and magnetic background from local film thickness and roughness variations from the frequency shift contrast [Fig. 4.37b] arising from the pattern of up/down domains $\Delta f_{dom}(x, y)$.

In a second step, the $\Delta f_{dom}(x, y, z)$ frequency shift data [Fig. 4.37b] is partially deconvolved from distortions arising from the tip oscillation with an amplitude A along the $\hat{n}$-axis (4.95), and the thickness and distance loss factors (4.42) and then discriminated to obtain a pattern of up/down domains. For the partial deconvolution, the Tikhonov method with a suitably chosen Tikhonov parameter is used. In the

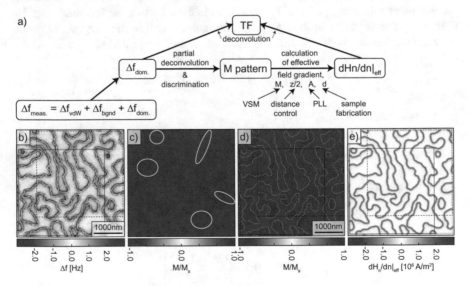

Fig. 4.37 a Principle of the tip calibration process. **b** Frequency shift arising from pattern of up/down domains, obtained from an MFM measurement at $z = 7$ nm, with an oscillation amplitude $A = 5$ nm on a Pt(10 nm)/ [Co(0.6 nm)/Pt(0.7 nm)]$_{\times 5}$/Pt(3 nm) magnetic multilayer sample with strong perpendicular magnetic anisotropy. **c** Pattern of up/down domains obtained by a partial deconvolution process from (**b**) followed by discrimination. The yellow ellipses highlight pixels with wrong magnetization orientation (**d**) corrected pattern of up/down domains obtained from (**b**). **e** Effective stray field derivative along the $\hat{n}$-axis for a finite oscillation amplitude $A = 5$ nm of the tip along the $\hat{n}$-axis. The insets in **d** and **e** highlight the symmetry and asymmetry of the domain walls along the y-direction, respectively. The squares highlighted by the solid and dashed black lines in (**b**), (**c**) and **e** indicate the data subset that has been used to obtain the tip transfer functions displayed in Fig. 4.38, respectively

center of larger domains (where the stray field becomes small because of the thickness loss factor, see bottom-most panel of Fig. 4.20b the discrimination procedure often leads to areas showing the wrong magnetization direction. Such areas (for example, those highlighted by the yellow ellipses in Fig. 4.37c are then manually corrected to obtain the pattern of up/down domains displayed in Fig. 4.37d. Note that the correct up/down-direction of the magnetization is apparent from the Δf_{dom} shown in Fig. 4.37b.

In a third step, the Fourier transform of the effective stray field derivative along the $\hat{\mathbf{n}}$-axis [Fig. 4.37e] is calculated from that of the domain pattern [Fig. 4.37d] using (4.95), the magnetization determined for example by vibrating sample magnetometry (here set to 1 A/m), the thickness of the magnetic layer d, the oscillation amplitude A, and $z/2$, half the tip-sample distance used in the experiment. According to (4.94), the Fourier transform of the tip transfer function $\text{TF}_{\text{tip}}(\mathbf{k})$ can then be obtained from the division of the Fourier transforms of the frequency shift [Fig. 4.37b] by that of the effective stray field derivative along the $\hat{\mathbf{n}}$-axis [Fig. 4.37e]. Such a division is however an ill-posed problem because the denominator can become very small (or even vanish) which would lead to a possibly extreme amplification of the measurement noise. Hence an appropriate regularization method must be used. In [290, 481], a pseudo-Wiener filter method was used. In our work, we use the Tikhonov regularization method to determine the transfer function from

$$\text{TF}_{\text{tip}}(\mathbf{k}, z/2) \approx \Delta f_{\text{dom.}}(\mathbf{k}, z/2) \cdot \frac{\left.\frac{\partial H_n^{A1}(\mathbf{r}, z/2)}{\partial n}\right|_{\text{eff}}}{\delta + \left|\left.\frac{\partial H_n^{A1}(\mathbf{r}, z/2)}{\partial n}\right|_{\text{eff}}\right|^2}, \quad (4.96)$$

where the Tikhonov parameter δ defines a limit for otherwise diverging solutions, effectively limiting the amplitudes of the transfer function for small effective stray field derivatives typically occurring at smaller spatial wavelengths.

The effect of the Tikhonov parameter becomes apparent when simulated MFM images are subtracted from the measured data and the least square differences are calculated. Subimages at the positions of the black squares were cut from the $\Delta f_{\text{dom.}}(x, y, z)$ frequency shift [Fig. 4.37b] and $\left.\frac{\partial H_n^{A1}(\mathbf{r}, z/2)}{\partial n}\right|_{\text{eff}}$ data [Fig. 4.37e] data to obtain a tip transfer function with a Tikhonov parameter $\delta = 10^{7.5}$, leading to a rather smoothened transfer function, i.e. with an cut-off at larger wavelengths. The real-space tip transfer function is displayed in Fig. 4.38a, together with the simulated MFM data [Fig. 4.38b] and the difference between the measured $\Delta f_{\text{dom.}}(x, y, z)$ frequency shift data [data inside black square of Fig. 4.37b] and the simulated MFM data [Fig. 4.38c]. In the difference data, the domain pattern is clearly visible indicating that the shorter spatial wavelengths required for an improved modeling of the domain walls are missing in the transfer function. The standard deviation of the difference data is 0.33 Hz. Using a Tikhonov parameter $\delta = 10^{3.5}$, the transfer function becomes sharper and has a higher signal at the center, but is also more affected by noise as pointed out by the yellow arrows in Fig. 4.38d. The simulated MFM data

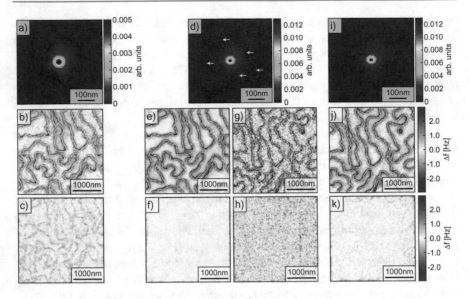

Fig. 4.38 Tip transfer function in real space, $\mathrm{TF_{tip}}(\mathbf{r})$, $z/2$ obtained by Tikhonov devonvolution with a Tikhonov parameter $\delta = 10^{7.5}$ (**a**), $\delta = 10^{3.5}$ (**d**), and after averaging 2916 individual transfer functions (**i**). The MFM data simulated with the transfer functions shown in **a**, **d**, and **i** are displayed in panels (**b**), (**e**)/(**g**), and (**j**) respectively. The corresponding difference data is displayed in panels (**c**), (**f**)/(**h**), and (**k**)

displayed in Fig. 4.38e has a larger contrast. Correspondingly, the difference image shown in Fig. 4.38f has a smaller contrast, and its standard deviation is only 0.09 Hz.

While MFM data simulated for the same domain pattern that has been used to obtain the tip-calibration function [Fig. 4.38e] agrees well with the measured data [Fig. 4.38b], a simulation performed at a different sample location [Fig. 4.38g] shows a strong granular up/down contrast in addition to the expected up/down domain pattern, leading to a correspondingly strong granular contrast of the difference image [compare Fig. 4.38h–f]. Trivially, the granular contrast must arise from a corresponding artifact contained in the tip transfer function. An improved transfer function can be obtained by averaging over many subimages cut from the original $\Delta f_{\mathrm{dom}}.(x, y, z)$ domain and $\left.\dfrac{\partial H_n^{A_1}(\mathbf{r}, z/2)}{\partial n}\right|_{\mathrm{eff}}$ data displayed in Fig. 4.37b and e, respectively. Here, 2916 tip transfer functions were averaged. This reduces the noise in the transfer function considerably, as visible in the real space image displayed in Fig. 4.38i. The MFM data simulated with the averaged transfer function is shown in Fig. 4.37j. The corresponding difference data depicted in Fig. 4.38k and the reduced standard deviation of 0.13 Hz reveal that this tip transfer function can now be used to simulate the MFM contrast arising from the stray field of an arbitrary magnetization pattern.

In order to select the Tikhonov parameter, or to assess to which spatial wavelength λ a transfer function obtained with a specific Tikhonov parameter is reliable, the decay of the transfer function with increasing wavenumber $k = 2\pi/\lambda$ is calculated. For this, the transfer functions obtained with different Tikhonov parameters are

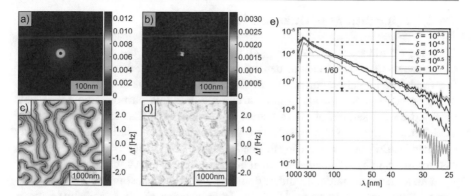

Fig. 4.39 a Tip transfer function in real space, $TF_{tip}(\mathbf{r}, z/2)$, obtained by circular averaging of the real space tip transfer function displayed in Fig. 4.38i. **b** Difference between the latter and the circularly averaged tip transfer function in real space. Note that the scale is a factor of 4 smaller to highlight the small differences. **c** and **d** Simulated MFM and difference data obtained with the circularly averaged tip transfer function. **e** Decay of the circularly averaged real space tip transfer functions with decreasing spacial wavelength for Tikhonov parameters of $\delta = 10^{7.5}$ (green curve) to $\delta = 10^{3.5}$ (thicker orange curve)

circularly averaged. Figure 4.39a displays the circularly averaged transfer function in real-space. The difference to the non-circularly averaged function is displayed in Fig. 4.39b with a color scale range that is four times smaller to make the small differences visible. The MFM image modeled with the circularly averaged transfer function, Fig. 4.39c, is smooth, but in the difference image [Fig. 4.39d], a faint contrast at the location of the domain walls, and a small contrast between the up and down domains is apparent.

The decay of the transfer function with increasing wavenumber $k = 2\pi/\lambda$ is plotted in Fig. 4.39e. Transfer functions with a Tikhonov parameter above $10^{5.5}$ (green, red, and blue curves) show a pronounced reduction of signal at smaller wavelenghts. At $\lambda = 30$ nm, the signal of the tip transfer function for $\delta = 10^{5.5}$ is about 30% smaller than that calculated for tip transfer function with $\delta = 10^{4.5}$. The difference between the latter transfer function and that calculated for $\delta = 10^{3.5}$ however remains below 10% at $\lambda = 30$ nm. A Tikhonov parameter $\delta = 10^{4.5}$ thus is ideal down to wavelengths of 30 nm, corresponding to a feature size of 15 nm. Note that for a spatial wavelength reduced from 300 to 30 nm, the signal of the tip transfer function is reduced by about a factor of 60. A further reduction of the MFM signal at small spatial wavelength then occurs because of the distance loss factor, which amounts to about a factor 6.6 for an MFM image recorded at 10 nm tip-sample distance. This demonstrates the need for sensitivity, as can for example be obtained by high-quality factor cantilevers discussed in Sect. 4.4.1.

4.4.6 Applications of Quantitative MFM

Once the tip transfer function $TF_{tip}(\mathbf{k})$, or with it, $\sigma_{tip}^*(\mathbf{k})$, the complex conjugate of the magnetic volume charge density of the tip (see (4.89)) has been obtained from an MFM tip calibration procedure (Sect. 4.4.5), it can be used to test the fitness of different candidate magnetization structures. For this purpose, the MFM data obtained on the sample of interest must be compared to that calculated from the model micromagnetic structures. The stray field patterns of the latter are conveniently obtained in 2D Fourier space using the methods described in Sect. 4.3. The calculation of the candidate MFM images can then be performed using (4.95).

In many cases, an initial candidate domain pattern can be estimated from the measured frequency shift pattern, Δf_{meas}, or from a *decanted* frequency shift pattern, Δf_{meas}^{dec}, where the distortions arising from the cantilever canting and the finite cantilever oscillation have been removed. Once such a candidate magnetization pattern is available and a suitable parametrization has been defined, the optimized parameters can be obtained from a least-squares minimization process. Such a *forward* modeling process is convenient, when the signal-to-noise of the measured data is limited.

Alternatively, components of the stray field (or its derivatives) at an arbitrary distance z above the sample, or in the best case the equivalent magnetic surface charge distribution of the sample of interest can be obtained from a measured MFM frequency shift pattern $\Delta f(x, y, z)$. Note that such a *backward* processing is more challenging, because the stray field decays rapidly with increasing tip-sample distance particularly for smaller spatial wavelengths and also the tip transfer function is less sensitive at smaller wavelengths. Consequently, the signal-to-noise ratio limits which part of the signal is lost in the noise and can thus not be recovered. The recovery process (or the *backward* process) is generally an ill-posed problem and thus typically performed by a deconvolution process. For this, it is necessary to test different deconvolution parameters (e.g. Tikhonov parameters) and analyze up to which spatial wavelength the deconvolution procedure remains within certain error limits. This procedure is similar to that used to determine the best Tikhonov parameter for obtaining the equivalent tip transfer function from MFM data measured for the calibration of the MFM tip (see Sect. 4.4.5). The tip calibration procedure also defines the minimal spatial wavelength that can be obtained by a successive deconvolution of measured MFM data in the best case. It is convenient to define a cut-off wavelength for the deconvolution process to avoid the amplification of noise.

4.4.6.1 Micromagnetism of the Cu/Ni/Cu System

Ferromagnetic multilayer thin film systems with perpendicular magnetic anisotropy (PMA) arising from interface anisotropy have been a subject of intensive studies in the late 1980's [98, 105, 143, 144] providing a basis for to date's perpendicular magnetic data storage systems as well as for thin films supporting skyrmions at room temperature (see Sect. 4.4.6.6). Cu/Ni/Cu trilayers show a strong PMA even for very large Ni thicknesses beyond 10 nm, where the effect of interface anisotropy becomes

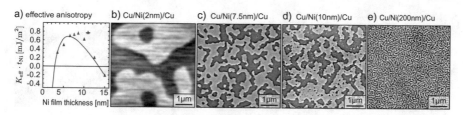

Fig. 4.40 a Dependence of the effective anisotropy $K_{eff} \cdot t_{Ni}$ of the Cu(100)/Ni(t_{Ni})/ Cu-system on the Ni film thickness t_{Ni}. **b** Static mode MFM image of a Cu(100)/ Ni(2 nm)/Cu film with large domains. **c, d,** and **e,** dynamic mode MFM images of Cu(100)/Ni(t_{Ni})/Cu-films with Ni thicknesses $t_{Ni} = 7.5, 10$, and 200 nm, respectively. Larger Ni thicknesses result in smaller domain sizes (Figure adapted from Bochi et al. [89], with permission from Americon Physical Society, and Marioni et al. [429], with permission from American Physical Society)

negligible. Instead, the PMA arises from the biaxial tensile strain of the Ni film that is epitaxially grown on the Cu(100)-substrate, and from the large magnetoeleastic coupling constant of Ni. Figure 4.40a displays the effective magnetic anisotropy values $K_{eff}(t_{Ni})$ measured by Bochi et al. [89] for Cu/Ni(t_{Ni})/Cu trilayers with Ni thicknesses, t_{Ni}, between 2 and 17.5 nm. The effective magnetic anisotropy values have been obtained from vibrating sample magnetometry (VSM). The MFM images, shown in Fig. 4.40b–e, acquired with a room temperature prototype of a UHV low temperature AFM [297], show a pattern of maze domains with a domain size that becomes smaller with increasing Ni film thickness, as can be expected from the increasing magnetostatic energy. From the dependence of the effective anisotropy on the film thickness, that becomes negative for thicknesses above about 13 nm, one would expect that the magnetization should be in-plane. Hence, large domains with straight domain walls are expected. However, even at a Ni film thickness of 200 nm, a pattern of maze domains with an positive/negative $\Delta f(x, y)$-contrast of ±60 Hz is observed, which is reminiscent of that expected for a pattern of domains with an up/down magnetization [Figs. 4.40e and 4.41].

Here, a quantitative analysis of the MFM data can be used to understand the micromagnetic state of the system. Figure 4.41a and b show the recorded MFM data and the simulated MFM contrast, respectively. The simulated image was obtained from a model up/down domain pattern with a shape as observed in Fig. 4.41a, the measured saturation magnetization M_s, and a calibrated tip equivalent magnetic charge distribution, (4.80) and (4.81) obtained from an MFM tip calibration [662]. The excellent match between the simulated and measured data is apparent from a comparison of the cross-sections shown in Fig. 4.41c taken at the locations of the black and blue lines in Fig. 4.41a and b, respectively. This confirms that a candidate magnetization structure consisting of a pattern of domains with an up/down magnetization of $\pm M_s$ is an excellent description of the micromagnetic structure. If the same procedure is performed for data taken on the Cu/Ni(200 nm)/Cu sample [Fig. 4.41d], the $\Delta f(x, y)$-contrast simulated for an up/down domain pattern is a factor of 1.58 times larger than the measured $\Delta f(x, y)$-contrast [compare Fig. 4.41d–e or the black to the blue cross-section in [Fig. 4.41f]. In [429], the reduced contrast is explained by closure domains with a magnetization canting angle of about 51° relative to the sur-

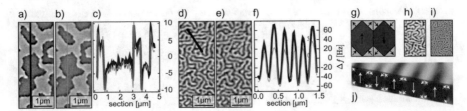

Fig. 4.41 **a** Measurement and simulation of the magnetization of Cu/Ni/Cu/Si(001) structures. **a** MFM image of a Cu(001)/Ni(7 nm)/Cu film, measured at 31 ± 1 nm tip sample distance. **b** Image simulated from perpendicular domains obtained from (**a**) with the calibration function $\sigma_{tip}^{*}(\mathbf{k})$. **c** Comparison of the sections indicated in (**a**) (black line) and (**b**) (blue line). **d** MFM image of a Cu(001)/Ni(200 nm)/Cu film. **e** Simulation assuming perpendicular domains obtained from d). f) comparison of the cross-sections indicated in (**d**) and (**e**). Figure adapted from Marioni et al. [429] with permission from American Physical Society. **g** Schematics of the magnetization texture inside the Cu(001)/Ni(200 nm)/Cu film. **h** MFM subimage cut from (**d**) for comparison with the domain structure obtained from micromagnetic simulations [displayed in (**i**)] using the experimental system parameters for the magnetization M_s, effective anisotropy, K_{eff}, and exchange stiffness A. **j** Magnetization texture of the Cu(001)/Ni(200 nm)/Cu film obtained from micromagnetic calculations. Panels **i** and **j** courtesy of Prof. Dr. D. Suess, Vienna University, Austria

face normal. Recent micromagnetic calculations performed by Prof. Dr. D. Suess, Vienna University, Austria confirmed the existence of such closure domains.

Note that this micromagnetic state of the sample leads to a linear magnetization loop with no hysteresis for a field applied along the perpendicular axis. Conventionally, such a magnetometry result would be interpreted as a hard axis loop with a sample magnetization that rotates from an in-plane to the perpendicular direction of the applied field. The domain pattern observed by MFM [Fig. 4.41d] however implies that most of the magnetic moments are aligned to a direction perpendicular to the sample plane. The absence of a hysteresis and the large linear range of the total magnetic moment with the applied field may make such systems useful for perpendicular field sensors in industrial applications.

4.4.6.2 Exchange Biased Samples

Coupling a ferromagnetic film (F) to an antiferromagnetic layer (AF) can significantly alter the hysteresis process and domain structure attributed to uncompensated spins occurring at the F-AF-interface (Fig. 4.42). This typically leads to a shift of the hysteresis loop along the field axis, H_{ex}, termed as *exchange-bias* (EB) effect, often accompanied with an enlarged coercivity and a vertical shift, m_{vert}, of the hysteresis loop. Exchange-biasing results from field cooling the sample through the Néel temperature T_N of the AF [Fig. 4.42a and b]. The ordered F magnetic moments then induce a specific orientation of uncompensated moments existing at the F/AF-interface that becomes locked or pinned if the sample is cooled sufficiently below T_N.

Alternatively, the sample can also be cooled with the ferromagnet in a demagnetized state [327]. Then no macroscopic exchange-bias effect occurs but the local order of the magnetic moments inside each domain will still imprint a specific orientation

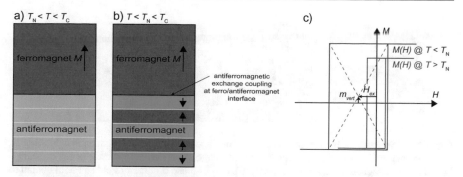

Fig. 4.42 a System consisting of a ferromagnet exchange-coupled to an antiferromagnet at a temperature T that is lower than the Curie temperature of the ferromagnet T_C but higher than the Néel temperature T_N of the antiferromagnet. The antiferromagnet is then in a paramagnetic state. **b** Once the sample is field-cooled below the Néel temperature of the antiferromagnet, $T < T_N$, the spins of the antiferromagnet order with a spin direction defined by the magnetic moment of the ferromagnet and the type of the ferromagnet/antiferromagnet exchange-coupling (here antiferromagnetic). **c** Hysteresis loops of the ferromagnet for $T_N < T < T_C$ (red) and for $T < T_N$ (blue). The latter loop is shifted along the field direction by the so-called exchange-field H_{ex} arising from the pinned uncompensated spins (pUCS) at the antiferromaget/ferromagnet interface and the exchange-coupling of the ferromagnetic moments to these pUCS. The ferromagnet thus has a preferred magnetization direction (unidirectional anisotropy). Because of the pinned uncompensated moments of the antiferromagnet, the latter has a remanent magnetic moment that gives rise to a vertical shift of the loop. Note that the sign of the vertical shift does not necessarily need to coincide with the sign of the exchange coupling at the ferromagnet/antiferromagnet interface. An additional phenomena often accompanying the exchange-bias effect is an increased hysteresis (wider loop)

of the uncompensated spins at the F/AF interface. First MFM work on exchange-biased samples revealing pinned uncompensated spins was performed by Kappenberger et al. [327]. The sample, consisting of a F/AF-heterostructure multilayer with the composition $CoO(1\,nm)\{[Co(0.4\,nm)/Pt(0.7\,nm)]_{\times 4}Co(0.6\,nm)/CoO(1\,nm)\}_{\times 10}$ [Fig. 4.43a], was demagnetized at room temperature with an oscillating in-plane field and zero-field cooled to 7.5 K below the $T_N \approx 250\,K$ of the 1 nm-thick CoO layers. Figure 4.43b shows the MFM data obtained after zero-field cooling. The contrast arises from stray fields generated by the stripe domains of the ferromagnetic layers, but also by the uncompensated spins imprinted by those domains during cool-down into the antiferromagnetic layers.

Figure 4.43c shows the MFM data recorded after applying of a field of 700 mT. The white domains with a magnetization opposite to the applied field have retracted and a pattern of much weaker stripe domains becomes apparent in the growing dark parts of the image. These stripe domains with a much fainter contrast become well visible in Fig. 4.43d, after saturation of the ferromagnetic layers in a field of 800 mT. Note that the range of the grey scale of Fig. 4.43d was reduced to match the $\Delta f = 1.5\,Hz$ peak-to-peak contrast. Note that the ferromagnetic layers are saturated, and that a thin film with a homogeneous magnetization does not generate a stray field (see Sect. 4.3.2). Hence, the MFM contrast visible in Fig. 4.43d arises from the imprinted pattern of pinned uncompensated spins. Note that the pattern remained even when the field was raised to 7 T [328], the maximum possible with the low temperature MFM for this

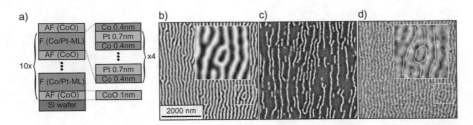

Fig. 4.43 a Structure of the ferromagnet/antiferromagnet (F/AF) multilayer sample exhibiting an exchange-bias effect at low temperatures. **b** MFM image of the domain structure of the exchange-bias sample demagnetized at room temperature by an in-plane oscillatory field and field-cooled to 7.5 K, well below the Néel temperature, T_N of the AF. The grey scale corresponds to a frequency shift contrast, $\Delta f = 29.2$ Hz. **c** MFM image of the domain structure at 700 mT displayed with $\Delta f = 10.3$ Hz. **d** MFM image obtained after saturation of the sample at 800 mT displayed with $\Delta f = 1.5$ Hz. The much weaker MFM contrast arises from the imprinted pattern of pinned uncompensated spins in the AF layers. Adapted from Kappenberger et al. [327] with permission of American Physical Society

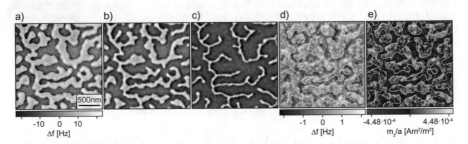

Fig. 4.44 a As-grown domain structure after zero-field cooling a Pt(2 nm)/CoO(1 nm)/Co(0.6 nm)/[Pt(0.7 nm)/Co(0.4 nm)]$_{\times 20}$ F/AF exchange-bias sample to 8.3 K. The domain walls are highlighted by the yellow lines. **b** Domain pattern at 100 mT. **c** Domain pattern at 200 mT, **d** MFM data obtained after saturation of the ferromagnetic layer in 300 mT. The granular contrast arises from the imprinted pattern of pinned uncompensated spins. To facilitate the comparison of the observed pattern with the ferromagnetic domain pattern obtained after zero-field cooling (**a**), the domain walls of the latter are again highlighted by yellow lines. Note that the Δf-contrast is much weaker than than observed in (**a**). **e** Pattern of the pinned uncompensated spins obtained from (**d**) by deconvolution of the Δf pattern by the tip transfer function and a transfer function relating the surface magnetic moment density to the stray field derivative. Adapted from Schmid et al. [579] with permission from American Physical Society

work [297]. An average pinned uncompensated moment density corresponding to about 7% of a full CoO monolayer was obtained [327]. While the density of the pinned uncompensated spins agreed with findings from other experimental techniques [498], and with the moment density expected from the lateral shift of the magnetization loop (the exchange field H_{ex}), the pinned uncompensated moments were found to be antiparallel to the F moments. This becomes apparent when comparing the zoomed areas in Fig. 4.43a and d highlighted by the yellow squares. This was an unexpected finding and started a debate on the existence of different pinned AF moments with only a part of them giving rise to the exchange bias effect [578].

A more refined quantitative analysis was performed by Schmid et al. [579]. In this work, a sample with a single F/AF-interface was designed to obtain the local areal density of the pinned uncompensated spins. The sample consisted of a Pt(2 nm)/CoO(1 nm)/Co(0.6 nm) AF/F bilayer deposited onto a ferromagnetic multi-layer consisting of [Pt(0.7 nm)/Co(0.4 nm)]$_{\times 20}$ on a 20 nm thick Pt seed on a silicon waver coated with its native oxide. Without the bottom, thick ferromagnetic multi-layer, the 0.4 nm-thick top F Co layer would show domains that are several microns in size [320]. The MFM image would then show a noticeable contrast only near the domain walls, while the domain centers would develop only negligible contrast (Sect. 4.3.2). While the F domains would still be recognizable [see for example Fig. 4.23b], the interpretation of the very granular pattern of pinned uncompensated spins would become challenging. It is thus convenient to engineer the thickness of the ferromagnetic Pt/Co multilayer or the number of Pt/Co-bilayer repetitions to obtain a domain size small enough to generate the up/down domain contrast shown in Fig. 4.44a. The domain patterns in fields of 100 and 200 mT are displayed in Fig. 4.44b and c, respectively. d) At 300 mT, the ferromagnetic layers are saturated. The weak, granular $\Delta f(\mathbf{r}, z_{ts})$-contrast displayed in Fig. 4.44d can thus be attributed to the stray field arising from the imprinted pattern of uncompensated spins that are responsible for the exchange-bias effect. Using the tip transfer function TF($\mathbf{k}$) determined from a calibration procedure (see Sect. 4.4.5), the local areal magnetic moment density of the pinned uncompensated spins could be determined [Fig. 4.44e]. This pattern of pinned uncompensated spins has later been used to model the evolution of the domain pattern and to obtain the macroscopic exchange field and increased coercivity observed experimentally for this sample [60].

4.4.6.3 Dependence of Magnetic Bubble Domain Diameter on the Applied Field

Vock et al. [672] used the transfer function method discussed in Sect. 4.4.5 to simulate MFM images from the stray field of candidate magnetization structures. Their work is one of the very few examples where MFM data acquired under ambient conditions has been used for a quantitative analysis. For the tip calibration, the maze domain pattern from Fig. 4.45a, occurring in a [Co(0.4 nm)/Pd(0.7 nm)]$_{\times 80}$ multilayer sample, has been used. In higher fields, Fig. 4.45b, isolated bubble domains appear. Using a micromagnetic model based on the work by Thiele [644], the expected radius of the bubble domains could be obtained within the field interval where bubbles are stable against expansion into stripe domains or annihilation in low and high fields, respectively.

4.4.6.4 Exchange Spring Media

The stabilization of a specific magnetization direction of a ferromagnetic (F) layer is of highest importance for various spintronic devices. Typically, such a stabilization is provided by the exchange-bias effect (Sect. 4.4.6.2). This technique enables applications such as spin valves [148,543] and magnetic tunnel junction devices [185,465] and it has also been considered in heat-assisted magnetic recording [444,505]. For

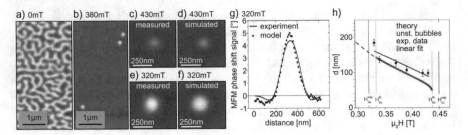

Fig. 4.45 **a** Maze domain pattern of the [Co(0.4 nm)/Pd(0.7 nm)]$_{\times 80}$ multilayer sample used for tip calibration. **b** Bubble domains. **c** and **d** Zoomed views of a bubble domain and simulated bubble domain contrast, respectively, in a field of 430 mT. **e** and **f** Zoomed views of a bubble domain and simulated bubble domain contrast, respectively, in a field of 320 mT. The bubble domains have a larger radius and therefore give rise to a stronger MFM contrast. **g** Cross-sections of experimental (blue line) and model data through the center of the bubble domain at 320 mT. **h** Theory of the expected dependence of the bubble domain diameter on the applied field (red line) is compared to data extracted from qMFM modeling (black points). (Figure adapted from Vock et al. [672]). Copyright (2011) IEEE

the latter example in particular [171,533,609], but likewise in general spintronics applications, exchange-coupled double-layer (ECDL) structures using rare-earth-based ferrimagnets (FIs) might provide useful design flexibility. Zhao et al. [728] have prepared a series of exchange coupled ferro/ferrimagnet double layer samples Pt(3)/[Pt(0.7)/ Co(0.4)]$_{\times 5}$/ Tb$_{26.5}$Fe$_{73.5}$(20)/Pt(10)/SiO$_2$/Si, where the numbers in the brackets indicate the thicknesses in nanometers, and the subscript number, the composition of the amorphous TbFe layer in atomic percent.

The magnetization process of the soft [Pt(0.7)/Co(0.4)]$_{\times 5}$ film was found to be surprisingly complex. Instead of the conventional lateral wall motion observed in the soft layer of a weakly exchange-coupled system (Sect. 4.4.6.2), a nucleation-dominated three-stage magnetization process showing a significant spatial variation occurs. Figure 4.46a–d show MFM data obtained after cooling the exchange-coupled bilayer system from room temperature to 10.5 K. In stage 1, a granular contrast pattern occurs inside the Co/Pt down domain which increases with the applied field [Fig. 4.46a and b]. In stage 2, the granular contrast becomes most intense and a strong evolution of the pattern occurs. Figure 4.46c shows MFM data taken at a field of 1500 mT characteristic for stage 2 of the magnetization process. Finally, in stage 3, the contrast of the granular pattern becomes weaker with increasing applied field, but the pattern no longer changes. Figure 4.46d, taken at 4 T, shows a typical image of the third magnetization stage. The characteristic magnetization structures that give rise to these observations can be inferred from matching MFM $\Delta f(\mathbf{r}, z_{ts})$-data simulated from candidate micromagnetic structures to measured MFM data. At zero field, the magnetization of the Co/Pt multilayer is opposite to that of the ferrimagnetic TbFe layer [Fig. 4.46e]. The MFM contrast simulated from this model magnetization pattern is displayed in Fig. 4.46i. Apart from the weak granular structure of the measured image, a good agreement between simulated and measured data is obtained. Note that the granularity arises from grain-to-grain variations of the magnetization

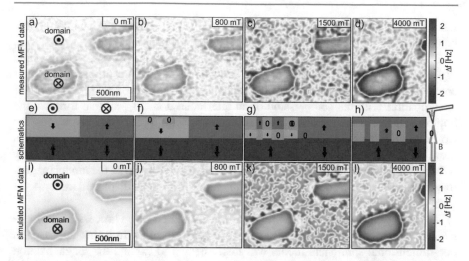

Fig. 4.46 a to **d** MFM images measured in fields of 0–4000 mT on an exchange coupled ferro/ferrimagnet double layer, Pt(3)/[Pt(0.7)/Co(0.4)]$_{\times 5}$/Tb$_{26.5}$Fe$_{73.5}$(20)/Pt(10)/SiO$_2$/Si sample. **e** to **h** Schematics of candidate magnetization structures used for modeling the MFM contrast of the different magnetization stages. **i** to **l** Modeled MFM images obtained from the candidate magnetization structures (see Zhao et al. [729] for details). Figure adapted from Zhao et al. [729] with permission from American Chemical Society

or the misalignment of the grains' easy axis with the z-direction which are both not considered in the candidate magnetization structure.

In stage 1, the gradual increase of the contrast is compatible with a rotation of the initially down magnetic moments of Co/Pt toward the field (up) direction. Because the magnetic moments of the Co/Pt multilayer at the interface are pinned to the magnetic moments of the high-anisotropy TbFe film, the Co/Pt multilayer magnetic moments at the top surface are expected to rotate more than those near the interface. For the modeling of the MFM contrast, the vertical structure of the spin chains at any given location on the image plane is modeled by a corresponding spatial distribution of "subdomain blocks" with zero magnetization, located at the top of the Co/Pt multilayer and reaching into different depths toward the interface [gray blocks in Fig. 4.46f]. The lateral distribution of these blocks can be inferred from the granular contrast observed in the MFM $\Delta f(\mathbf{r}, z_{ts})$-data displayed in Fig. 4.46b. With optimized sublayer depths, again an excellent agreement between the modeled and measured MFM $\Delta f(\mathbf{r}, z_{ts})$-data could be obtained [compare, for example, Fig. 4.46k–c].

In stage 2, a strong increase of the contrast is accompanied by a substantial change in the appearance of the granular pattern. These observations are compatible with isolated Co/Pt grains switching their magnetization from a canted down to a canted up state (as was used in the modeling). This is reminiscent of a Stoner–Wohlfarth magnetization process with a field applied away from the easy axis where an instability of the magnetization state occurs. The candidate magnetization structure for stage 2 is depicted in Fig. 4.46f. With the latter, again an excellent agreement of the modeled [Fig. 4.46i] with the measured MFM $\Delta f(\mathbf{r}, z_{ts})$-contrast [Fig. 4.46c] could be obtained.

In the third magnetization stage [2000 mT $< B \le$ 7000 mT see for example Fig. 4.46d], the pattern of the granular contrast inside the up domain no longer changes, and a reduction of its contrast in increasing fields is observed. This behavior is compatible with the compression of an interfacial domain wall that forms at the bottom and top of the ferromagnetic and ferrimagnetic layers, respectively. As the Co/Pt multilayer approaches saturation, its magnetic moment distribution becomes more homogeneous and its stray field thus decays (note that the stray field of a layer with a homogeneous magnetization vanishes). Figure 4.46h shows the model magnetization distribution used to obtain the simulated MFM image displayed in Fig. 4.46l with the measured data shown in Fig. 4.46d.

To better understand the nucleation-governed magnetization process, micromagnetic modeling was performed. Two approaches were used. In a first approach, a one-dimensional spin-chain model was implemented to explain the magnetization behavior of isolated grains. The 3-stage magnetization process inferred from the MFM observation and modeling of the MFM contrast could be confirmed. In a second modeling approach, a full 3D-micromagnetic calculation was performed. The result showed that the magnetization behavior can be well understood with the more simple one-dimensional spin chain model and confirmed that the even simpler candidate magnetization structures used for the simulation of the observed MFM images are justified. A theoretical study revealing the full complexity of the magnetization process and further experimental work with thicker F layers have recently been published by Vogler et al. [674] and Heigl et al. [266], respectively.

4.4.6.5 Magnetic Vortices in Magnetic FeCo Nanowires

In their work, Vock et al. [673] have electrochemically grown FeCo nanowires in anodized aluminum oxide templates. Figure 4.47a shows an SEM topview of the $Fe_{52}Co_{48}$-filled Al_2O_3 template. An MFM overview image is displayed in Fig. 4.47b. For the subsequent analysis, Vock et al. used the transfer function theory as described in Sect. 4.4.5. The tip was calibrated on the maze domain pattern obtained from a $[Pt(0.9nm)/Co(0.4nm)]_{\times 100}$ multilayer [Fig. 4.47c and d].

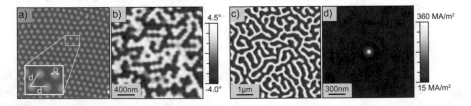

Fig. 4.47 a SEM image of the $Fe_{52}Co_{48}$ filled Al_2O_3 nanowire template. **b** MFM image of the nanowire array in the as-grown, zero-field state. **c** MFM image of the band domain state of the $Pt(5 nm)/[Pt(0.9 nm)/Co(0.4 nm)]_{\times 100}/Pt(2 nm)$ used for the tip calibration. **d** Circularly averaged $\frac{\partial H_z^{tip}}{\partial z}$ obtained from the tip calibration procedure. (Figure adapted from Vock et al. [673] with permission from AIP Publishing)

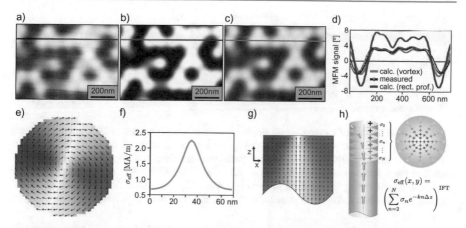

Fig. 4.48 **a** Measured MFM image. **b** Simulated MFM image for a homogeneous magnetization. **c** Simulated MFM image for a vortex at the top end of the nanowire. **d** Cross-sections of the measured MFM data (black line and dots), simulated MFM data with a homogeneous magnetization (red line), and simulated MFM data for a vortex (green line). **e** Vortex structure, **f** cross-section through the effective magnetic surface charge generated by the vortex, **g** side-view of the vortex structure, and **h** schematics displaying how the effective magnetic surface charge, $\sigma_{\text{eff}}(x, y)$. (Figure adapted from Vock et al. [673] with permission from AIP Publishing)

Because of the high aspect ratio, $L_z/d = 93$, of the nanowires and the thus result-ing high shape anisotropy, $K_{\text{sh}} = 1130\,\text{kJm}^3$, their magnetization is predominately aligned along the wire axis and a homogeneous effective surface charge distribution given by M_s is expected. In a demagnetized sample, two wire magnetization direc-tions will occur leading a bright/dark MFM contrast. An exemplary MFM image is displayed in Fig. 4.48a together with a simulated MFM image in Fig. 4.47b. This reveals that the model of a homogeneous surface charge distribution is not valid. Micromagnetic simulations were then performed, revealing that a magnetic vortex is formed at the ends of the nanowire such that most magnetic moments at the sur-face are in-plane, considerably reducing the magnetic surface charge [Fig. 4.48e]. From the 3-dimensional distribution of the magnetic moment distribution obtained from the micromagnetic calculation, an effective surface magnetic charge density, $\sigma_{\text{eff}}(x, y)$ could be calculated [Fig. 4.48f]. The MFM contrast obtained from the lat-ter, depicted in Fig. 4.48c matches the measured contrast [Fig. 4.48a] well [compare also the green and black cross-sections in Fig. 4.48d].

4.4.6.6 Skyrmions

Skyrmions are whirls of magnetic moments with dimensions ranging from a few hundred nanometers down to the atomic scale. Skyrmions appear in ferromagnetic bulk or thin film materials with a broken inversion symmetry where, apart from the colinear ferromagnetic exchange interaction, a Dzyaloshinskii–Moriya interaction (DMI) occurs. The DMI interaction favors spatially-modulated magnetization states like spin-spirals and skyrmions that can occur in ordered hexagonal lattices or, if skyrmion pinning is dominant, also in strongly disordered arrangements. Skyrmions

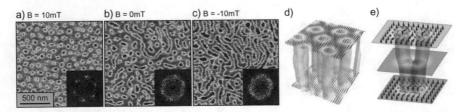

Fig. 4.49 a MFM data of a $Fe_{0.5}Co_{0.5}Si$ single crystalline sample cooled to 10 K in 20 mT shown together with the Fourier transform highlighting the hexagonal order of the skyrmion lattice. **b** and **c** Data acquired at 0 mT and −10 mT, respectively, and corresponding Fourier transforms. **d** Sketch of the skyrmion lattice. The skyrmion walls are chiral Bloch walls. **e** Sketch of a magnetic configuration describing the merging of two skyrmions. Figure adapted from Milde et al. [457]. Copyright (2013) of the American Association for the Advancement of Science

possess a nontrivial topology, and carry a finite topological charge. It is therefore not possible to continuously transform a skyrmion into a uniform magnetization state. The skyrmions thus are topologically protected, and hence a finite amount of energy is needed to annihilate them. Skyrmions respond to electric currents via spin-transfer and spin-orbit torques and can be moved through device-like geometry structures. This makes them attractive for future spintronic devices.

The first magnetic force microscopy results have been obtained by Milde et al. [457] on $Fe_{0.5}Co_{0.5}Si$ single crystalline samples with a B20 crystal structure and bulk DMI. Figure. 4.49a shows MFM data acquired after field cooling in 20 mT to 10 K. A (slightly irregular) hexagonal skyrmion lattice is visible. A reduction of the magnetic field to 0 mT [Fig. 4.49b] and finally to −10 mT [Fig. 4.49c] leads to the coalescence of neighboring skyrmions and to the formation of elongated structures. The images show how the number of skyrmions and with them the topological charge of the visible sample area is reduced. Note that in these B20 single crystalline materials exhibiting bulk DMI, the skyrmion spin texture has Bloch walls as indicated schematically in Fig. 4.49d. The hexagonal order gradually disappears as the skyrmions coalesce in higher fields parallel to their core magnetization. The magnetic configuration describing the merging of two skyrmions is depicted in Fig. 4.49e. At the merging point the magnetization vanishes at a singular point (arrow). This defect can be interpreted as an emergent magnetic antimonopole, which acts like the slider of a zipper connecting two skyrmion lines. Note that the existence of DMI in these material lifts the degeneracy of the Bloch wall chirality and all walls must show the same chirality according to the specific sign of the DMI.

Skyrmions at room temperature were first observed by XMCD x-ray microscopy [466] in $SiO_x/Pt(10)/[Ir(1)/Co(0.6)/Pt(1)]_{\times 20}$-multilayers. The same system but with smaller numbers of repeats was also studied at room temperature by MFM by the same group under ambient conditions. Figure 4.50a and b show $3 \times 3\mu m^2$ MFM images of the demagnetized state obtained under ambient conditions of the an $[Ir(1)/Co(0.6)/Pt(1)]_{\times n}$-multilayers with $n = 3$ and 20, respectively. Although the signal-to-noise ratio of the three-repeats sample is clearly reduced, MFM images showing isolated skyrmions or chiral bubble domains could also be obtained Fig. 4.50c and d for multilayers with $n = 5$ and 10, respectively. Note that in these

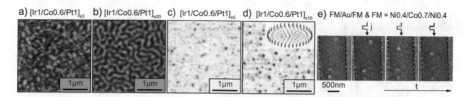

a) [Ir1/Co0.6/Pt1]$_{\times3}$ b) [Ir1/Co0.6/Pt1]$_{\times20}$ c) [Ir1/Co0.6/Pt1]$_{\times5}$ d) [Ir1/Co0.6/Pt1]$_{\times10}$ e) FM/Au/FM & FM = Ni0.4/Co0.7/Ni0.4

Fig. 4.50 a and **b** MFM data of a [Ir(1 nm)/Co(0.6 nm)/Pt(1 nm)]$_{\times n}$-multilayer with 3 and 20 repeats, respectively, with interfacial DMI arising from the Ir/Co and Co/Pt interfaces. The MFM data shows the zero-field state. **c** and **d** Skyrmions or chiral bubble domains in Ir/Co/Pt-multilayers with 5 and 10 repeats. The inset in **d** shows the spin texture of a skyrmion with a clock-wise Néel wall. **e** Transport of skyrmions in a Ni(0.4 nm)/Co(0.7 nm)/Ni(0.4 nm)/Au(3 nm)/Ni(0.4 nm)/Co(0.7 nm)/Ni(0.4 nm) structure. Current pulses of 3 ns with $j = 3.9 \cdot 10^{11}$ A/m^2. Figure adapted from Cros et al. [2] and Hrabec et al. [289], distributed under the terms of the Creative Commons Attribution License (http://creativecommons.org/licenses/by/4.0)

thin film systems the DMI arises from spin-orbit coupling occurring predominantly at the Co/Pt-interface giving rise to domains, and skyrmions with clock-wise Néel walls [see inset in Fig. 4.50d]. Improved MFM images of skyrmions and their current-induced generation and dynamics were then reported by the same group, again using an MFM operated under ambient conditions, in [289] [Fig. 4.50e]. However, from the MFM images it remains unclear whether the visible roundish objects are bubble domains with chiral Néel walls but a homogeneous magnetization region in their center, or "true" skyrmions with the spins rotating away from the z-axis away from the skyrmion central axis [like shown in the inset of Fig. 4.50d]. To distinguish these distinct but topologically equivalent spin textures, a quantitative analysis of MFM data acquired with improved signal-to-noise ratio is required.

A first quantitative analysis of MFM data acquired on thin film multilayer samples was reported by Yagil et. al. [709] for a [Ir(1)/Fe(0.5)/Co(0.5)/Pt(1)]$_{\times20}$ multilayer (where the numbers in brackets are the thicknesses of the individual layers in nanometers). Figure 4.51a displays an MFM image recorded at 5 K in a field of -0.3 T using a cantilever with a spring constant $c_L \approx 1$ N/m, a free resonance $f_0 \approx 75$ kHz, an oscillation amplitude $A \approx 30$ nm, and a scan height $h \lesssim 50$ nm. The tip and the sample were saturated together in a positive field, such that the skyrmions appear with a magnetization opposite to that of the tip and hence lead to a positive frequency shift or a repulsive tip-sample interaction force. Note also that the low temperature, the relatively soft cantilever, and the relatively large oscillation amplitude, together with a presumably high quality factor in vacuum, an excellent signal-to-noise ratio was obtained. The magnetic scalar potential of the skyrmion was described by a multipole expansion with a dipole and quadrupole moment. The sign of the quadrupole moment indicates whether the chirality is clockwise or anti-clockwise and hence describes the helicity of the Néel skyrmions which is defined by the sign of the interfacial Dzyaloshinkii-Moriya interaction. To fit the MFM data, the tip was modeled as a thin shell defined by $g(z) = \alpha(z^4 + \beta z)^{1/4}$ with axial symmetry with $\alpha = 0.24$, and $\beta = 2.7 \cdot 10^6$ nm^3. The results of the fit are displayed in Fig. 4.51b, where each skyrmion was fitted with individual dipole and quadrupole moments. The fit of an

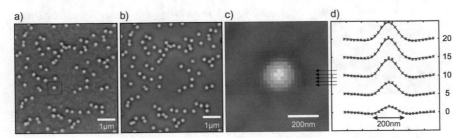

Fig. 4.51 a MFM data obtained on a Ir(1)/Fe(0.5)/Co(0.5)/Pt(1)]$_{20}$ multilayer at 5 K and in a field of -0.3 T showing skyrmions. **b** Fitted skyrmions using a multipole model to describe the magnetic scalar potential of the skyrmion magnetization field, and a simple, geometry-based model for the MFM tip. **c** Zoomed view of the skyrmion highlighted by the red frame in (**a**). **d** Cross-sections though the skyrmion displayed in (**c**). The solid lines show the fitted cross-sections. Figure adapted from Yagil et al. [709] with permission from AIP Publishing

individual skyrmion and the corresponding cross-sections are shown in Fig. 4.51c and d. Fitting all skyrmions visible in Fig. 4.51a allowed to determine that the Néel skyrmions have a helicity $\gamma = 0$, and a radius $r_{sk} = 34 \pm 27$ nm.

Modeling of the MFM contrast arising from skyrmions in an [Ir(1)/Co(0.6)/ Pt(1)]$_{\times 5}$ multilayer system with interfacial Dzaloshinskii-Moriya interaction using a calibrated tip was first performed by Bacani et al. [38]. The MFM was operated at room temperature but under vacuum conditions to obtain a quality factor of about 40'000, and with it, a superior sensitivity (see Sect. 4.4.1). Note, however that in more recent work, cantilever that have a quality factor up to 1'000'000 can be prepared; see Sect. 4.4.1 for example images. Bimodal magnetic force microscopy with capacitive tip-sample distance control [593] was used to measure a large series of images in different fields with a reproducible average tip-sample distance. However, due to an increase of the cantilever quality factor with time, the tip-sample distance feedback had to be re-calibrated repeatedly. Here, the use of the frequency-modulated capacitive tip-sample distance control mode [727] developed later would have been beneficial for an even better reproducibility of the tip-sample distance. The appearance of the skyrmions at zero-field after saturation in negative fields and their expansion into larger (bubble) domains in increasing positive fields, are displayed in Fig. 4.52.

Higher-resolution images of isolated skyrmions, their annihilation in increasing fields, and the background frequency shift signal are displayed in Fig. 4.52a, b, and c, for a pair of skyrmions. Using differential imaging techniques as described in Sects. 4.4.3.3 and 4.4.3.4, the MFM signal observed after skyrmion annihilation Fig. 4.52b and c could be attributed to the spatial variations of the average Co-layer thickness. Using Tikhonov deconvolution, the spatial variation of the magnetization, corresponding to a spatial variation of the magnetic moment density if the film thickness were constant could be obtained, and explained by an average film thickness variation of ± 1 atomic layer of Co. Such a spatial variation of the Co layer thickness then leads to a strong variation of the local DMI as predicted from ab-initio theory [711]. To analyze the MFM contrast arising from the skyrmions, a parametrized,

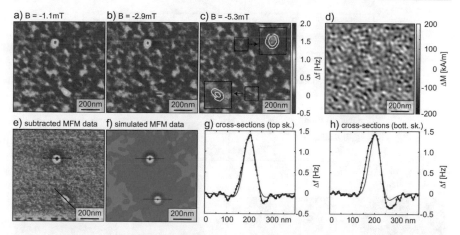

Fig. 4.52 **a** Measured MFM image at $B = -1.1$ mT. **b** The bottom skyrmion is repeatedly annihilated by the additional field from the tip at $B = -2.9$ mT. **c** At $B = -5.3$ mT both skyrmions are deleted. The background contrast arises from variation of the areal magnetic moment arising from a corresponding variation of the total Co-film thickness. **d** shows the corresponding local variation of the magnetization when the Co-film thickness is assumed to be constant at the nominal value obtained by deconvolution of (**c**) using quantitative MFM methods as described in Sect. 4.4.4 and also in [38]. **e** Background subtracted MFM data obtained by subtracting (**c**) from (**a**). **f** Simulated MFM $\Delta f(x, y)$-image using a skyrmion model magnetization pattern with $D = 3.45$ mJ/m^2 and $D = 3.43$ mJ/m^2, for the top and bottom skyrmion, respectively. The agreement obtained between simulated and measured data is excellent; see simulated (red lines) and measured (black lines with points) cross-sections displayed in (**g**) and (**h**), respectively. (Figure adapted from Bacani et al. [38], distributed under the terms of the Creative Commons Attribution License (http://creativecommons. org/licenses/by/4.0))

semi-analytical model was used to describe the skyrmion spin textures and for the calculation of the MFM contrast expected from it (Sect. 4.4.4.2) using the calibrated tip transfer function (Sect. 4.4.5). Excellent agreement between the simulated and measured skyrmion MFM contrast images could be obtained from fitting only the DMI-value and keeping the film anisotropy, K_u and its exchange stiffness, A fixed [Fig. 4.52e–h].

As already pointed out by Tetienne et al. [643], the stray field above a counter-clock wise magnetic moment texture is amplified, whereas that for a clock wise texture is attenuated. This is because the Néel wall type skyrmions arising from an interfacial Dzyaloshinkii-Moriya interaction (iDMI) generates a non-vanishing divergence of the magnetization field, i.e. magnetic volume charges in the inside of the thin film multilayer structure. The stray field emanating from the magnetic volume charges then either amplifies or attenuates the one arising from the magnetic surface charges (see (4.44)).

Figure 4.53a shows that the measured MFM image of a skyrmion in a Pt(10)/ Co(0.6)/Pt(1)/[Ir(1)/Co(0.6)/Pt(1)]$_{\times 5}$/Pt(3) multilayer sample. The iDMI in this multilayer favors a clock wise spin texture such that the stray field of the magnetic volume charges attenuates that of the magnetic surface charges. Figure 4.53b and d display the MFM contrast modeled for a clock wise and counter-clock wise skyrmion spin

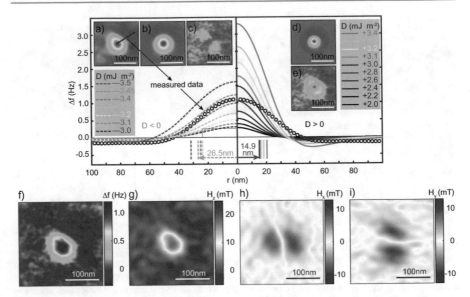

Fig. 4.53 **a** MFM image of a skyrmion. **b** Simulation of the MFM image using an iDMI < 0 chosen to match the experimental peak contrast as per panel (**a**). **c** The difference image (**a**–**b**) reveals that the simulated skyrmion MFM image matches the measured image well. **d** Simulation of the MFM image using a iDMI > 0 chosen to match the experimental peak contrast as per panel (**a**). **e** The difference image reveals that the modeled skyrmion MFM image matches the center contrast of the measured skyrmion but not the width of the measured skyrmion. The cross-section calculated for iDMI = $-3.45 \, \text{mJm}^{-2}$ nicely match that from the measurement, while all cross-sections for iDMI > 0 fail to match the measured contrast. **f** Larger view of the MFM data of the skyrmion already depicted in (**a**). **g** Pattern of the z-component of the stray field obtained from a deconvolution of the MFM data with the tip-transfer function. **h** and **i** Display the x- and y-components of the stray field calculated from (**g**). Figure adapted from Marioni et al. [430] with permission from American Chemical Society

texture, respectively. The difference images obtained from subtracting the simulated from the measured MFM data that is displayed in Fig. 4.53c and e reveal that only the simulation based on the clock-wise spin texture matches the experimental data. The matching of simulated MFM data from chiral candidate spin textures to measured MFM data can thus by used to determine the type of domain wall (Bloch or Néel-type) and for the case of Néel-type walls the chirality of the domain wall with it the sign of the DMI. Similar work was also performed by Meng et al. [446] to determine the chirality in a complex oxide multilayer structures generating DMI and nanoscale skyrmions.

Note that it is generally convenient to fit MFM data from a stray field arising from candidate magnetization structures to measured MFM data versus performing a deconvolution of the measured MFM signal to obtain the stray field. The latter is generally more challenging because the quality of a deconvolution process is typically limited by signal-to-noise ratio of the measured MFM data. Nevertheless, such a deconvolution process was possible for the skyrmions measured by Bacani et al. [38,430] using an MFM operated under vacuum conditions. Figure 4.53f shows that the measured MFM image of a skyrmion in a Pt(10)/Co(0.6)/Pt(1)/[Ir(1)/Co(0.6)/Pt(1)]$_{\times 5}$/Pt(3) multilayer sample, while the z-component of the stray field obtained from a deconvolution process of the calibrated tip-transfer function is depicted in Fig. 4.53g. Because the stray field outside a magnetic sample is given by a scalar magnetic potential (see Sect. 4.3.1), (4.26) can be used to obtain the x- and y-component of the stray field displayed in Fig. 4.53h and i, respectively.

4.4.6.7 Deconvolution of the Stray Field from Measured MFM Data

The renewed interest in quantitative MFM methods becomes apparent in two recent publications addressing the calibration of MFM tips [481] and the evaluation of the precision of stray field patterns obtained from the deconvolution of MFM data by different groups [290]. All data was measured with scanning force microscopes operated under ambient conditions resulting in low quality factors Q^{air} around 200, with Nanosensor PPP-MFMR cantilevers having stiffnesses c_L^{air} between 2.2 and 9.8 N/m and a nominal resonance frequencies between about 50 and 120 kHz. The tip radius is typically below 50 nm.

Because of the much smaller measurement sensitivity obtained under ambient conditions, the tip calibration was performed on a Pt(2 nm)/[Co(0.4 nm)/Pt(0.9 nm)]$_{\times 100}$/Pt(5 nm)/Ta(5 nm)/SiO$_x$/Si(100) multilayer sample. Such a sample generates a substantially stronger stray field than thinner samples with a smaller number of Co/Pt repetitions. At zero field, the sample shows a pattern of maze domains with a domain width of about 170 nm. The MFM data measured at $z = A + z_{lift}$, where $A = 20$ nm is the oscillation amplitude of the cantilever, and $z_{lift} = 30$ nm is the lift height is shown in Fig. 4.54a. A magnetization pattern M_z^0 is then obtained by a discrimination procedure applied to the measured $\Delta\phi$ data. A cross-section taken at the location of the red line in Fig. 4.54a is shown as the magenta line in Fig. 4.54b. To take into account the finite width of the Bloch domain walls, the M_z^0 map is convoluted with a kernel operation describing the domain wall resulting in a M_z^{ref} (Sect. 4.3.3) map from which the z-component of the stray field H_z can be calculated [the blue line in Fig. 4.54b shows a cross-section of H_z at the location of the red line in Fig. 4.54a]. The transfer function $G_\alpha(\mathbf{k}, z)$ describing the MFM imaging process is then obtained using a pseudo-Wiener filter method as

$$G_\alpha(\mathbf{k}, z) = \Delta\phi^{ref}(\mathbf{k}, z)\frac{\hat{H}_z^{ref}(\mathbf{k}, z)}{|H_z^{ref}(\mathbf{k}, z)|^2 + \alpha} , \tag{4.97}$$

where $\Delta\phi^{ref}(\mathbf{k}, z)$ is the 2D Fourier transform of the measured phase shift $\Delta\phi^{ref}(\mathbf{r}, z)$ pattern, $H_z^{ref}(\mathbf{k}, z)$ is the 2D Fourier transform of the z-component of the stray field calculated from the magnetization pattern M_z^{ref}, and α is the Wien-filter regularization parameter. One-dimensional representations of the transfer function $G_\alpha(k_x)$ in Fourier and $G_\alpha(x)$ in direct space for a fixed z are displayed in Fig. 4.54c and d, respectively, for three different deconvolution parameters α. Note that this process is equivalent to that of van Schendel et al. [662], described in more detail in Sect. 4.4.5, where a Tikhonov regularization was used.

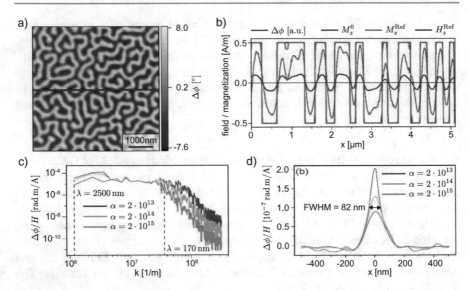

Fig. 4.54 a Typical MFM phase image acquired on the Pt(2 nm)/[Co(0.4 nm)/Pt(0.9 nm)]$_{\times 100}$/Pt(5 nm)/Ta(5 nm)/SiO$_x$/Si(100) multilayer sample used for the tip calibration. **b** The magenta curve shows the phase signal of the MFM image at the red line in (**a**). The red and green curves show the magnetization cross-sections for a magnetization patterns M_z^0 and M_z^{ref} for zero-domain wall width and a domain wall width of 16 nm, respectively. The blue line shows the z-component of the cross-section of the stray field H_z^{ref} calculated from the magnetization pattern M_z^{ref}. **c** The blue, orange, and green curves shows one-dimensional representations of the transfer functions $G_\alpha(k, z)$, obtained with Wien filter parameters $\alpha = 2 \cdot 10^{13}$, $2 \cdot 10^{14}$, and $2 \cdot 10^{15}$, respectively. The dashed vertical lines indicate spatial wavelenghts λ equal to the image size (5000 nm) and domain size (170 nm). **d** Real-space representations of the transfer functions $G_\alpha(k, z)$. The curve obtained for the best Wien-filter parameter $\alpha = 2 \cdot 10^{13}$ has a FWHM of about 82 nm. Adapted from Hu et al. [290]

The work of Hu et al. [290] reports that such a tip calibration [determination of the transfer function $G_z^{ref}(\mathbf{k}, z)$] was performed by three different research groups using the same Co/Pt multilayer calibration sample. All three groups then used their transfer function $G_\alpha(\mathbf{k}, z)$ to obtain the stray fields from $\Delta\phi^{ref}(\mathbf{k}, z)$, the 2D Fourier transform of the measured phase shift patterns, again using a pseudo Wien-filter approach for the deconvolution process,

$$H_{z,\alpha'}(\mathbf{k}, z) = \Delta\phi^{ref}(\mathbf{k}, z) \frac{\hat{G}_z^{ref}(\mathbf{k}, z)}{|G_z^{ref}(\mathbf{k}, z)|^2 + \alpha'} \,, \tag{4.98}$$

where α' denotes the Wien-filter parameter used for the deconvolution that is different from that used for the determination of the transfer function $G_\alpha(\mathbf{k}, z)$.

To test the deconvolution process, two different samples were fabricated: Sample S1 is an epitaxial Ta(3 nm/SmCo$_5$(12 nm)/Ru(9 nm)/Al$_2$O$_3$(0001) grown by pulsed laser deposition. Sample S2 is a sputter-deposited Pt(2 nm)/[(Co(0.53 nm)/Pt(1.32 nm)]$_{\times 10}$/Pt(5 nm)/Ta(15 nm)/SiO$_x$/Si(100) multilayer sample that was pat-

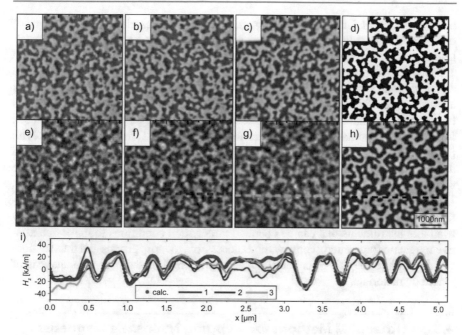

Fig. 4.55 **a** to **c** MFM phase contrast data obtained by three different research groups at the same spatial location on an epitaxial Ta(3 nm/SmCo$_5$(12 nm)/Ru(9 nm)/Al$_2$O$_3$(0001) grown by pulsed laser deposition. **d** Magnetization pattern obtained by a discrimination process of the measured MFM phase contrast data. **e** to **g** $H_z^{ref}(\mathbf{r}, z)$ data obtained by deconvolution of the measured MFM phase contrast images displayed in (**a**) to (**c**) with the tip transfer function obtained from a calibration measurement [see Fig. 4.54]. **h** $H_z(\mathbf{r}, z)$ data calculated from the magnetization pattern displayed in (**d**). **i** comparison of the cross-sections of the $H_z^{ref}(\mathbf{r}, z)$ data displayed in **e** to **g** to the cross-section of the calculated $H_z(\mathbf{r}, z)$ data shown in (**h**)

terned into randomly arranged rectangular structures ranging from $2\,\mu$m down to 250 nm.

The $\Delta\phi^{ref}(\mathbf{r}, z)$-data obtained by the three groups at the same locations of sample S1 are displayed in Fig. 4.55a–c together with the stray field patterns [Fig. 4.55e–g] obtained from the measured $\Delta\phi^{ref}(\mathbf{r}, z)$-data using the deconvolution process described by (4.98). The magnetization pattern obtained from discriminating one of the $\Delta\phi^{ref}(\mathbf{r}, z)$-maps and the $H_z^{ref}(\mathbf{r}, z)$ pattern calculated from it are displayed in Fig. 4.55d and h, respectively. The comparison of the calculated stray field map [Fig. 4.55h] with the maps obtained from the deconvolution process [Fig. 4.55e–g] reveals that the latter patterns appear less sharp, i.e. do not contain sufficient high-spatial frequency components to match the sharp transitions arising at the location of the domain walls visible in the calculated stray field pattern [Fig. 4.55h]. Nevertheless, the cross-sections taken at the dashed lines displayed in Fig. 4.55e–h reveal that the spatial variations of the stray field obtained by the three groups at the same sample location agrees within about 20%.

The results displayed in Figs. 4.54 and 4.55 reveal the state of the art of quantitative magnetic force microscopy techniques obtained using data acquired with instruments working under ambient conditions. Section 4.4.1 describes the advantages of operating a magnetic force microscope in vacuum discussing the improved signal-to-noise ratio possible with cantilevers reaching quality factors between 200'000 and 1'000'000. The operation in vacuum requires the use of more advanced tip-sample distance control modes described in Sect. 4.4.2. These operation modes, however allow the use of cantilevers with a stiffness around 0.3 N/m which is typically 10 times softer but have about the same resonance frequency as the cantilevers used for intermittent-contact/lift-mode operation (see Sect. 4.1.2) in the work of Hu et al. [290]. Consequently, according to (4.50), the sensitivity of the MFM operated with the high-quality factor cantilever in vacuum compared to the instruments used in the work of Hu et al. [290] is about 100–200 times better.

This demonstrates the advantages MFM operated with highest quality factor cantilevers using advanced operation modes, but also the necessity for commercially available instruments that can be operated under such conditions to make quantitative magnetic force microscopy methods accessible to more groups and thus to more widely explore the full potential of such experimental methods for the analysis of magnetic materials.

4.5   Other SPM Methods for Mapping Nanoscale Magnetism

4.5.1   SPM Methods Mapping the Magnetic Field

Various scanning probe methods used to map nanoscale magnetism have been developed. Here, only methods that have been widely applied by different groups to study various materials are reviewed. For this reason, spin-polarized tunneling scanning tunneling microscopy (SP-STM) and exchange force microscopy (MExFM) are discussed in Sects. 4.5.2.1 and 4.5.2.2, while other STM-based techniques that can address atomic scale magnetism like spin-flip tunneling [268,278,346] and single atom magnetic resonance imaging [699] are not described here.

MFM is a robust lab tool to image stray fields emanating from magnetic sample surfaces with high spatial resolution at various temperatures and in fields of up to several Tesla and can even perform quantitative field measurements. However, the ferromagnetic tip used in MFM inevitably generates a stray field that can perturb the micromagnetic state of the sample (see Sects. 4.2.2 and 4.2.3), particularly if magnetically-soft samples are imaged with large magnetic moment tips (which are often used to compensate, for example, for the lack of sensitivity if an MFM is operated under ambient conditions). Further, the MFM tip is a finite-size nanoscale object, inherently limiting the lateral resolution to about 10 nm [446]. In order to overcome these limitations of the MFM, other scanning probe techniques sensitive to the magnetic stray field that are less invasive, including one with an atomic-scale field sensor, have been developed. All methods mapping the stray field have in common the fact that from the measured stray field only limited information on the magnetization distribution inside the sample can be obtained. This is because the stray field emanates from magnetic surface and volume charges (see Sect. 4.3.1) arising from the divergence of the magnetization field. Different magnetization fields can have the same divergence and, trivially, any divergence-free magnetization field does not generate a stray field. Furthermore, because the stray field outside a magnetic sample is a conservative field, i.e. is generated from a scalar potential, all vector components of the stray field can be calculated from a single component of the stray field measured in a plane above a sample (see (4.26) and Fig. 4.53). The stray field decays exponentially into the outside space with a decay constant given by the spatial wavelength of the stray field. It is hence advantageous to maximize the sensitivity of the field sensor, and to scan it at the smallest possible tip-sample distance. While for MFM substantial efforts have been undertaken to obtain a robust tip-sample distance control (see Sects. 4.1.2 and 4.4.2), corresponding experimental techniques have unfortunately not yet been explored extensively for other SPM methods mapping the stray field.

4.5.1.1 Scanning Hall Probe Microscopy

Scanning Hall probe microscopy [103,501] is often described as a non-invasive (the magnetic field generated by the Hall sensor current can be neglected) and quantitative technique to map stray field emanating from the sample surface. While the first is correct, the latter statement can not be made for fields varying over the spatial wavelengths of the same order or smaller than the size of the Hall bar structure. Then, the part of the stray field component perpendicular to the Hall bar sensor passing through it must be assessed before a quantitative analysis of the stray field at the surface of the sample can be performed. Further, a quantitative mapping of the stray field can only be compared to that calculated from model-magnetization structures, if the distance between the Hall sensor and the surface of the sample has been determined.

First SHPM data with sub-micron spatial resolution were demonstrated by Chang et al. [103]. A submicron Hall structure fabricated from a GaAs/ $Al_{0.3}Ga_{0.7}As$ heterostructure was used to obtain a spatial resolution of about $0.35\,\mu m$ and a field sensitivity of about 0.01 mT. The Hall bar was structured close to the edge of a chip, and the chip edge was used as a tunneling tip to control the distance between the Hall bar structure and the surface [Fig. 4.56a]. Images of vortices in a c-axis oriented $La_{1.85}Sr_{0.15}CuO_4$ of $0.8\,\mu m$ thickness are displayed in Fig. 4.56b and c. Vortices in a 350 nm-thick $YBa_2Cu_3O_{7-\delta}$ film on MgO [Fig. 4.56d] were later imaged by Oral et al. [501] also using a GaAs/$Al_{0.3}Ga_{0.7}$ as Hall bar, albeit with a slightly larger dimension (Hall bar wire width of $1\,\mu m$).

In principle, a smaller Hall bar structure would be beneficial for the spatial resolution. The minimum size of the GaAs/$Al_{0.3}Ga_{0.7}$ is however limited to about $1.5\,\mu m^2$ because of surface depletion effects of the 2D electron gas. Much smaller Hall bar structures could be fabricated using single crystalline InSb [Fig. 4.57a] and polycrystalline Bi [Fig. 4.57c]. A field sensitivity of about $0.08\,mT/\sqrt{Hz}$ was obtained. However, the Hall bar cross is about $5\text{--}10\,\mu m$ away from the metal edge used as a tunneling tip for distance control [see arrow in Fig. 4.57a] such that the distance between the Hall bar and the sample surface is about 80 nm for a tilt angle of $1.2°$

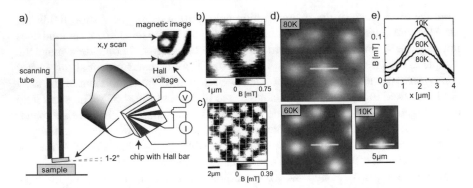

Fig. 4.56 a Schematics of the scanning Hall probe microscope (SHPM). **b** 5 μm × 5 μm scan of the magnetic field above a c-axis oriented $La_{1.85}Sr_{0.15}CuO_4$ of 0.8 μm thickness imaged in 0.5 mT and at 4.2 K. The white circular features are the vortices. **c** 9.3 μm × 9.3 μm SHPM image of a 100 nm-thick Nb film with a square-grid pattern of 0.5 μm circular holes measured in a field of 0.39 mT. About 1/3 of the holes are occupied by vortices. Figure adapted from Chang et al. [103] with permission from AIP Publishing. **d** SHPM images of vortices in a 350 nm-thick $YBa_2Cu_3O_{7-\delta}$ film acquired at 80, 60, and 10 K. **e** Corresponding cross-sections through the center of one vortex. Figure adapted from Oral et al. [501] with permission from AIP Publishing

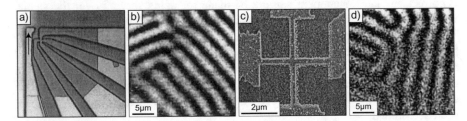

Fig. 4.57 a Optical image of a 1.5 μm² InSb Hall probe with the metallization (arrow) used for STM control of the Hall probe to surface distance. The Hall cross is about 10 μm away from the chip edge. **b** Room temperature SHPM image of a 5.5 μm-thick bismuth-substituted garnet film measured with the Hall sensor depicted in (**a**) operated at a current of 50 μA. **c** 200 nm × 200 nm Bi Hall probe. **d** SHPM image of the garnet film acquired with the Bi Hall probe at a current of 40 μA. Figure adapted from Sandhu et al. [571]. Copyright (2004) IEEE

of the chip. For a field with a spatial wavelength of 100 nm, the distance loss for $z = 80$ nm is 0.0066 leading to a much lower field sensitivity of only 12.2 mT/$\sqrt{Hz}$ instead of the 0.08 mT/$\sqrt{Hz}$ determined for homogeneous fields. A smaller sensor-to-surface distance would hence clearly be advantageous. The limited "real-world" sensitivity becomes apparent from the data recorded on a Bi-substituted iron garnet film displayed in Fig. 4.57b and d measured by the InSb and Bi Hall probes, respectively.

More recently, SHPM has been used for the mapping of stray fields with larger lengths scales. Figure 4.58a shows the setup of the SHPM instrument implemented by Shaw et al. [600]. For distance control, a tuning fork setup was used which is more reliable than the tunnel current feedback for larger area samples. Figure 4.58c,

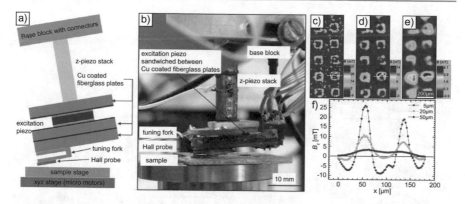

Fig. 4.58 a Schematics of the microscope. **b** Optical image of the microscope showing the probe and sample stages. **c**, **d**, and **e** SHPM images of the B_z distribution above a imprinted structure consisting of $100\,\mu m \times 100\,\mu m$ squares of NdFeB powder acquired at a Hall probe to surface distance of 5, 20, and $50\,\mu m$, respectively. **f** Corresponding cross-sections of the B_z distributions. Figure adapted from Shaw et al. [600] with permission from AIP Publishing

d and e show SHPM images of hard magnetic powder (NdFeB) based micro-flux sources acquired at 5, 20, and $50\,\mu m$ Hall probe to sample distance.

Imaging stray fields at wavelengths considerably larger than the Hall probe sensor size also permits quantitative field measurements. The traceability of calibrated Hall probes fabricated close to the tip of AFM cantilevers has recently been investigated by Gerken et al. [202]. This work also reviews different materials for the Hall sensors and points out under which conditions these are best used.

4.5.1.2 Scanning SQUID Microscopy

A superconducting quantum interference device (SQUID) is a superconducting ring typically containing two weak links, each having the same critical (super) current I_0 for the case of a symmetrical SQUID [Fig. 4.59a]. The critical current through the SQUID (through both weak links) then becomes periodic in the applied magnetic flux ϕ_a with a period given by the flux quantum $\phi_0 \approx 2.067 \cdot 10^{-15}$ Wb. If an applied current is biased just above the critical current [Fig. 4.59b], the voltage drop across the SQUID varies sinusoidally [Fig. 4.59c]. In order to determine the magnetic flux penetrating the SQUID, typically a flux-closed loop with an ac-flux modulation scheme is applied [121,351]. A feedback then applies a flux $\delta\phi$ to the SQUID loop such that the dc component of the total flux (external flux ϕ_a from the applied magnetic field and flux $\delta\phi$ generated by the feedback) is set such that the SQUID voltage is at an extremum, and the first harmonic arising from the flux modulation vanishes [Fig. 4.59d]. For microscopy the SQUID loop or a pick-up loop integrated into the SQUID must be made small and brought close to the surface of the sample [351]. The first two-dimensional scanning SQUID microscope was built by Rogers and Bermon at IBM research [554] to image superconducting vortices in devices designed for the IBM Josephson computer program. As reviewed by Kirtley [351], there are different competing strategies for improving the spatial resolution of a

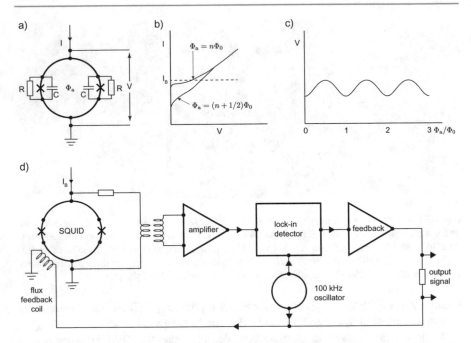

Fig. 4.59 a Schematics of a dc SQUID with two identical weak links (crosses). **b** I-V-characteristics. **c** V versus ϕ_a/ϕ_0 at constant bias current I_B. $\phi_0 \approx 2.067 \cdot 10^{-15}$ Wb is the flux quantum. **d** Schematics of the flux-closed loop. Figure adapted from Clarke et al. [121]

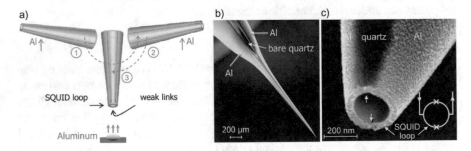

Fig. 4.60 a Schematic description of three self-aligned deposition steps for fabrication of SOT on a hollow quartz tube pulled to a sharp tip (not to scale). In the first two steps, aluminum is evaporated onto opposite sides of the tube forming two superconducting leads that are visible as bright regions separated by a bare quartz gap of darker color in the SEM image (**b**). In a third evaporation step, Al is evaporated onto the apex ring that forms the nanoSQUID loop shown in the SEM image (**c**). The two regions of the ring between the leads, marked by the arrows in (**c**), form weak links acting as two Josephson junctions in the SQUID loop. The schematic electrical circuit of the SQUID is shown in the inset of (**c**). Figure adapted from Finkler et al. [180] with permission from American Chemical Society

SQUID microscope sensor: First, the SQUID loop can be fabricated very small with narrow and thin constructions of the Josephson weak links. Hao et al. [256] reported a SQUID sensor with a diameter of about 370 nm and a noise level of $0.2\,\mu\phi_0/\sqrt{\text{Hz}}$.

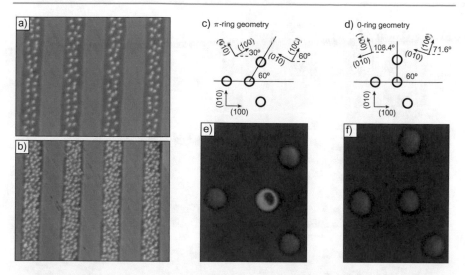

Fig. 4.61 Scanning SQUID microscope images of $35\,\mu$m-wide and $200\,$nm thick YBCO strips grown on a SrTiO$_3$ substrate by pulsed laser deposition cooled in magnetic field $\mu_0 H$ equal to 20 [(**a**)] and $50\,\mu$T [(**b**)]. Figure adapted from Kuit et al. [373] with permission from American Physical Society. **c** π-ring geometry of three fused (100) single crystalline SrTiO$_3$ substrates with four superconducting YBCO rings predicted to show a half-integer flux quantization for the YBCO ring at the intersection point of the three substrates for the case of d-wave superconductivity. **d** Zero-ring geometry: all superconducting YBCO rings should show integer flux quantization values. **e** Experimental data revealing zero flux in the three superconducting YBCO rings running across a crystalline boundary and a half-interger flux quantum for the YBCO ring at the intersection point of the three substrates. **f** All superconducting YBCO rings show zero flux. Figure adapted from Kirtley et al. [353] with permission from Springer Nature

A second approach was reported by Finkler et al. [180]. The SQUID was fabricated by three evaporations of Al onto a quartz tubes that have been pulled to form a sharp tip with apex diameter down 100 nm [Fig. 4.60]. For their SQUID with an active area of $0.34\,\mu$m^2, Finkler et al. reported a flux sensitivity of $1.8 \cdot 10^{-6}\,\phi_0/\sqrt{\text{Hz}}$. A third strategy is to use a more conventional SQUID with a larger area, but to have shielded superconducting leads to a small loop picking up the field emanating from the sample [352]. Similar to Hall sensors (see Sect. 4.5.1.1), SQUID sensors are non-invasive sensors offering an even better magnetic flux sensitivity at the cost of a lateral resolution limited by finite area of the SQUID loop. Unlike Hall sensors, which can also be applied for room temperature work, the use SQUID sensors is restricted to low temperatures. For this reasons, SQUID sensors have first been applied to study vortices in superconductors [554]. Figure 4.61a and b display SQUID microscope data recorded on $35\,\mu$m wide stripes of YBa$_2$Cu$_2$O$_{7-\delta}$ (YBCO) strips cooled in different fields. Figure 4.61c and d show the schematics of a sample used to test the type of superconducting order parameter in YBCO and the corresponding experimental results in Fig. 4.61e and f, respectively. The half-integer flux quantum appearing at the superconducting YBCO ring running across the three crystalline boundaries proved the existence of a d-wave superconducting wave function in this

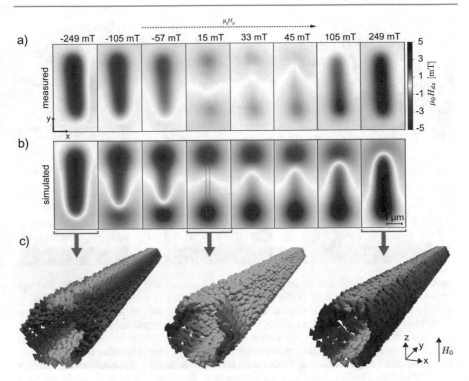

Fig. 4.62 a Reversal of a 4 μm long ferromagnetic nanotube as measured by the scanning SQUID and **b** as generated by numerical simulations of the equilibrium magnetization configuration. The dashed line inidcates the position and size of the FNT. **c** Simulated configurations corresponding to three values of H_0. The middle configuration, nearest to zero field, shows a mixed state with vortex end domains of opposing circulation sense. Arrows indicate the direction of the magnetization, while red (blue) contrast corresponds to the magnetization component along $\mathbf{e}_z$ or $-\mathbf{e}_z$. Figure adapted from Vasyukov et al. [663] with permission from American Chemical Society

high-temperature superconducting material. Further examples of work performed on vortices can be found in the review article by Kirtley [353].

SQUID microscopy has also been used to image current distributions in integrated circuits [106,603] and, more recently, to image the stray field of individual ferromagnetic nanotubes [663]. Figure 4.62 shows the magnetic reversal of a 4 μm-long ferromagnetic nanotube (FNT). The FNT consisted of a nonmagnetic GaAs core with a hexagonal cross-section surrounded by a 30 nm-thick magnetic shell of CoFeB. The data displayed in Fig. 4.62 were acquired with a SQUID-on-tip sensor [180] attached to a tuning fork used for controlling the sensor-to-surface distance. Here, the sensor was scanned about 300 nm above the FNT. Such SQUID-on-tip sensors could be fabricated with diameters as small as 46 nm and sensitivities of $50\,n\Phi_0/\sqrt{\text{Hz}}$ corresponding to spin sensitivities down to $0.38\,\mu_B/\sqrt{\text{Hz}}$ were obtained [664], clearly proving the unparalleled sensitivity that can be reached with a SQUID sensor.

4.5.1.3 NV Center Microscopy

To improve the lateral resolution of microscopes mapping the magnetic field, Chernobrod and Bergman [113], based on earlier ideas of Sekatskii and Letokov [596], proposed the use of single spin nanoscale quantum sensors. The main advantages are that the sensor has atomic scale dimensions and can offer excellent field sensitivity. An excellent embodiment of such an atomic-sized field sensor is a nitrogen-vacancy (NV) center in diamond [234,313]. The proposals and first proof-of-concept experiments, as well as various experimental and theoretical studies are reviewed in [557].

Figure 4.63a shows a sketch of an NV center, consisting of a substitutional nitrogen atom (N) and a vacancy (V) at one of the nearest neighbor sites of the diamond crystal lattice. The energy level diagram of an NV center is depicted in Fig. 4.63b. The $^3A_2$ ground level is a spin triplet state, whose sublevels are split by spin-spin interaction into a singlet state of spin projection $m_s = 0$, and a doublet with $m_s = \pm 1$. The latter is separated by $D = 2.87$ GHz in the absence of a magnetic field [indicated by the blue arrow in Fig. 4.63b]. A magnetic field B_{NV} applied along the quantization axis of the NV center $\mathbf{n}_{NV}$ (the direction of the NV-axis), leads to a splitting of the

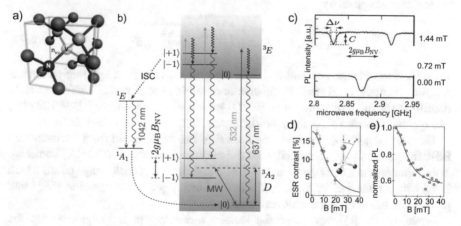

Fig. 4.63   a Atomic structure of the NV defect in diamond. **b** Energy level scheme. The notation $|i\rangle$ denotes the state with spin projection $m_s = i$ along the NV defect axis $\mathbf{n}_{NV}$. Spin conserving optical transitions from the $^3A_2$ spin triplet ground state to the ^3E excited state are shown with the green solid arrows. Such transitions are efficiently excited through non-resonant green illumination on the phonon sidebands. The dashed arrows indicate spin selective intersystem crossing (ISC) involving the singlet states ^1E and $^1A_1$. The infrared (IR) transition occurring at 1042 nm between the singlet states is also shown. **c** Optically detected electron spin resonance (ESR) spectra recorded for different magnetic field magnitudes applied to a single NV defect in diamond. The ESR transitions are shifted owing to the Zeeman effect, thus providing a quantitative measurement of the magnetic field projection along the NV defect quantization axis. These spectra are recorded by monitoring the NV defect PL intensity while sweeping the frequency of the microwave (MW) field. Spectra for different magnetic fields are shifted vertically for clarity. **d** ESR contrast and **e** normalized PL intensity as a function of magnetic field amplitude applied with an angle $\theta = 74 \pm 1°$ with respect to the NV defect axis $\mathbf{n}_{NV}$. The solid line is the result of a rate equation model developed in [642]. Figure adapted from Rondin et al. [557]. Copyright (2012) IOP Publishing and courtesy of P. Maletinsky

$m_s = \pm 1$ states by $2g\mu_B B_{NV}$, where $g \approx 2$ is the electron g-factor, and B_{NV} is the field along the NV center quantization axis. The defect can be optically excited by green laser light leading to spin-conserving transitions to an ^3E excited level, which is also a spin triplet. Relaxation can then either occur through the same radiative transition generating a broadband red photoluminescence (PL), or through a second path involving a non-radiative intersystem crossing (ISC) to singlet states. The non-radiative ISCs to the ^1E singlet state are found to be strongly spin selective, with rates from the $m_s = \pm 1$ states being much higher than from the $m_s = 0$ states. Conversely, the lowest $^1A_1$ singlet state to the ground state preferentially decays into the $m_s = 0$ state. Consequently, since the ISCs are non-radiative, the red PL is much higher, if the $m_s = 0$ state is populated. This spin-dependent PL response is then used for the detection of the electron spin resonance (ESR) signal of the NV center by optical methods. Applying a resonant microwave (MW) field to excite the NV center from its $m_s = 0$ to the $m_s = \pm 1$ states then leads to a drop in the PL signal [lowest graph in Fig. 4.63c]. An external field B_{NV} then leads to a splitting of the $m_s = \pm 1$ states and consequently to the appearance of two resonance peaks [upper graphs in Fig. 4.63c].

At sufficiently weak fields and a transverse field smaller than 5 mT, the ESR frequencies are given by

$$\nu_\pm(B_{NV}) = D \pm \sqrt{\left(\frac{g\mu_B}{h} B_{NV}\right)^2 + E^2}, \tag{4.99}$$

where h is the Planck constant and E is the off-axis zero-field splitting parameter resulting from the local strain in the diamond matrix (and $D = 2.87$ GHz as already described above). For high-purity (CVD)-grown diamond samples $E \approx 100$ kHz, and typically about 5 MHz for nano-diamonds, but $E \ll D$ is always fulfilled.

Hence, using (4.99), the magnetic field B_{NV} can be obtained from the measured ESR frequencies [optical detection of the magnetic resonance (ODMR) contrast]. This either requires a measurement of the ESR spectrum at each image point, which is very time intensive, or the use of a lock-in technique as described in [585] that permits the tracking of the ESR resonance.

Appel et al. [34] have used the lock-in measurement technique to map the stray fields above an epitaxial Cr_2O_3 antiferromagnetic thin film on a single crystalline Al_2O_3 substrate with an NV center-to-surface distance h_{NV} of about 100 nm [Fig. 4.64a]. Cr_2O_3 is a collinear magnetoelectric antiferromagnet with relevance for antiferromagnet spintronics applications [366]. The surface of the (0001)-oriented Cr_2O_3 thin film consists of a layer of Cr atoms of one of the AF sublattices, and hence all have the same magnetic moment direction. This surface termination thus leads to a surface magnetic moment density corresponding to a few μ_B/nm^2 linked to the AF order parameter that is insensitive to the local roughness [Fig. 4.64b]. Figure 4.64c–e show consecutive field maps obtained during cooling the sample through its antiferromagnetic ordering temperature. The data reveals that a magnetic field sensitivity of a few μT was obtained.

Note, however, that the acquisition of such images where the ESR resonance is tracked is time intensive. For the data displayed in Fig. 4.64a a single-pixel integration time of 7 s has been used. For this reason, a simpler, contour imaging method

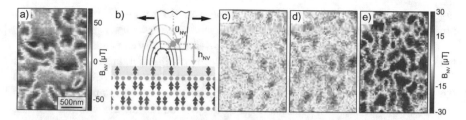

Fig. 4.64 a Map of the measured stray magnetic field B_{NV} above the Cr_2O_3 film. **b** Measurement geometry and relevant experimental parameters. **c, d,** and **e** Consecutive magnetic field maps obtained by cooling through the paramagnet antiferromagnet phase transition at a temperature corresponding to 301.5 K, 300.5 K, and 299.5 K, respectively. Figure adapted from Appel et al. [34]. Copyright 2020, American Chemical Society

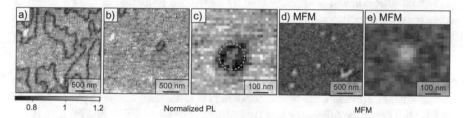

Fig. 4.65 Domains and isolated skyrmions in a Pt(5 nm)/F/Au(3 nm)/FM /Pt(5 nm), where the ferromagnetic layer, F, was a Ni(0.4 nm)/Co(0.7 nm) /Ni(0.4 nm) multilayer. **a** and **b** Photoluminescence (PL) quenching images recorded at $B_z = 0$, 3 mT. **c** PL quenching image of a single skyrmion. Figure adapted from Gross et al. [230]. Copyright 2020, American Physical Society. **d** and **e** MFM images acquired by Hrabec et al. [289] in a field of $B_z = 14$ mT of skyrmions in the same system shown for comparison. Figure adapted from Hrabec et al. [289]. Copyright 2020, Springer Nature

[557] is often used for a more rapid characterization of the stray fields emanating from magnetic structures. For this, the microwave frequency remains fixed, but its amplitude is square-wave modulated (ON/OFF) at kHz frequencies, and the corresponding NV fluorescence rate modulation is measured. Dark contours in the measured image then correspond to locations where the applied microwave frequency is in resonance with the $m_s = 0$ to $m_s = \pm 1$ transitions and thus (in a first order approximation) are contours of constant magnetic field along the NV center quantization axis. An example for this contrast mechanism measured by Gross et al. [230] is shown in Fig. 4.65a–c that display domains and isolated skyrmions in a Pt(5 m)/F/Au(3 nm)/FM/Pt(5 nm), where the ferromagnetic layer, F, was a Ni(0.4 nm)/Co(0.7 nm)/Ni(0.4 nm) multilayer. For comparison of the signal-to-noise ratio obtained with NV center microscopy, MFM data acquired by the same group (by Hrabec et al. [289]) is displayed in Fig. 4.65d and e. It may seem surprising that the SNR of the MFM data (although not performed with high quality factor cantilevers) is comparable to that acquired by NV center microscopy. However, one should keep in mind, that NV center microscopy relies on the interaction of the magnetic field with an atomic size sensor, while the MFM maps the interaction of the field with a large number of magnetic moments of the tip. Moreover, the stray field of the MFM tip may influence the micromagnetic state of the sample, particularly if tips with a

high magnetic moment are used, i.e. to compensate for a non-optimized sensitivity arising from lower-quality factor cantilevers.

For stronger sample stray fields, NV center magnetometry can become more problematic. The ESR frequencies then depend strongly on the orientation of the magnetic field with respect to the NV center axis, and obtaining quantitative information on the stray field becomes increasingly challenging. Moreover, the optically-induced spin polarization and spin dependent PL of the NV center decays and the contrast of the ESR signal vanishes [Fig. 4.63d]. As outlined in the review article of Rondin et al. [557], this is an important limitation for using NV magnetometry to perform nanoscale imaging of ferromagnetic structures, where the magnetic fields (and especially transverse fields) can easily exceed tens of millitesla, particularly at small distances above the surface. Rondin et al. [557] however, also point out, that the decreased ESR contrast is accompanied by an overall reduction of the PL intensity when the off-axis component of the field increases [Fig. 4.63e] and that this effect can be used to perform all-optical magnetic field imaging, albeit without the possibility for a quantification of the measured signal.

The fundamental limit for the sensitivity of spin-based magnetometers is given by the spin projection resulting from the intrinsic statistical distribution of quantum measurements. However, the spin-readout in NV center magnetometry is performed by PL with detection efficiency that typically is about $\varepsilon \sim 10^{-3}$. Photon shot-noise thus limits most of the NV center magnetometry data reported so far. An improved photon collection is thus beneficial for obtaining a high field sensitivity. A successful approach has been presented by Maletinski et al. [420], who integrated NV centers in cylindrical diamond nanopillars and demonstrated a field sensitivity of about $6\,\mu T/\sqrt{Hz}$ for a linewidth of about 10 MHz [Fig. 4.63c] and a PL count rate of about 200'000.

The fundamental limit of the ESR line width $\Delta v \sim 1/T_2^*$ is however determined by the inverse of the inhomogeneous dephasing time of the NV defect electron spin, T_2^* [557]. It has been pointed out that it is very challenging to reach this limit experimentally, and particularly to reach a long T_2^* time, e.g. 100 μs, which can be obtained only in ultrapure diamond samples for NV center defects not too close to the surface and kept at sufficient distance from other surfaces such as, for example, that of the magnetic sample. In the best case, a dc-field sensitivity of $40\,nT\,Hz^{-1/2}$ can be obtained. A further increase of the sensitivity can be obtained by ac-techniques. For ultrapure, isotopically engineered CVD diamond films, Balasubramanian et al. [40] obtained $T_2 = 1.82$ ms and a sensitivity of $4.3\,nT/\sqrt{Hz}$. Such high field-sensitivities make NV centers well-suited for single spin experiments, i.e. to detect the magnetic field of only a few nT emanating from a single electron spin [228].

Clearly, NV center microscopy can obtain extremely high sensitivities allowing the detection of magnetic fields arising from single spins. Moreover, an NV center-based field sensor does not influence the micromagnetic state of the sample making NV center microscopy an ideal technique for the study of magnetically soft samples. However, in spite of the excellent field-sensitivities obtainable with NV centers having long coherence times, the application of such sensors for quantitative imaging of stray fields emanating from ferromagnetic samples still remains limited and the obtained signal-to-noise ratio is in many cases worse than that obtained with state-of-the-art MFM techniques. One of the key problems with NV center microscopy is the placement of the magnetic field sensor in close proximity to the sample surface. The most wide-spread method to fabricate NC centers in controlled locations is the use of ion implantation followed by subsequent annealing procedures [480]. Unfortunately, it has been discovered that NVs implanted at shallow depths, i.e. $\lesssim 20$ nm have inferior spin coherence optical properties. To overcome this problem, Ohno et al. [500] proposed to δ-doping of diamond with nitrogen, i.e. to controllably incorporate N atoms during diamond growth. They achieved an NV layer thickness of 2 nm at a depth of 5 nm with a $T_2 > 100 \mu s$. However, the implantation depth of the NV center in the diamond crystal is only one part of total distance between the sample surface and the NV center. For a microscopy application, the diamond crystal containing the NV defect must be integrated into the tip of an AFM. In addition, a confocal microscope and an MW antenna must be combined with the AFM system to measured the spin-dependent PL signal. Two important issues must be simultaneously addressed: first, the photon collection efficiency must be maximized, and second, the AFM must allow a reasonably good control of the diamond apex to surface distance during imaging. An excellent experimental approach to increase the photon collection efficiency has been presented by Maletinsky et al. [420]. Diamond chips with integrated cylindrical nanopillars having diameters of about 200 nm and a lengths of 1 μm are microfabricated from a [001]-oriented diamond crystal containing implanted NV centers. Because of the [001]-orientation, the NV axis orientation is tilted by 54.7° from the nanopillar direction. The nanopillar can then act as a waveguide for the PL light towards the back side of the diamond chip from where the light can be collected with the optical microscope. Typically, the diamond chip with the nanopillar structure is attached to a tuning fork that then acts as an AFM cantilever, and is used for controlling the distance from the nanopillar's front most point to the surface. This however imposes a further limitation: a tuning fork based AFM is not well suited for scanning larger, i.e. micron-sized areas reliably with reasonably short scan times, because the force sensitivity of the tuning fork is limited and its resonance frequency is small. Reliable scanning at a well-controlled tip-sample distance with sub-nanometer precision as, for example, required for differential imaging techniques or to compare magnetic field data acquired, e.g. at different external fields, becomes challenging. Magnetic stray fields of periodic structures or the different Fourier components of an arbitrary stray field pattern decay exponentially with the distance from the surface of a sample. Hence, a precise and reliable control of the distance between the sensor and the surface of the sample is a necessary condition for calibration measurements, e.g. to calibrate the z-position of the NV center within the nanonpillar, and then to quantitatively map the stray field of the sample. Note that for a quantitative MFM work, considerable experimental effort has been undertaken to obtain an optimized control of the tip-sample distance to make quantitative imaging [728] and differential imaging possible [446].

4.5.2   SPM Methods Mapping Magnetism at the Atomic Scale

4.5.2.1   Spin-polarized Tunneling Microscopy

Since its invention, the scanning tunneling microscope (STM) [81,83] has become an established surface science tool. In an STM, a metallic tip is brought into a close proximity to a conducting sample. At a sufficiently small tip-sample distance, typically below 1 nm, a tunnel current can flow, which depends on the applied bias, U, on the electronic states of the tip and sample, and exponentially on the tip-sample distance. The tunnel current decays by about one order of magnitude for an increase of the tip-sample distance by 1Å. This rapid decay ultimately permits to image surfaces with atomic resolution, in spite of the radii of several tens of nanometers typical for STM tips, because most of the tunneling current flows through the apex atom.

An STM can be operated in an imaging mode where the tip-sample distance is adjusted by a feedback such that the measured tunnel current remains constant. Alternatively, the tip can be scanned at constant average height, or with a slow distance feedback, and the variation of the tunneling current arising from the local topography or spatial variations of the local density of states (DOS) can be mapped, provided that the topography is sufficiently small to avoid a tip-sample crash. To explore the electronic states, the dependence of the tunneling current I on the sample bias U can be explored. For this, either the dependence $\frac{dI}{dU}$ on the sample bias U at a selected tip position $\mathbf{R}_t$, or its dependence $\frac{dI}{dU}$ on the tip position $\mathbf{R}_t$ at a selected bias U are recorded to either locally map the electronic states or acquire a spectroscopic image of the sample. Using magnetic tips, the current can become spin-polarized [704] and surface states with different spin-polarization can be distinguished [Fig. 4.66a–c]. Like normal STM, SP-STM can achieve atomic resolution of spin-textures. Examples are shown in Fig. 4.66e and f.

Heinze et. al. [269] used non-polarized and spin-polarized STM tips to image a single Mn monolayer grown pseudomorphically on a W(110) substrate [inset in Fig. 4.66d]. With a non-magnetic STM tip, an atomic resolution image showing the pseudomorphic growth of the Mn on the W(110) substrate is obtained [Fig. 4.66d]. Using an Fe-coated W-tip having an in-plane magnetic moment orientation, the theoretically predicted c(2×2) AFM superstructure [Fig. 4.66e] was imaged using the constant current mode with a current setpoint of 40 nA and a bias of -3 mV. Generally, SP-STM with atomic resolution of non-periodic structures is however best performed using a spectroscopic imaging mode at a well-selected bias to maximize the spin-polarized term in

$$\left.\frac{dI}{dU}\right|_U \propto n_t n_s(\mathbf{R}_t, \varepsilon_F + eU) + \mathbf{m}_t \mathbf{m}_s(\mathbf{R}_t, \varepsilon_F + eU) \,, \qquad (4.100)$$

where ε_F is the Fermi energy, n_s, n_t, $\mathbf{m}_s$, $\mathbf{m}_t$ are the local density of states of the sample, tip, magnetization density of states of the sample, and tip, respectively [704].

An example of SP-STM with atomic resolution using such as spectroscopic technique was reported by Meier et al. [445] and shown in Fig. 4.66f. The image shows a 3d-representation of data acquired on a sub-monolayer of Co deposited on Pt(111).

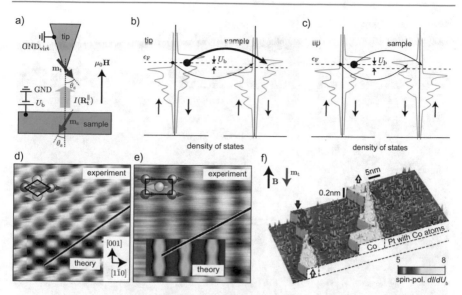

Fig. 4.66 a Schematics of a spin-polarized tip above a magnetic sample in a field $\mu_0\mathbf{H}$. **b** and **c** Electronic states for spin-up and spin-down electrons for a parallel and antiparallel arrangement of $\mathbf{m}_t$ with $\mathbf{m}_s$, respectively. Figure adapted from Phark et al. [519] Copyright 2020, Springer Nature. **d** and **e** $2 \times 2\,\mathrm{nm}^2$ constant-current STM images of one monolayer of Mn on W(110) imaged with an unpolarized W-tip (**d**) and spin-polarized Fe-coated W-tip (**e**), respectively. While panel **d** reveals the Mn atomic lattice, the data displayed in panel **e** shows the c(2×2) antiferromagnetic ground state. **f** Shows data obtained on a partial Co layer on a Pt(111) substrate recorded at 0.3 K. The STM topograph is color-coded spectroscopic dI/dU-data. Figure adapted from Wiesendanger et al. [694]. Copyright 2020, American Physical Society

Because of the perpendicular magnetic anisotropy arising from the Co/Pt interface, the Co nanowire along the Pt(111) step-edges has a magnetization that is either up or down [yellow or red arrow in Fig. 4.66f], respectively. The STM topography image recorded at $I = 0.8\,\mathrm{nA}$ and $U = 0.3\,\mathrm{V}$ is color-coded with $\frac{dI}{dU}$-data acquired simultaneously using a bias-modulation of 20 mV. Apart from the magnetic nanowires, single Co-atoms with a magnetization along the external up-field are visible on the Pt-terraces. The magnetic moment of these isolated Co-atoms could be imaged as a function of the applied field, documenting the atomic resolution of spin-textures.

Spin-polarized tunneling [Fig. 4.67a] can image the spin-texture of a sample with atomic resolution. Figure 4.67d displays SP-STM results of skyrmions in Pd(1 ML)/Fe(1 ML)/Ir(111) obtained with a Cr spin-polarized tip having a perpendicular magnetization. A model spin-texture of these skyrmions is shown in Fig. 4.67e. A spin-polarized Fe coated W-tip with an in-plane magnetization was used to obtain the SP-STM data of chiral domain walls existing in 1.75 ML of Fe on W(110) [Fig. 4.67g].

A magnetic contrast can however also arise if non-spin-polarized STM tips are used, from a local tunneling anisotropic magnetoresistance (TAMR) [91,482,678, 683] [Fig. 4.67b]. Data showing the contrast arising from chiral domain walls existing

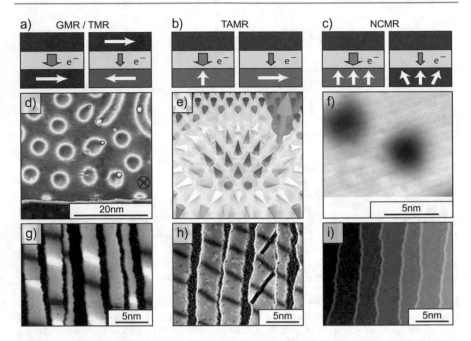

Fig. 4.67 Magnetoresistive effects in tunneling junctions. **a** Sketch of the giant magnetoresistance (GMR) or tunneling magnetoresistance (TMR) effects, in which two magnetic electrodes are separated by a metal or insulator/vacuum, respectively. **b** The tunneling anisotropic resistance (TAMR) effect does not require a magnetic sensor electrode (tip). The effect arises from the intrinsic spin-orbit coupling (SOC) within the magnetic layer giving a different conductance for an out-of-plane and in-plane magnetization of the sample. **c** The non-collinear magnetoresistance (NCMR) effect arises from a sample with a non-collinear spin texture giving rise to a contrast if the degree of the non-collinearity changes. **d** Skyrmions in Pd(1 ML)/Fe(1 ML)/Ir(111) imaged by SP-STM using a Cr spin-polarized tip. **f** Spin texture of the skyrmions in (**d**). **f** Skyrmions in Pd(1 ML)/Fe(1 ML)/Ir(111) imaged with a non-spin-polarized W tip by the NCMR contrast. **g** Domain walls in 1.75 ML of Fe on W(110) imaged by SP-STM with an Fe coated W-tip revealing the in-plane magnetic orientation. **h** Same sample as **g** but imaged with a non-spin-polarized W-tip using the TAMR contrast. **i** Topography image of 1.75 ML of Fe on W(110). **a** to **c** Adapted from Hannecken et al. [251] with permission from Springe Nature. **d** and **e** Adapted from from Romming et al. [556] with permission from American Physical Society. **f** Adapted from Hanneken et al. [251] with permission from Springer Nature. **g** to **i** Adapted from from Bode et al. [91] with permission from American Physical Society

in 1.75 ML of Fe on W(110) are displayed Fig. 4.67h. The typical topography of such a sample is displayed in Fig. 4.67i.

A change of the local differential tunnel conductance can further arise from the non-collinearity of spin textures in the sample [non-collinear magnetoresistance (NCMR) effect], which can induce a mixing between the spin channels and consequently result in a change of the local electronic state [251,371] [Fig. 4.67c]. An NCMR contrast can solely occur in the presence of non-colinear spin structures. Nanoscale skyrmions are therefore ideal candidate spin textures leading to such a contrast. Figure 4.67f shows a $\frac{dI}{dU}$-image of two skyrmions in Pd(1 ML)/Fe(1 ML)/

Ir(111) recorded at $B = -2.5\,\mathrm{T}$, $U = +0.7\,\mathrm{V}$, $I = 1\,\mathrm{nA}$, and $T = 4\,\mathrm{K}$ obtained with a non-spin-polarized W-tip.

SP-STM can thus image surface magnetism with atomic resolution and also provide spectroscopic information on an atomic scale. Using alternative contrast mechanisms such as the TAMR and NCMR, STM can further address the influence of the local spin-orbit coupling on the band structure and map the local degree of non-colinearity of the spins and thus provide data that can be directly compared to that obtained from atomistic simulations. SP-STM, however delivers a magnetic contrast only, if the electrons can tunnel into the magnetic layer of the sample. Hence, SP-STM methods can provide information on the spin texture solely if these and thus the magnetic layer containing them are located at the surface of the sample. This excludes samples where the magnetic layers are covered by other layers, e.g. for oxidation protection or by electrodes to introduce electrical currents or perform a read-out of the spin texture. Also, disentangling magnetic and topographical contrast becomes challenging on samples with a roughness on the scale of a few nanometers, which is typical for polycrystalline samples.

4.5.2.2 Magnetic Exchange Force Microscopy

With the invention of the atomic force microscope (AFM) [86], or more generally, the scanning force microscope (SFM), a scanning probe microscopy tool to image insulating sample surfaces with highest lateral resolution became available. In their publication, Binnig et al. [86] presented scanlines acquired on an Al_2O_3 surface displaying features having a width of about 3 nm. Images showing structures with atomic-scale periodicities were presented a few years later by various groups [22, 87,449,450] and it was soon recognized that the images show atomic periodicity. However, atomic-scale defects or unit cell steps with atomic extension perpendicular to the step-edge were never observed. The first images with true atomic resolution were obtained almost a decade after the invention of the AFM by Giessibl [203], Kitamura and Iwatsuki [304], and by the Morita group by Ueyama et al. [657] and Sugawara et al. [628]. For this, the tip is brought into close vicinity to the sample surface such that short-range forces arising from incipient chemical bonds between the tip apex atom and surface atoms occur (see [210] for a review of the earlier work on AFM with atomic resolution). Since 1995, various semiconducting, metallic, and insulating samples have been imaged with atomic resolution. More recently, the controlled functionalization of the tip, either by a CO molecule [231] of by an O atom [464], has become a popular technique to image organic molecules on surfaces.

In case the tip is covered with a magnetic material, the instrument becomes sensitive to the magnetic stray field emanating from the sample surface. However, if the apex atom of an AFM tip coated with a ferromagnetic or antiferromagnetic material is approached sufficiently close to the surface of a magnetic sample, the inter-atomic chemical bonding energy can depend on the relative spin-orientation of the tip apex and surface atom.

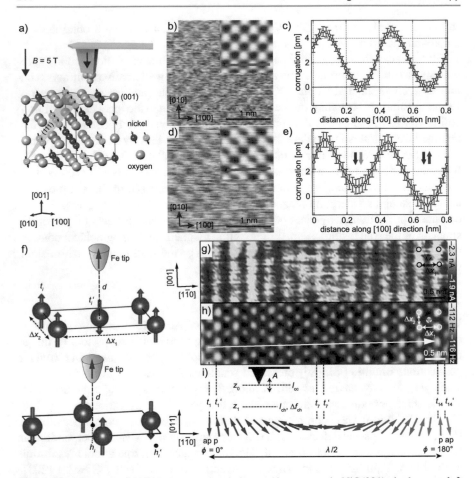

Fig. 4.68 a Concept of MExFM on the insulating antiferromagnetic NiO(001) single crystal. **b** Raw topography data recorded at $T = 7.6$ K, with a frequency shift kept constant at -22 Hz and unit cell averaged data of the chemical unit cell (inset at the top right). A cantilever with a spring constant of ≈ 34 N/m and a free resonance frequency $f_0 = 159$ kHz was used with an oscillation amplitude $A = 6.65$ nm at $U = -1.2$ V. **c** Line section of the spatially averaged magnetic unit cell along the [001]. **d** Raw data measured at $\Delta f = -23.4$ Hz showing the magnetic unit cell and unit cell averaged data of the chemical unit cell (inset at the top right). **e** Cross-section: an additional apparent height difference between nickel atoms of opposite spin orientations due to the magnetic exchange interaction with the spin of the iron tip is about 1.5 pm. Figure adapted from Kaiser et al. [323]. Copyright 2020, Springer Nature. **f** Sketches of the antiferromagnetic c(2×2) unit cell which locally approximates the spin spiral. **g** Current I_{ch} and **h** frequency shift Δf_{ch} images of one half of the spin-spiral period [see (**i**)], measured at constant height (z_1) that is by 0.29 nm closer to the surface than the height z_0 at which the current feedback loop was opened ($U = -10$ mV and $I_{cc} = -2$ nA). Parameters: oscillation amplitude $A = 50$ pm, $U = -0.1$ mV, tip magnetization normal to the surface. The arrow in **g** depicts the contrast variation due to the reversal of a single spin along the [1$\bar{1}$0] direction. **i** Side-view sketch of one half of the cycloidal spin spiral along the [1$\bar{1}$0] direction in one monolayer Mn on W(110) together with the experimental measurement scheme. Figure adapted from Hauptmann et al. [261]. Copyright 2020, Springer Nature

Consequently, a magnetic exchange force can be measured and in the best case a magnetic image with atomic resolution can be obtained. The concept of magnetic exchange force microscopy (MExFM) was first demonstrated by Kaiser et al. [323] who used an Fe-coated cantilever tip in a field of 5 T to image rows of antiferromagnetically ordered Ni spins of opposite orientation on the surface of a NiO(001) single crystal [Fig. 4.68a–e]. In a later work, Schmidt et al. [582] also resolved the antiferromagnetic structure of an Fe monolayer on W(001). While Kaiser et al. [323] and Schmidt et al. [582] both relied on an AFM equipped with a microfabricated cantilever, MExFM contrast on the NiO(001) surface has also been demonstrated by tuning fork AFM [522]. Note, that chemical interaction forces (including the spin-dependent part) are much larger than typical magnetic dipole forces measured by MFM such that the inferior force or force gradient sensitivity of the tuning forks compared to that of cantilevers is not relevant. On the contrary, when operated with small oscillation amplitudes comparable to the range of the interatomic forces, tuning fork based AFMs can easily obtain a signal-to-noise ratio that can surpass that of cantilever AFM operated with nanometer-sized oscillation amplitudes [260,262,522]. Moreover, the high spring constant of a tuning fork and the mesoscopic metal tip attached to it, considerably facilitated measurements that simultaneously acquire tunnel current and frequency shift signals, either by operation at constant tunnel current (STM-type feedback scheme) or at constant frequency shift (AFM-type feedback scheme).

It has been pointed out by Hauptmann et al. [262], that force-based detection of magnetic structures by MExFM can be compared to current-based measurements by SP-STM (see Sect. 4.5.2.1): MExFM can address magnetic insulators [323,522]. Furthermore, force-based magnetic detection can be combined with simultaneous spin-polarized current measurements (SPEX) [260] providing additional information facilitating the disentanglement of magnetic and topographic contrast contributions [261] [Fig. 4.68f–i]. Moreover, magnetic exchange force spectroscopy (MExFS) can directly quantify exchange forces [260,583], which, combined with first-principles calculations, allows for determining the interplay of various exchange mechanisms [634] as well as the role of chemical functionalization of the tip [387,388]. However, like SP-STM, exchange force microscopy techniques can solely be applied to systems where the magnetic states of the samples are at the surface of the samples. This excludes most technologically relevant samples, where the magnetic layers are typically covered for example by an oxidation protection layer. In this case, microscopy techniques mapping the magnetic stray field such as MFM and NV center microscopy are the methods of choice, although providing only indirect information on the magnetic moment distribution of the magnetic layer of the sample.

Other Members of the SPM Family

5

Abstract

Members of the family of scanning probe microscopes, be the scanning near-field optical microscopy or electrochemical scanning tunneling microscopy are introduced and some application examples are discussed.

In this chapter, some other members of the scanning probe microscopy family are briefly described. All probe microscopes are based upon probing tips, but some tips are rather different from standard STM tips. Methods such as scanning near-field optical microscopy (SNOM), scanning ion conductance microscopy (SICM) or photoemission microscopy with scanning aperture (PEMSA), are based on tips with apertures, where light, ions or electrons can pass through. The scanning near-field acoustic microscope (SNAM) is the acoustic analogue of the SNOM. Other methods essentially depend on a standard STM feedback and measure outcoming radiation, as in the case of STM with inverse photoemission (STMiP), or measure the influence of incoming light on the tunneling current, as in the case of laser STM (LSTM), or measure the temperature of the tip, as in the case of scanning thermal microscopy (SThM), or perform measurements in an electrolyte, as in the case of electrochemical STM (ECSTM). In scanning noise microscopy (SNM), the noise of the tunneling current is measured at a compensated thermovoltage. In scanning capacitance microscopy (SCM), the capacitance between probing tip and sample is measured. In scanning potentiometry microscopy (SPotM), the electrical potential, which depends on the resistivity of the sample, is measured. In scanning spreading resistance microscopy (SSRM), the spreading resistance is monitored. Scanning tunneling atom probe (STAP) is an example of the combination of STM with a time-of-flight mass spectrometer, where the mass of single ions from the probing tip can be analyzed. TERS is tip enhanced Raman scattering, where plasmons in the nanocavity enhance the electromagnetic fields, where molecular vibrations at the single molecule level can be detected. PIFM means photon induced force microscopy, where light

© Springer Nature Switzerland AG 2021
E. Meyer et al., *Scanning Probe Microscopy*, Graduate Texts in Physics,
https://doi.org/10.1007/978-3-030-37089-3_5

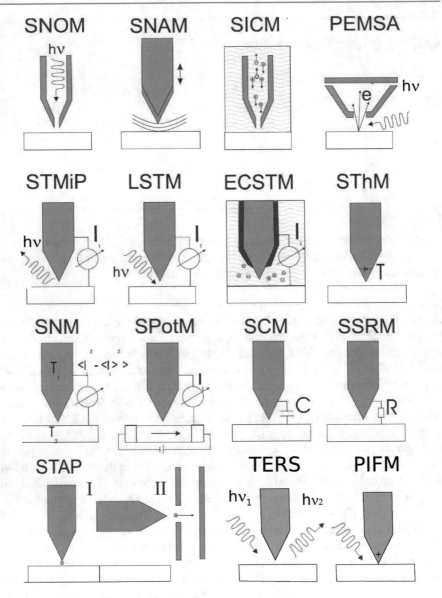

Fig. 5.1 Scanning probe microscopes (see text)

irradiation leads to the creation of image charges, which can be detected by force microscopy. All these modes are schematically depicted in Fig. 5.1.

5.1   Scanning Near-Field Optical Microscopy (SNOM)

The scanning near-field optical microscope (SNOM) probes the rapidly decaying optical near field in close proximity to objects at surfaces. In this way, the resolution limit of classical optical microscopy at one half of the wavelength is overcome, in the same way as the STM overcomes the resolution limits of scanning electron microscopy. Furthermore, SNOM allows a straightforward local spectroscopy for chemical identification and other optical techniques like polarization analysis or second harmonic generation. The drawback of this very attractive type of probe microscopy is the difficulty in reproducibly manufacturing optical probes for nanometer-scale studies.

The term 'optical near-field' describes the non-radiating part of the electromagnetic field close to a light emitter or scatterer. A well-known example is the evanescent field outside a glass prism, where a light beam is totally reflected at the inner surface. When a second body is brought into the range of the evanescent wave, the electromagnetic field can couple to this body and the total reflection is weakened.

The near field and radiative far field are always coupled. Whenever an object is interacting with an optical near field, the far-field radiation is changed. This is exploited in SNOM: the near field evanescing from an optical probe is used to excite or scatter surface objects, or the optical probe collects or scatters near-field radiation emitted or scattered from objects at the surface. By scanning the probe over the surface, lateral variations in the optical properties can be recorded. The optical near field exhibits variations on the same length scale as the size of the emitting structure. In contrast, the propagating far field cannot vary on a scale smaller than half of the wavelength. Figure 5.2 shows schematically the example of two Hertzian dipoles in close proximity. SNOM can overcome the resolution limit of classical optical microscopy, however, only by bringing the probe within a distance from the surface which is of the same order as the distance of the objects to be resolved.

Small apertures at the apex of tapered optical fibers are the most widely used optical near-field probes. The fibers can be tapered by heating the glass locally and pulling the fiber apart. Alternatively, the sharpening can be realized by an etching process [281]. The fiber has to be covered by a metal coating to confine the light and to prevent light leaking out of the tapered end. Then a sub-wavelength aperture has to be made in the metal coating. The aperture can be produced by shadow effects during

Fig. 5.2 Radiation of two Hertzian dipoles in close proximity. The near field is depicted by *grey disks* and the propagating far field by *dotted* and *solid circles*

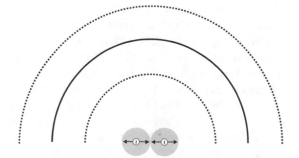

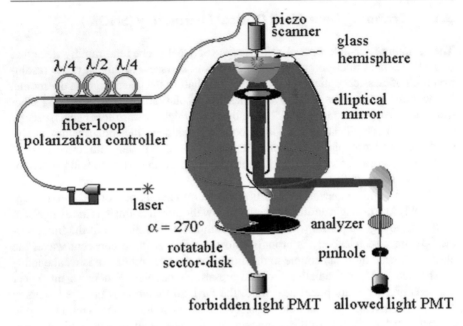

Fig. 5.3 SNOM setup for the detection of allowed and forbidden light. Reprinted from [300], http:// josaa.osa.org/abstract.cfm?URI=josaa-16-1-141, Copyright 1999 from Optical Society of America

the metal evaporation, by cutting the fiber apex after evaporation with a focused ion beam [377], or by an electrochemical etching process [473]. As an alternative to tapered fibers, tetrahedral glass tips covered with a gold film have been used to confine the light [358].

There are different configurations for SNOM using optical fiber tips. The sample can be illuminated from the rear side in a geometry where total internal reflection is found. The lateral variation of the evanescent field in front of the surface is probed by detecting the light which couples into the fiber while scanning in close proximity. The intensity of the detected light decays exponentially with the tip–sample distance and this mode has therefore been called scanning tunneling optical microscopy. The more widely used configuration is coupling of light into the fiber and illumination of the sample by the near field emerging from the nano-aperture at the fiber apex. The resulting light radiation can be detected by far-field detectors. In reciprocity to the evanescent field of a totally reflected beam, the propagating field produced by the interaction of the near field with the sample surface travels into directions of high deflection angles which are forbidden for propagating fields arriving at the surface. By detecting only these forbidden light contributions, near-field interactions can be selectively imaged with high lateral resolution (see Fig. 5.3) [300].

The resolution of this illumination mode is primarily limited by the size of the aperture and the aperture–sample distance. Correspondingly, the main difficulties of the method are the reproducible manufacturing of optical probes and the tip–sample distance control. Apertures formed by shadowing during the metal evaporation often

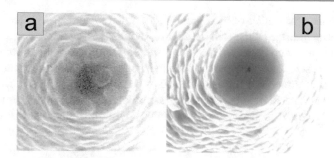

Fig. 5.4 Scanning electron micrographs of apertures at the apex of tapered optical fibers produced by **a** shadowed evaporation (aperture diameter 200 nm), **b** electrochemical etching (aperture diameter 50 nm). Note the improved flatness and regular shape of the aperture processed by etching. Courtesy of D. W. Pohl and A. Bouhelier

have grainy rims and exhibit a strong variation in diameter. Etching processes controlled by the transmission of light through the emerging aperture are a more promising technique [473]. For typical results see Fig. 5.4. The distance control in SNOM is more difficult than in other SPM methods due to the fact that the light intensity versus distance curve is very complicated and often non-monotonic. Therefore, it cannot be used for the distance control. Instead, the damping of a lateral oscillation of the fiber due to shear forces between apex and sample is exploited. The fiber is excited into oscillation by means of a small quartz tuning fork attached to it, and the change in resonance parameters of the oscillation (usually the phase between excitation and oscillation) is used for the distance feedback controller [229]. Unfortunately, the separation of controlled signal and measured signal involves numerous artifacts. Even small topographic features which change the overall tip–sample distance can appear as optical inhomogeneities [265].

Recently, some new concepts have been proposed to overcome the difficulties described above. The integration of the optical aperture into the tip of a typical SFM cantilever allows one to use the well-established SFM techniques for distance control and eventually to produce the optical probes in a batch microfabrication process [156]. A semiconductor laser has even been integrated into the cantilever as light source [270]. The idea of an active optical scanning probe has been miniaturized by exciting a single molecule attached to the apex of a tapered optical fiber and studying its fluorescence variations due to near-field interactions with objects on the sample [455]. A completely different approach is the so-called apertureless SNOM. Here, the variation in light intensity scattered in the near field between a standard SPM tip and surface objects is recorded. The problem in this technique is the poor signal-to-noise relation, and the discrimination of scattering in the near-field region and scattering in the rest of the setup. Suggested examples for well-defined applications of apertureless SNOM are the change in fluorescence lifetime in the near field of a tip or the generation of second harmonic light in the confined area.

The applications of SNOM are numerous and only a few examples can be given here. For an overview see the proceedings of the conference series entitled NFO [11]. The ability of SNOM to study the optical activity of single molecules is of great importance for studies of biologically functional systems [661]. Technological applications include the analysis of optical waveguides [41]. The electronic structure of semiconductor nanostructures has been studied with femtosecond time-resolved SNOM [16]. A scientific application is the measurement of the propagation of plasmons excited in structured metal films [93]. The combination of an SFM tip as scattering probe and Raman spectroscopy exploits the surface enhancement of the Raman effect in order to study surface chemistry with the highest lateral resolution [182].

5.2   Scanning Near-Field Acoustic Microscopy (SNAM)

In contrast to the scanning acoustic microscope [539], no lenses are used in scanning near-field acoustic microscopy (SNAM). Instead, quartz resonators with resonance frequencies in the range from 30 kHz [242,244] up to MHz [456] were used to measure the interaction between the surface and the probing tip, which is attached to the quartz (see Fig. 5.5). The distance dependence of the damping signal in Fig. 5.6 is shown for a 1 MHz oscillator. The periodic interference is due to the formation of standing acoustic waves between the tip and the surface. The radiation losses exhibit a periodicity of half the wavelength of sound. In the near-field range, at distances smaller than the wavelength, radiation losses are expected to decrease to zero. However, a strong increase is observed, which is related to the hydrodynamic

Fig. 5.5 Principle of SNAM. A tuning fork sensor is brought towards the surface. The damping signal of the resonator is related to radiation losses of the acoustic waves and to hydrodynamic damping in the near-field regime. Reprinted from [456], https://doi.org/10.1016/0040-6090(95)05853-2, with permission from Elsevier

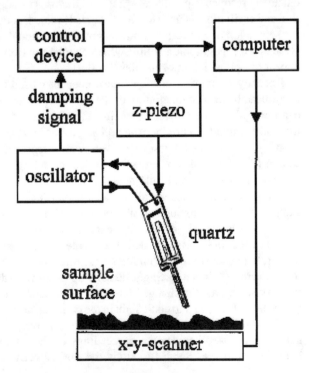

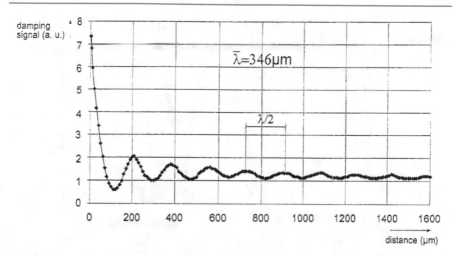

Fig. 5.6 Distance dependence of the damping signal of the SNAM, with a resonance frequency of 1 MHz. The periodic signal is related to radiation losses. The near-field regime is related to hydrodynamic damping in air. Reprinted from [456], https://doi.org/10.1016/0040-6090(95)05853-2, with permission from Elsevier

damping force. The lateral resolution of the order of several microns was attributed to this hydrodynamic damping force. Lateral resolution below 0.1 μm, which is smaller than the mean free path of the gas molecules in air, is attributed to other interactions. The reader is also referred to Chap. 3 about conventional SFM, where damping at these distances is discussed in more detail.

Related experiments are the investigation of ferroelectric films by force microscopy. Ac voltages are applied between the probing tip and the surface and these lead to the creation of acoustic waves. The acoustic waves are detected with a piezoelectric transducer, mounted at the rear side of the sample. This geometry is reminiscent of the original SAM geometry, where piezoelectric films were used as acoustic detectors. Therefore, the method is also called SNAM [395]. If the thickness variations of the piezoelectric film are directly detected by SFM, the method is called piezoresponse force microscopy (PFM) [164,636].

5.3   Scanning Ion Conductance Microscopy (SICM)

The scanning ion conductance microscope (SICM) uses a micropipette which approaches the surface of a sample immersed in an electrolyte [255]. The ion conductance between the electrode in the solution and the electrode in the pipette is measured. As the pipette approaches the sample surface, the cross-section of the ion path from one electrode to the other is reduced and the conductance decreases. The resolution is essentially given by the size of the aperture. Compared to glass pipette tips, microfabricated probes [532] have the advantage of smaller aperture size and higher stability. Structures of the order of 100 nm can be resolved. The distance is controlled by the ionic current.

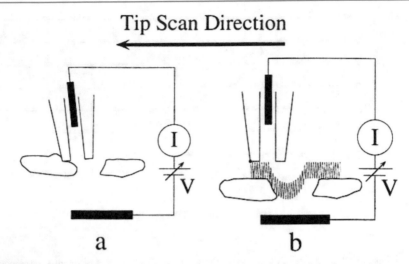

Fig. 5.7 Scanning ion conductance microscopy combined with SFM. **a** The microscope is operated in contact mode. **b** The microscope is operated in tapping mode SFM, where the interaction force is reduced and the disturbance of the sample is smaller. Reprinted from [535], https://doi.org/10.1016/S0006-3495(96)79416-X, with permission from Elsevier

Alternatively, a combination of SICM and SFM can be used [535]. In this case, the pipettes are made of pulled quartz glass tubes, where the front-end portion is bent. The deflection of the pipette can be measured with a conventional laser beam deflection method. As shown in Fig. 5.7, the microscope can be operated either in contact mode or in tapping mode. In both cases, the ionic current is measured during imaging. However, the resolution on the pores is found to be improved in tapping mode [535]. The main application of SICM is to measure the distribution of ionic currents through the pores in a porous surface. Examples are nucleopore membranes or polymer membranes. Promising applications of SICM on living cells were presented by Korchev et al. [364, 365]. So far, the lateral resolution is still rather poor, in the micron range. Recently, several groups [474, 723] have suggested using SICM as a lithographic tool, where metallic wires are deposited on a surface.

5.4   Photoemission Microscopy with Scanning Aperture (PEMSA)

In photoemission microscopy with a scanning aperture (PEMSA) the photoemission of electrons escaping from a surface and passing through a small aperture is measured (cf. Fig. 5.8). Using polarized light, the circular polarization dependence of this photoemission current can be measured. McClelland and Rettner introduced this method and achieved a resolution of 30 nm [442]. Magnetic domains of a Co/Pt film were imaged by polarization-dependent PEMSA. The magnetic contrast of PEMSA is related to scanning Kerr microscopy. The photons which contribute to the signal must be absorbed very close to the surface. The created photoelectrons have low

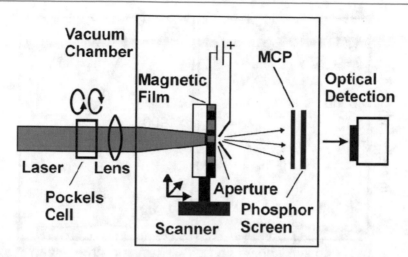

Fig. 5.8 Diagram of photoemission microscopy with scanning aperture (PEMSA) apparatus. The Pockels cell modulates the laser polarization and the sample is scanned relative to the aperture in order to image variation of the photoemission current. Reprinted from [442], https://doi.org/10.1063/1.1290721, with permission from AIP Publishing

energies (a few eV), which limits the mean free path to a few nanometers. The apertures used by McClelland et al. were 30–45 nm in diameter. The sample was first biased at +1 V to suppress photoemission and was brought towards the tip until the tunneling current rose. Then the sample was retracted 30 nm and the bias switched to −10 V to detect photoemission. With 2 mW of light, a 100 nA photocurrent was generated and 1 pA was passed through the 30 nm aperture. The photoemission was recorded as the light polarization was alternated. Then the sample was moved to the next position and the cycle repeated. Since the resolution is approximately given by the mean free path of the photoelectrons and the aperture size, McClelland et al. suggest that the resolution would surpass SNOM resolution. The potential of PEMSA has not yet been fully explored. Energy analysis may provide useful information about the local band structure. Experiments analogous to ultraviolet photoemission spectroscopy (UPS) or X-ray photoemission spectroscopy (XPS) may be performed, where the local chemistry can be determined with superb lateral resolution.

5.5 STM with Inverse Photoemission (STMiP)

The excitation of photon emission by tunneling electrons in an STM was observed by Gimzewski et al. [214]. In the case of metallic films, the emission corresponds to surface plasmons in the film that are inelastically excited by the tunneling electrons and then decay by the emission of photons [72,515]. Enhanced photoemission is found on small metallic islands. Contrasts of the photon maps on a scale of some tens of nanometers were attributed to local variations in the field strength of the tip-induced plasmon modes, which are determined by the surface geometry of the

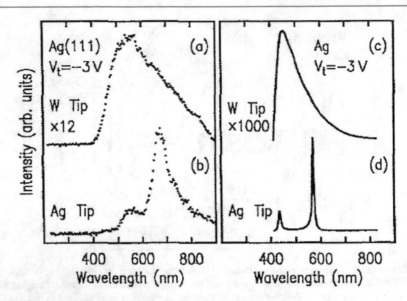

Fig. 5.9 Effect of tip material from STM junction with Ag(111) surfaces on fluorescence spectra. **a** and **b** are experimental results at a tip voltage of −3 V for W and Ag tips. **c** and **d** show the corresponding theoretical findings. Reprinted from [73], https://link.aps.org/doi/10.1103/PhysRevLett. 71.3493, with permission from AIP Publishing

junction and the dielectric properties. As shown in Fig. 5.9, the photoemission spectra have a main peak corresponding to an interface plasmon which fits into the cavity. The additional peaks correspond to higher photon energies from higher-order modes having nodes inside the cavity. The Ag tip shows a redshift with respect to the curve with a W tip. This is explained by the fact that the W tip has no well-defined plasmon mode and its role in the formation of the interface plasmon is passive. On a sub-nanometer scale, photon intensity variations are related to variations in the probability for inelastic tunneling current due to variations in the final density of states [75]. The method was also applied to study the luminescence spectra of semiconductors [29].

To use electrons to induce light emission has made significant progress and became a powerful technique, because it combines the high resolution of STM with the valuable information of optical spectroscopy. To apply this technique to molecules has been hindered for a long time by the presence of the metallic substrate, which quenches the light emission of the molecules. One possibility to overcome this problem is to partially detach polymeric molecules from the surfaces. By detaching a polythiophene wire from a Au(111) surface, it became possible to decouple it from the metal and to observe its electroluminescence [547]. An alternative approach is to deposit the molecules on thin insulating films, such as Al_2O_3 or NaCl. In this case, the molecules were found to be optically active and spectra from single molecules were acquired [107,538,726]. Doppagne et al. [151] were able to identify vibronic levels with submolecular resolution by comparison with Raman spectra from bulk samples (cf. Fig. 5.10).

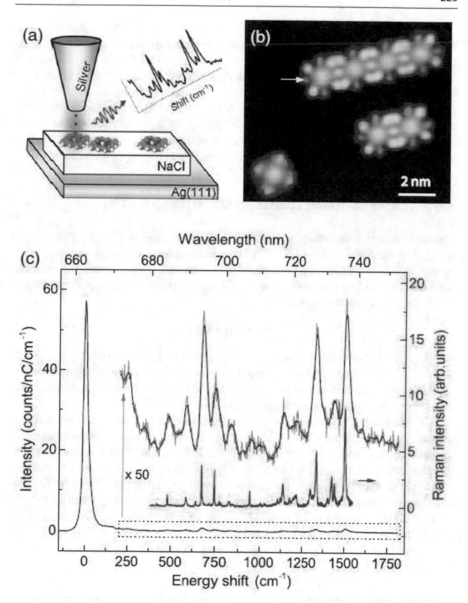

Fig. 5.10 The STM injects electrons into single molecules of zinc-phthalocyanine (ZnPc) to induce light emission. The molecules were deposited on thin NaCl-films to decouple them from the metallic silver substrate. The detected light shows an intense emission line at 1.9 eV, but also weaker lines, which can be attributed to vibrational motions of the molecules. The vibronic lines can be identified by comparison with Raman spectroscopy from bulk molecular crystals. **a** Schematics of the STM induced light emission. **b** STM image of a few ZnPc molecules. **c** Light intensity emitted from a single ZnPc compared with a Raman spectrum from a bulk molecular crystal. Reprinted from [151], https://link.aps.org/doi/10.1103/PhysRevLett.118.127401, with permission from AIP Publishing

In this context, the plasmon-enhanced Raman scattering has to be mentioned. It is well known that the presence of metallic tips enhances the intensity of Raman scattered light, also called tip enhanced Raman scattering (TERS). The combination of low temperature STM with a Raman scattering can even yield resonance conditions, where the intensity is large enough to probe TERS spectra of single molecules [724]. It is essential to tune the plasmon frequency by changing the separation between tip and sample in order to match excitation frequency with the sum of the Raman frequency plus plasmon frequency in order to meet the resonance conditions. Then, it is possible to compare TERS spectra on different parts of porphyrin molecules.

5.6   Laser Scanning Tunneling Microscopy (LSTM)

The laser scanning tunneling microscope (LSTM) combines STM with laser light [676]. The main difficulty of LSTM is to exclude thermal expansion due to heating by the laser light and to measure a signal in a large optical background. A possible solution is to mix two laser sources with different frequencies ω_1, ω_2 and to measure the difference frequency signal

$$\Delta\omega = \omega_1 - \omega_2 \,,$$

which is generated by the nonlinear properties of the STM junction. The junction is a rectifier, which means that an additional current is generated by the light exposure. Thus, the rectified current and the radiated difference signal can be measured. The latter signal is also used in apertureless SNOM.

For infrared laser light, the coupling is dominated by the antenna properties of the tunneling junction. In the visible regime, tip-induced plasmons and propagating surface plasmons can become important. Effects such as thermal expansion, thermoelectric effects and photoconductivity have to be taken into account. Atomic resolution images of graphite were achieved with the difference signal [676]. The most promising application is the combination of laser spectroscopy and STM.

Related techniques are the ac-STM, which was introduced by Kochanski et al. [357]. In this approach microwave radiation was irradiated on the tip and higher harmonics of the tunneling current were created. Kochanski was able to demonstrate that even insulators could be imaged by the ac-STM.

5.7   Electrochemical Scanning Tunneling Microscopy (ECSTM)

STM in electrolytes can be used to study in situ corrosion or electrochemical deposition processes. Both inorganic and organic systems, including biological systems, may be investigated. The main experimental problem is to reduce Faraday and capacitive currents between tip and sample to a low level. Therefore, tunneling tips are covered by some insulating material (e.g., epoxy resin, nail varnish, wax, silicon polymers). Only the front end of the tip remains uncovered. Using the potentiostatic STM concept [527,606], the electrochemical tip currents can be minimized by

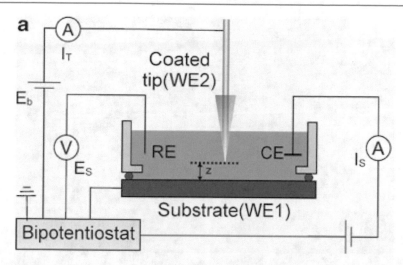

Fig. 5.11 Principle features of an electrochemical STM (ECSTM). The sample voltage E_S and tip voltage E_T are referenced relative to the reference electrode (RE). The tip voltage is selected in a regime where electrochemical currents are negligible. The potential on the sample is freely adjustable in order to deposit material on or remove material from the sample during STM operation. Reprinted from [527], https://doi.org/10.1007/978-90-481-9751-4_46, with permission from Springer Nature

adjusting the tip potential E_T relative to a currentless reference electrode (RE). The material of this reference electrode depends on the electrolyte composition. If the electrolyte contains a metal cation Me^{z+} in equilibrium with the corresponding metal surface, the electrode can be a simple metal wire Me in the STM cell (e.g., Ag, Cu, Pb-coated Pt reference electrodes). A versatile reference electrode is the Ag/AgX electrode, where X stands for Cl or Br. This can be used in solutions containing Cl^- or Br^-, respectively.

As shown in Fig. 5.11, the sample voltage (E_S) can be adjusted relative to the reference electrode. Thus, material can be deposited on or removed from the sample during STM operation. The tip potential is kept in a regime where the electrochemical current is practically negligible. Therefore, the tunneling current dominates and can be used as feedback parameter to probe the sample topography. The bipotentiostatic control is completed by a current-carrying counterelectrode (CE), where the sum of substrate (I_s) and tip (I_t) currents is measured. A state-of-the-art ECSTM requires high standards, such as the use of a gas purification system, in order to achieve the necessary cleanliness of the surfaces [700]. A series of ECSTM images in a sulfuric acid electrolyte is given in Fig. 5.12, where the formation of a sulfate adlayer is observed. Since the image contrast is given by the tunneling current, the ECSTM can achieve atomic resolution.

A further probe microscope which should be mentioned in this context is the scanning electrochemical microscope (SECM). In SECM, the feedback input is the Faraday current flowing into a micron-sized electrode probe at separations of a few microns. Typical resolutions were obtained in the micrometer range [48].

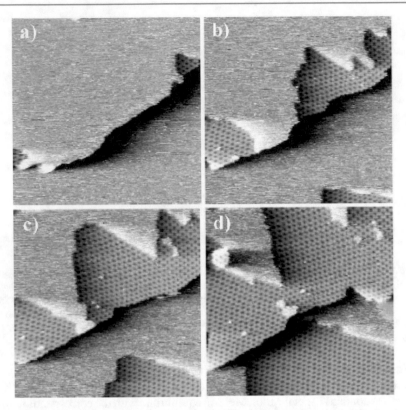

Fig. 5.12 Series of STM images of Cu(111) recorded at time intervals of 2 min, showing the formation of a Moiré structure ($101 \times 101\,nm^2$, $I_t = 1\,nA$, $E_T = 169\,mV$ and $E_S = -725\,mV$ versus Hg/Hg_2SO_4 electrode). The Moiré structure is related to the incommensurate structure of the adsorbate film (sulfate and bisulfate adlayer) on Cu(111) in a sulfuric acid electrolyte. Reprinted from [700], https://doi.org/10.1063/1.1149971, with permission from AIP Publishing

5.8   Scanning Thermal Microscopy (SThM)

Scanning thermal microscopy (SThM) is based upon a miniaturized thermal sensor at the apex of the probing tip. Therefore, local temperature variations are measured. These temperature inhomogeneities may arise from heat sources, such as electrical currents in wires, or may reflect local variations in the thermal conductivity of the sample. In the latter case, either the tip or sample has to be heated in order to create a temperature gradient between tip and sample. The first scanning thermal microscope was introduced by Williams and Wickramasinghe [696]. It consisted of a thin film thermocouple at the end of an STM tip. The thermocouple was heated and, as it was brought closer to the surface, the tip was cooled due to heat transfer between tip and sample. Using the thermovoltage as a feedback input, even insulating surfaces could be imaged. In this mode topography and temperature variations can hardly be distinguished. Instead of using a thermocouple, other temperature-sensitive properties such as the electrochemical potential [686] or the contact potential were measured

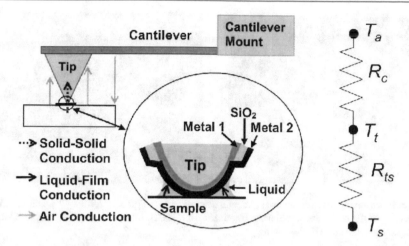

Fig. 5.13 Schematic diagram of a cantilever probe used for scanning thermal microscopy. The heat transfer mechanisms between tip, cantilever and sample are indicated. In ambient pressure, the heat conduction through the liquid meniscus dominates. Reprinted from [602], https://doi.org/10.1063/1.1334658, with permission from AIP Publishing

[488]. However, these properties also depend on the electronic structure of the sample and make an unambiguous interpretation difficult. Therefore, we will restrict the following discussion to microscopes which include a small temperature sensor in the probing tip and have an independent force sensor to probe the topography.

The most systematic work was performed with SThM, which contains a thermocouple junction at the tip end of a cantilever. After the first attempts with etched wire thermocouples [418], several papers were presented where batch fabrication processes were realized [389,409,458]. Figure 5.13 shows the key elements of an SThM. A thermocouple junction is integrated on the apex of a probing tip of an SFM sensor (cantilever with integrated tip). In SThM, several tip–sample heat transfer mechanisms exist, such as solid–solid conduction through the contact, liquid conduction through the liquid film bridging tip and sample and air conduction. In ambient pressure, where a meniscus is formed between tip and sample, liquid conduction is dominant. In dry conditions or on hydrophobic surfaces, air conduction becomes important. In vacuum, solid–solid conduction and radiation losses have to be considered.

Generally, the tip temperature is given by

$$T_t = T_s + \frac{T_a - T_s}{1 + \Phi} , \tag{5.1}$$

where T_s is the local surface temperature, T_a is the ambient temperature, $\Phi = R_c/R_{ts}$, R_{ts} is the tip–sample thermal resistance and R_c is the total thermal resistance of the cantilever and tip relative to ambient. Taking into account the fact that $R_{ts} \approx 10^5$ K/W is rather large, it becomes clear that R_c should also be large in order to achieve a high sensitivity. Therefore, it appears favourable to use cantilever materials with low thermal conductivity, such as silicon nitride or silicon oxide. The thermocouple

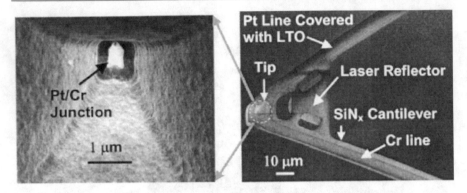

Fig. 5.14 Scanning electron micrographs of a batch-fabricated probe. *Left*: Pt–Cr junction. *Right*: Overview of the probe and cantilever. Reprinted from [602], https://doi.org/10.1063/1.1334658, with permission from AIP Publishing

itself should consist of materials yielding a high thermopower and a reasonably low thermal conductivity, e.g., Pt/Cr [602] and Pd/Au thermocouples [458] (cf. Fig. 5.14).

According to (5.1), variations of the tip temperature ΔT_t are determined by

$$\Delta T_t = \left(\frac{1}{1+\Phi}\right)\Delta T_s - \frac{T_s - T_a}{(1+\Phi)^2}\Delta\Phi\ . \tag{5.2}$$

Assuming that the ambient temperature T_a and the cantilever thermal resistance R_c are constant, only changes in the sample temperature T_s and the tip–sample thermal resistance R_{ts} can change the tip temperature T_t. T_s can only vary due to local heat sources, such as an electrical current or light exposure. Variations in R_{ts} can be due to variations in the thermal conductivity of the sample. However, one also has to take into account tip–sample distance variations and geometrical effects (see also Chap. 6). Remarkably, the method has the potential to probe subsurface features with an information depth in the micron range [250].

So far, most applications have been dedicated to the failure analysis of integrated devices with spatial resolution in the sub-100 nm range. Fiege et al. have investigated hot spots of electronic devices with a sensitivity of 5 mK [179]. Other examples are 30–500 nm wide metal lines, single transistors and vertical-cavity surface-emitting quantum well lasers, where the temperature distribution can be mapped [419].

Apart from the above-described integrated thermocouples, other successful designs have been realized, including the use of resistance thermometers [250] and and bimetallic cantilevers [478]. Recently, the piezoresistance of microfabricated cantilevers was used to probe temperature variations with high accuracy [665]. It was found that the cooling of the probing tip is extremely sensitive and can be used to regulate the probing tip above the surface. Surprisingly, a rather fast response time in the μs range was achieved with this setup. In this case, the probing tip is heated and it is found that conduction through air is the dominant heat transfer channel.

Interesting extensions of SThM have been suggested by Oesterschulze et al. [496], where the thermal probe is heated periodically. This results in an amplitude and phase

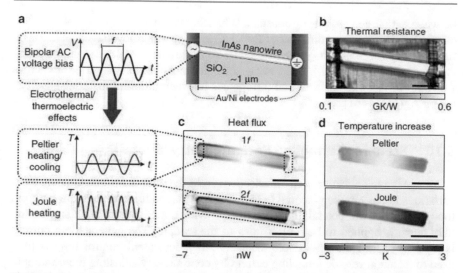

Fig. 5.15 An elegant way to separate the contributions from variations of the tip-sample resistance from real sample-related heat source variations is to use an AC-modulation of the sample heat source, which is a InAs nanowire (**a**). Then the DC-component (**b**) and the f-component and $2f$-component (**c**) can be used to separate the Joule dissipation in the sample and the Peltier-component (**d**). Figure from [447], https://doi.org/10.1038/ncomms10874, distributed under the terms of the Creative Commons Attribution License (http://creativecommons.org/licenses/by/4.0)

signal. In addition to the thermal conductivity, known from static scanning thermal microscopy, this dynamic method may yield information about thermal diffusivity.

A very elegant way to separate different contributions to the heat flow is to modulate heat source with frequency f. The DC-component and the two AC-components, f and $2f$, can be measured separately. Then, the different contributions can be separated: A map of the Joule heating of the sample and the Peltier component can be plotted (cf. Fig. 5.15), where variations of the the tip-sample thermal conductance were eliminated [447, 448].

5.9   Scanning Noise Microscopy (SNM)

A mean-square noise current $\langle I_t(t)^2 - \langle I_t(t) \rangle^2 \rangle$ is measured in an STM. Möller et al. [461, 462] found that the current noise at zero dc tunneling current ($\langle I_t \rangle = 0$) is white noise. Under these conditions, the current noise is Johnson noise, given by the formula

$$ S_I = \frac{4k_B T}{R_T} , \tag{5.3} $$

where S_I is the power spectral density of the current noise, k_B is the Boltzmann constant, T the temperature and R_T the tunneling resistance. Möller et al. measured values of the order of $40 \, \text{fA}/\sqrt{\text{Hz}}$ for $R_T = 10^7 \, \Omega$ at $T = 290 \, \text{K}$. The current noise is proportional to the tunneling resistance and depends exponentially on the distance. Therefore the current noise can be used as the feedback signal to keep the separation between sample and probing tip constant. When the tip is heated by a laser beam ($P \approx$ 40 mW), the temperature difference ($T \approx 5$ K) leads to a thermovoltage which can be measured by a second voltage feedback. The input parameter of this second feedback is the normal dc current $\langle I_t \rangle$. Local maps of the thermovoltage show variations across step sites of metallic surfaces and differences between different materials, such as Cu on Ag(111) [461].

5.10   Scanning Tunneling Potentiometry (SPotM)

A voltage is applied between two electrodes across the sample surface, causing local variations in the surface potential [475]. The STM is modified to measure both the topography and the potential. An ac voltage with a modulation frequency in the kHz range is applied to both electrodes of the sample. The generated ac current is used to control the tip–sample distance. An independent control loop is then used to maintain zero dc tunneling current by continuously adjusting the dc sample voltage in order to monitor local variations in the potential across the sample. This technique is useful for measuring local potential variations on Schottky barriers, p–n junctions and heterojunctions. In close analogy, the local potential can also be measured by Kelvin force microscopy [489] (see Chap. 3), which is called scanning force potentiometry, with the advantage that the method can be applied to insulating as well as conductive surfaces. A combination of contact SFM with conductive probing tips was demonstrated by Hersam et al. [274]. In this case the SFM tip is grounded through a limiting resistor. The local potential is directly related to the current through the SFM tip. Hersam et al. have shown that electrical failure of nanowires can be observed by the discontinuity in the potential gradient [274]. The nanowire can then be repaired by the SFM tip, which was used to move gold into the gap.

5.11   Scanning Capacitance Microscopy (SCM)

Scanning capacitance microscopy (SCM) is an important method for characterizing dopant profiles of semiconductor devices (see Fig. 5.16) [695]. Comparison with SIMS data have shown qualitative agreement, but quantitative dopant concentrations can only be achieved in combination with advanced modeling [147].

The original SCM used an insulating stylus, which was moved across a surface [440]. The stylus was used to maintain a gap between the electrode and the sample. The corresponding capacitance between electrode and sample was measured on a local scale without feedback. Later, Williams et al. combined the capacitance measurement with an STM [697]. The sample is scanned with a feedback at constant electrode–sample capacitance. Barrett and Quate [51] introduced the combination of a contact SFM with a capacitance measurement, which turned out to be very successful. Contact SFM is a common tool for imaging topography on the nanometer scale and the capacitance measurement gives additional information about localized

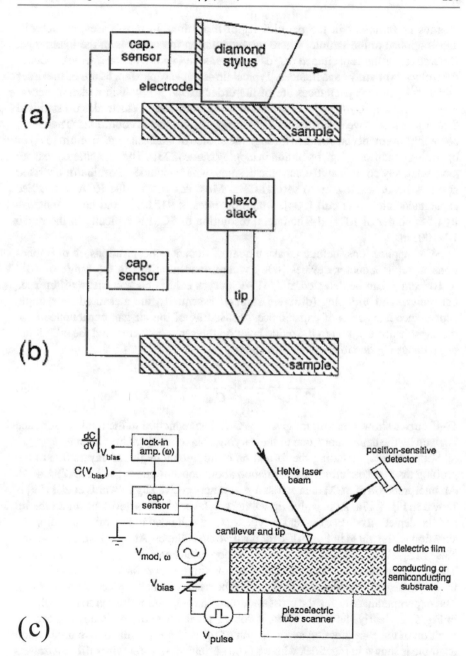

Fig. 5.16 Schematic diagrams of scanning capacitance microscopes. **a** An electrode is mounted on an insulating diamond stylus, which is scanned without feedback in contact. **b** Capacitance sensor combined with an STM tip. The probing tip can be scanned at constant electrode–sample capacitance. **c** Combination of contact SFM and a capacitance sensor [51]. The surface is imaged at constant repulsive normal force. A modulation of the applied voltage is used to measure derivatives of the capacitance, which reduces the influence of stray capacitances. **a** Reprinted from [440], https://doi.org/10.1063/1.334506, with permission from AIP Publishing. **b** Reprinted from [698], https://doi.org/10.1063/1.102312, with permission from AIP Publishing. **c** Reprinted from [51], https://doi.org/10.1063/1.349388, with permission from AIP Publishing

charges or variations in the permittivity of thin films. In most cases, an ac voltage is applied to the sample, where the modulation frequency is in the kHz range. Derivatives of the capacitance dC/dV are measured with a lock-in, which reduces the influence of stray capacitances. Typical tip–sample capacitances are of the order of 10^{-19} F. Stray capacitances are of the order of 10^{-13} F, which makes it necessary to measure dC/dV with modulation techniques. Metal electrodes on standard SFM tips undergo wear processes due to the repulsive contact conditions. Therefore, some improvement can be made through the choice of material (e.g., iridium [51]) or operating conditions (e.g., operation in tapping mode [223]). The capacitance sensors have relatively complicated electronic designs, which include an oscillator operated at a few hundred MHz up to several GHz. Most designs use the RCA VideoDisc capacitance pickup circuit [504], which operates at 915 MHz and has a sensitivity of the order of 10^{-20} F. The lateral resolution of SCM is typically in the range 10–100 nm.

SCM applications include measurement of dopant concentrations of p–n junctions in semiconductor devices [698], where concentrations in the range of 10^{15} to 10^{20} cm^{-3} can be detected. dC/dV–V curves exhibit characteristic differences between n- and p-doping (different slopes). Essentially, the measured tip–sample capacitance is a series of capacitances, consisting of the air gap capacitance C_{air}, the capacitance C_{ox} due to the oxide layer on the semiconductor, and the additional capacitance C_D due to the depletion layer:

$$\frac{1}{C_{total}} = \frac{1}{C_{air}} + \frac{1}{C_{ox}} + \frac{1}{C_D} . \tag{5.4}$$

C–V curves show a maximum given by C_{ox}. The reduction in the total capacitance is given by the capacitance due to the varying size of the depletion layer.

Some models, including the formation of additional depletion layers due to the probing tip, yield quantitative information about concentrations [117, 159, 291, 494]. An interpretation of SCM data across a p–n junction given by Edwards et al. [159] is shown in Fig. 5.17a. At negative tip voltages, a layer of holes develops under the tip and the depletion edges of n- and p-type carriers will bend to the left. An analogous situation is illustrated in Fig. 5.17c for positive tip voltages. At the so-called flat-band bias voltage V_{FB} the width of the observed depletion layer is approximately correct. The corresponding C–V curves exhibit the maximum capacitance C_{ox} (for simplicity C_{air} is neglected), which is reduced for the cases of small C_D according to (5.4). The experimental data for the cross-section of a MOS field-effect transistor shown in Fig. 5.18 clearly show the different doping levels. A variation is observed in the position of the p–n junction as a function of tip voltage. An illustrative application example is shown in Fig. 5.19, where a (100)-oriented Si wafer with different dopant concentrations was characterized by SCM. 3D information about the impurity concentration was determined by Kang et al. [325] by position-dependent dC/dV–V measurements, where the image contrast was compared with simulations.

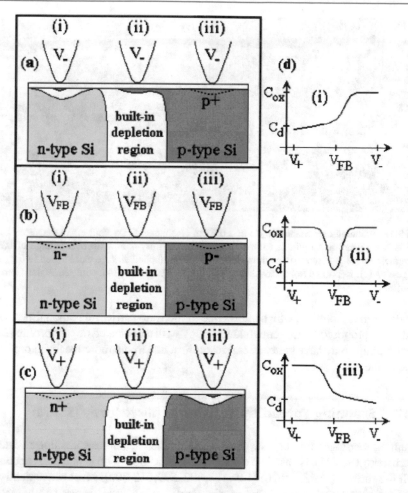

Fig. 5.17 Schematics of the carrier movement in silicon near a *p–n* junction. Due to the applied tip voltage, an additional depletion layer may be formed, which extends or reduces the apparent width of the built-in depletion layer (**a**)–(**c**). **d** Qualitative summary of the tip–sample capacitance versus voltage curves for different positions (i)–(iii). Reprinted from [159], https://doi.org/10.1063/1. 120849, with permission from AIP Publishing

Another application is charge storage on nitride-oxide–silicon (NOS) systems, where information densities of 27 Gbits/in^2 were demonstrated by Barrett et al. [51]. Some instrumental improvements and other proposals concerning SCM are mentioned briefly here. Zavyalov et al. proposed a low temperature treatment of cross-sections of *p–n* junctions, which produced insulating layers with consistent quality and reproducibility [722]. Improved capacitance sensors by Tran et al. [651] lowered the capacitance sensitivity to the zeptofarad (10^{-21} F range). Yamamoto et al. [710] proposed a design with an insulating SFM cantilever and a metallic tip, which may significantly reduce the stray capacitance. Experiments by Hamers et al. [249] show that SCM combined with irradiation of short laser pulses can be used to achieve

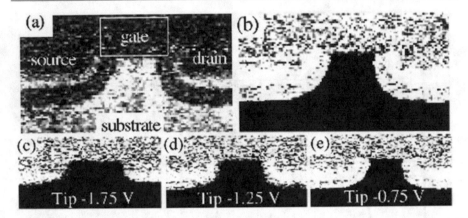

Fig. 5.18 Scanning capacitance image of a 0.6 μm N-channel MOS field-effect transistor. Image size is $2 \times 1 \, \mu m^2$. **a** Amplitude (dC/dV) and **b** phase images at a tip bias voltage of 0.25 V. Movement of the *p–n* junction contrast as a function of applied voltage can be seen in the phase images (**c**)–(**e**). Reprinted from [159], https://doi.org/10.1063/1.120849, with permission from AIP Publishing

ultrafast time resolution, which allows one to measure carrier relaxation times. Goto et al. have proposed to use semiconductor tips with a dedicated dopant profile [223]. Even dull tips may have better lateral resolution due to the more localized depletion zone of the tip.

5.12   Scanning Spreading Resistance Microscopy (SSRM)

Scanning resistance microscopy (SRM), also called scanning spreading resistance microscopy (SSRM), is based on a conducting SFM tip which is in repulsive contact with the surface (see Fig. 5.20) [139,173,492,599,612,702]. Most conveniently, an SFM probing tip with boron-doped diamond coating [484] is used to adjust the forces ($\approx$50–200 μN). In addition, a metallic layer of tungsten can improve contact formation [173]. The local spreading resistance is related to the local resistivity. Maps of the spreading resistance show a lateral resolution of the order of the contact radius (typically 10–15 nm). The technique is applied to semiconductor surfaces with emphasis on silicon. According to Eyben et al. [173], the local pressure can be large enough ($\approx$11 GPa) to create a β-tin phase transition of silicon [Si(II) phase], which is a metallic phase of silicon. The interface between the silicon (II) phase and the normal silicon (I) phase gives a contact, which allows one to draw reliable conclusions about the local resistivity to a depth of some 10 nm. The spreading resistance R in the diffusive limit is approximately given by the Maxwell formula

$$R = \rho/4a \, , \tag{5.5}$$

where ρ is the resistivity of the sample (the resistivity of the tip is neglected) and a the contact radius. With a typical value of 1 Ω cm and $a = 50$ nm one finds a resis-

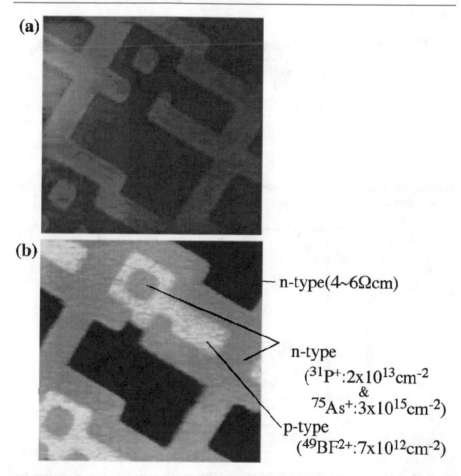

Fig. 5.19 a SFM image of an Si surface with different dopants. **b** dC/dV image of the same area. *Dark contrast* corresponds to the *n*-type silicon wafer (4–6 Ω cm). *Bright contrast* corresponds to the *p*-type doped areas (BF$_2$ 7 × 10^{12} cm^{-2}). The scan area is 20 × 20 μm^2. Reprinted from [710], https://avs.scitation.org/doi/abs/10.1116/1.589397, with permission from American Vacuum Society

tance of 50 kΩ. In practice, calibration curves are performed with samples of well-known resistivity, where an empirical, nonlinear relation $R = f(\rho)$ is assumed (see Fig. 5.21). Experimentally, a dynamic range of dopant concentrations of 10^{14} to 10^{19} cm^{-3} is accessible. The calibrated tips can then be used to measure complex semiconductor structure, such as cross-sections of CMOS transistors or bipolar transistors, where the different dopant concentrations and corresponding resistivities can be determined. Comparative SCM and SSRM measurements of test structures demonstrated similar performance, but also showed the need for calibration procedures for both techniques [122].

Most SRM experiments involve rather large forces (>50 μN), which are necessary to pierce through contamination layers and to establish reliable contacts under

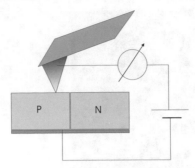

Fig. 5.20 Schematic view of scanning spreading resistance microscopy. A conductive SFM tip is scanned over a heterogeneous surface at relatively large repulsive forces

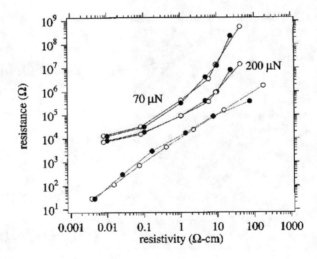

Fig. 5.21 Calibration curve of scanning spreading resistance microscopy. A tungsten-covered, ion-implanted diamond tip was loaded on *n*-type (*open circles*) and *p*-type (*filled circles*) (100) silicon at loads of 70 and 200 μN. Calibration curves for a conventional W/Os spreading resistance probe (at a load of 50 mN) are given for comparison (*dotted lines*). Reprinted from [139], https://avs. scitation.org/doi/abs/10.1116/1.588478, with permission from American Vacuum Society

ambient conditions. Some work has also been carried out at lower forces and correspondingly smaller contact diameters. Due to the reduced forces, not only diamond-coated tips can be used, but also metallic coatings, such as platinum have been found to be reliable [78, 645]. For the nanometer-sized contacts, which are comparable to the mean-free path $L \approx 100\,\text{Å}$ in metals, the Maxwell formula (5.5) is not applicable. Ballistic transport dominates the transport through nm-sized contacts and the resistance is given by the Sharvin expression

$$R_{\text{Sharvin}} = \frac{h}{2e^2} \frac{\lambda_{\text{F}}^2}{\pi^2 a^2} , \qquad (5.6)$$

where h is Planck's constant, e the electron charge and λ_F the Fermi wavelength. With typical parameters $a = 3\,\text{nm}$, $h = 6.626 \times 10^{-34}\,\text{J s}$, $e = 1.602 \times 10^{-19}\,\text{C}$ and $\lambda_F = 0.5\,\text{nm}$, a Sharvin resistance of about $37\,\Omega$ is found. Similar values were found by Lantz et al. for UHV experiments using Pt/Ir-coated tips on graphite [382,383]. Lantz et al. were able to demonstrate that the conductance is proportional to the tip–sample contact area. Therefore, the Sharvin equation seems to be applicable for these small contacts. However, Lantz et al. also observed that the metallic coatings are easily worn, which makes a reliable determination of the contact resistance difficult. Alternatively, clean silicon tips (with the oxide removed) were investigated and the conductance was found to be proportional to the contact area. However, the quantitative analysis is more difficult than with metals. Essentially, a Schottky contact is formed with the metal surface. Transport mechanisms such as thermoemission and tunneling have to be taken into account. Lantz et al. also performed experiments in UHV at very low loads, of the order of nanonewtons [383]. In contrast to ambient conditions, contamination layers can be removed by appropriate sample and tip preparation. Therefore, there is no need to penetrate through the contamination layer by the application of high loads. Low load experiments and correspondingly small contact diameters are interesting for high resolution applications.

Bietsch et al. investigated gold nanowires with conducting SFM in air. The structures were imaged in tapping mode [78]. At selected positions, the tip was lowered and loads of the order of 700 nN were applied in order to penetrate and form reliable contacts with gold. Under these conditions, both the Maxwell and the Sharvin resistance contribute to the contact resistance. Values of $\approx 11\,\Omega$ indicate that transmission is not perfect, but some additional scattering occurs at the Pt/Au interface. Having established a reliable contact formation, the resistance of gold nanowires was investigated. Two-point probing and potential probing were established (see Fig. 5.22). With nanowire cross-sections of $1.2 \times 10^{-15}\,\text{m}^2$, resistivities of $\rho_n = 12\,\mu\Omega\,\text{cm}$ were determined, which is more than 5 times the bulk resistivity of gold ($\rho_n = 2.2\,\mu\Omega\,\text{cm}$). The granular structure and small size are the reasons for these high resistivities.

Another application of conducting SFM is the investigation of dielectric breakdown of gate oxides [476,495]. Murrell et al. determined the threshold voltage required to generate a small tunneling current in the oxide. Relatively large variations in these threshold voltages indicate that the breakdown strength in conventional metal-oxide–silicon capacitors may not be limited by the intrinsic dielectric strength, but by imperfections in the Si/SiO$_2$ structure.

In conclusion, conducting SFM has a lot of potential to characterize the conductance of nanometer-sized structures. Extensions such as the four-point probe by Petersen et al. [516] are most promising. SRM is already a useful tool for determining the local doping concentrations of semiconductor devices.

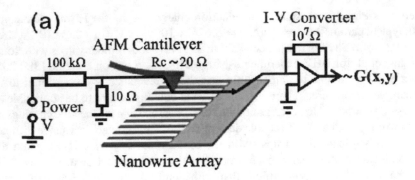

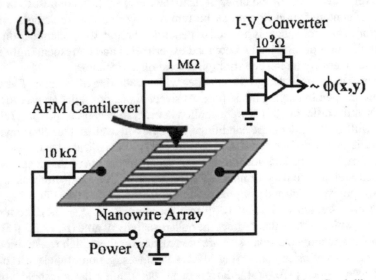

Fig. 5.22 Sensing an array of nanowires. **a** Two-point conductivity probing. Comb-like wire array with a common electrode and the SFM tip as the second electrode. The current is proportional to the conductivity between tip and sample electrode. **b** Potential probing. Array of wires in a closed electrical circuit. The conductive SFM measures the local potential. Reprinted from [78], https://avs.scitation.org/doi/abs/10.1116/1.591353, with permission from American Vacuum Society

5.13   Scanning Tunneling Atom Probe (STAP)

The combination of STM with mass spectroscopy appears promising, since high lateral resolution of STM topography can be combined with chemical identification [604, 687]. The principle of the method is shown in Fig. 5.23. A sharp STM tip is scanned over a surface. A short, positive voltage pulse is applied to the tip above the region of interest, causing some atom transfer from the sample to the tip. Then the STM tip is retracted from the surface and positioned in front of a time-of-flight mass spectrometer. In close analogy to atom probe technology, a large, negative dc

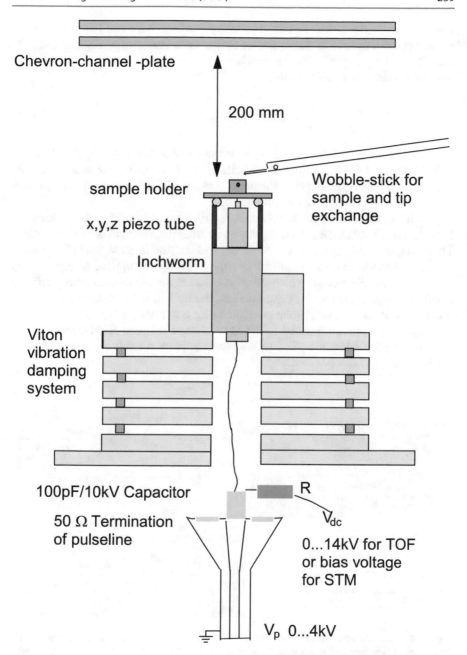

Fig. 5.23 Design of a scanning tunneling atom probe (STAP) instrument. Reprinted from [687], https://avs.scitation.org/doi/abs/10.1116/1.591353, with permission from Elsevier

voltage (e.g., $V_{dc} \approx 10\,kV$) and short voltage pulse (e.g., $4\,kV$ and $10\,ns$) are applied to the tip, leading to field desorption of atoms from the tip. The flight time of the ions from the tip to the detector is measured. The mass-to-charge ratio m/n can be determined according to the formula

$$\frac{m}{n} = \alpha(V_{dc} + \beta V_p)\frac{(t - t_0)^2}{d^2} ,\qquad (5.7)$$

where V_{dc} is the dc voltage on the tip, V_p is the pulse height, t the flight time, t_0 the delay time and d the drift distance. Weierstall et al. [687] were able to acquire mass spectra from STAP probing tips. Elements such as W, Si, Ge were detected, which were previously picked up on a sample surface.

The main problem with STAP concerns the diffusion of atoms on the probing tip. In particular, metal atoms diffuse rapidly away from the region of highest curvature. Therefore, the probing tip may have to be cooled. Tips in STM operation tend to become blunt, which would require very large fields for desorption. A solution may be to use local electrodes, which allow one to create large fields with relatively small voltages. In the previous experiments, the tip had to be transferred from the STM position to the atom probe position using a normal UHV manipulator. The combination of microfabricated switches to change between STM operation and atom probe position within a few milliseconds appears suitable for acquiring local maps of chemical composition.

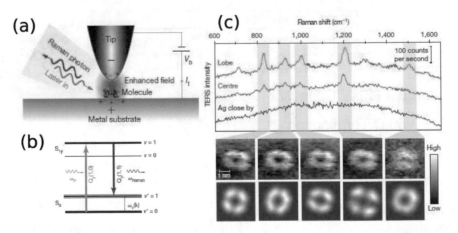

Fig. 5.24 a Tip enhanced Raman scattering (TERS) uses locally enhanced fields to study single molecules. **b** Schematic of the optical transitions involved in the Raman process. **c** The Raman spectrum shows characteristic differences for the different positions on the H_2TBPP molecule on Ag(111). The mapping of the peaks of the Raman spectrum, shows different appearances of the molecule, which are comparable to the theoretical calculations at the bottom. Reprinted from [724], https://doi.org/10.1038/nature12151, with permission from Springer Nature

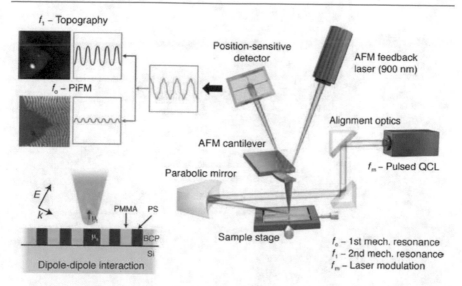

Fig. 5.25 Photon induced force microscopy (PIFM). Schematic diagram of PIFM, where light is focused on the apex of the AFM tip. Due to photoinduced polarizability an additional force component can be detected by suitable modulation techniques. Figure from [491], https://advances. sciencemag.org/content/2/3/e1501571, distributed under the terms of the Creative Commons Attribution License (http://creativecommons.org/licenses/by/4.0)

5.14 Tip Enhanced Raman Scattering (TERS)

To get the topographical as well as the chemical information about a molecule is one of the big challenges in SPM. Molecular vibrations are suitable to get a fingerprint to identify subgroups in a molecule. The tip apex in close vicinity to the metallic surface gives a strong enhancement of the electromagnetic field. By matching the resonance of the nanocavity plasmon to the molecular vibrations one can increase the sensitivity. This matching is achieved by changing the distance with the STM feedback. Under these conditions it becomes possible to image a single porphyrin molecule on silver with submolecular resolution [724] (Fig. 5.24).

5.15 Photo Induced Force Microscopy (PIFM)

The light illumination of the probing tip in close vicinitiy to the sample leads to photoinduced polarizability. If the light intensity is modulated at a suitable frequency, photoinduced forces can be detected by the force microscope. By changing the wavelength to meet the absorption of some specific material, one can spatially map the chemical components of heterogeneous samples, such as block copolymers films, on the nanometer scale. The photoinduced force is due to dipole-dipole interaction between the probing tip and the sample, when illumiinated with a monochromatic, coherent light source [491, 544] (Fig. 5.25).

Artifacts in SPM

6

Abstract

Recurring artifacts in all modes of operation are addressed. Examples are related to the convolution with the tip geometry and the influence of piezo scanner non-linearities.

6.1 Introduction to Artifacts in SPM

In this chapter artifacts of scanning probe microscopy are discussed. The tip artifact, where the sample topography is convoluted with the tip geometry is the most common artifact. A second class of artifacts are topography images, which are influenced by local variations of properties, such as conductance, elasticity, adhesion or friction. The third class of artifacts are local measurements, such as SNOM, STM-induced photoemission or lateral force measurements, which are influenced by local topography. The fourth class of artifacts are instrumental artifacts.

6.2 Tip Artifact: Convolution with Tip Shape

The most common artifact in scanning probe microscopy is the tip artifact. It has been observed with STM [213] or SFM [26,238,271,426]. Topographic features, which have a large aspect ratio compared to the probing tip are not correctly reproduced. The acquired image is a convolution between the probing tip shape and the sample feature. A blunt tip will broaden topographic features and reduce corrugation amplitudes. Multiple tips can create "shadows" or repeated features in the images ("ghost images").

A simple criterion is given by the curvature of the probing tip. All sample features that have a smaller radius of curvature than the radius of curvature of the probing tip are not completely imaged. Commercial manufacturers guarantee a radius of curvature of about 15 nm as an upper bound. In practice, the tip geometry can be

© Springer Nature Switzerland AG 2021
E. Meyer et al., *Scanning Probe Microscopy*, Graduate Texts in Physics,
https://doi.org/10.1007/978-3-030-37089-3_6

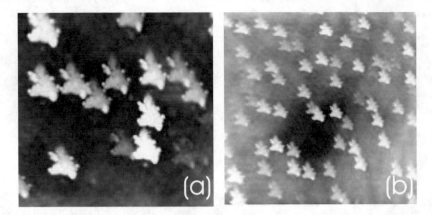

Fig. 6.1 Observation of tip artifacts with non-contact force microscopy on the $Al_2O_3(0001)$-surface, where the probing tip is imaged by needle-like structures of the surface. Nanometer-sized mini-tips are observed. The probing tip is a commercially available silicon tip, covered by its native oxide [10]. **a** $459 \times 459\,nm^2$-area, **b** $918 \times 918\,nm^2$-area

rather complicated, including nanometer-sized mini-tips. Therefore, it is advisable to characterize the probing tip with an scanning electron microscope or to image some standard samples, which have needle-like structures. The imaging of these needle-like structures gives a direct image of the probing tip. In Fig. 6.1 sharp needle-like structures are imaged with a non-contact force microscope on the $Al_2O_3(0001)$-surface.[1] The images show a collection of "islands" that are all identical, which is a clear indication that the probing tip is imaged instead of the needle-like surface feature. Remarkably, the probing tip has several nanometer-sized tips, which yield most probably high resolution on flat parts of the sample.

In the above case, the tip artifact is rather obvious and the tip geometry can be directly determined from the observed image. In this case, the radius of curvature of the tip is much larger than the radius of curvature of the needle-like surface feature ($R \gg R_s$). In addition the spacing between the front-most needles is relatively large. The situation is schematically drawn in Fig. 6.2. Westra et al. could show that SFM images of thin metallic films with columnar structure are often dominated by this tip artifact [689].

In the case, where the tip radius R is comparable with the spacing between needles or the dimensions of holes ($R \approx w$ or $R < w$), the situation, as shown in Fig. 6.3, is more complex. Again, the curvature of the SPM profile is given by the radius of curvature of the probing tip R and can be calculated with the formula of a truncated sphere with the same height h and width w as the observed profile:

$$R = \frac{h^2 + (w/2)^2}{2h} \tag{6.1}$$

---

[1]The sample was heated in ultrahigh vacuum, which led to the formation of needle-like structures.

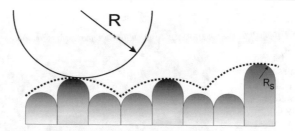

Fig. 6.2 The model surface, consisting of asperities with radius of curvature, R_s, is imaged by a probing tip of radius of curvature, R. Since $R \gg R_s$, the profile of constant interaction (dotted line) is not representative of the surface. The observed features correspond to the inverted probing tip with radius R

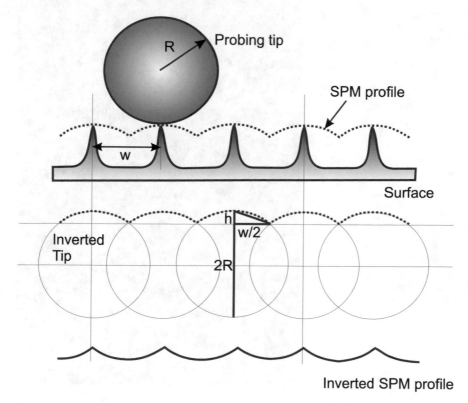

Fig. 6.3 The SPM profile of narrowly spaced holes is dominated by the inverted tip images, which are spaced by the hole dimension, w. The observed corrugation height is reduced, because the tip can not reach the bottom of the holes. Also, the SPM profile does not appear as holes, but as hillocks with a curvature given by the radius of curvature of the probing tip, R. The inverted SPM profile resembles more closely to the real surface with the correct spacing between the holes. However, this procedure is not generally applicable (see text)

Fig. 6.4 **a** SFM image of
polyhexamethylene (nylon)
spin coated onto glass with a
sharp probing tip. **b** SFM
image of the same sample
with a blunt tip, which is
terminated by a wedge.
Reprinted from [238],
https://doi.org/10.1063/1.
106862, with permission
from AIP Publishing

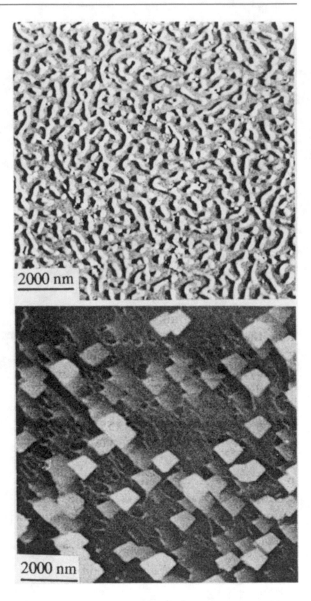

It becomes evident that the SPM profile has a reduced corrugation amplitude
compared to the real surface. Also, the narrow holes appear as hillocks. For ion-
bombarded surfaces, where small holes are present, Chen et al. [111] found that
inverted images resemble to the real topography more closely (cf. Fig. 6.3). However,
this procedure is not generally applicable. It may only be helpful for surfaces with
equally spaced holes, where the width, w, is smaller than the radius of curvature of
the tip.

Even more striking tip artifacts were presented by Grütter et al. [238]. They
found that some tips of microfabricated force sensors do terminate in wedge-like or

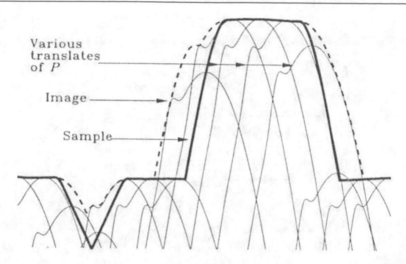

Fig. 6.5 Image construction (surface topography $s(x, y)$ and tip geometry $t(u, v)$ are known): The tip is reflected at the origin and scanned along the sample. The image $i(x, y)$ is the envelope of the translates of the reflected tip. Reprinted from Villarrubia [670], https://pubmed.ncbi.nlm.nih.gov/27805154, Public domain

rectangular shape, e.g., rectangles of 20–600 nm were observed. Images of polyhexamethylene showed strong tip artifacts, which did not resemble the topography at all (cf. Fig. 6.4). A possible way to overcome this problem is to deconvolute the acquired image by numerical simulations. For this purpose, the tip geometry has to be determined by some means, e.g., SEM-images or imaging of needle-like test structures. Such a deconvolution procedure has been presented by Grütter et al. [238] for SFM images of diamond films. Some commercial "tip characterizers" are available [97], which consists of arrays of sharp silicon tips.

A systematic description of tip deconvolution algorithms for SPM has been given by Villarrubia [670]. The tip geometry is described by a function $t(x, y)$ and the sample topography is given by a function $s(x, y)$. The tip is lowered at a position $(x\prime, y\prime)$ until it touches the surface. The apex marks the height of the image at the position $(x\prime, y\prime)$. The imaging equation is then given by:

$$i(x\prime, y\prime) = -min_{(x,y)}[t(x - x\prime, y - y\prime) - s(x, y)] \tag{6.2}$$

where the minimum is taken over all (x, y)-positions. This image equation can be reformulated into the form:

$$i(x, y) = max_{(u,v)}[s(x - u, y - v) + p(u, v)] \tag{6.3}$$

where $p(u, v) = -t(-u, -v)$ is the tip function reflected through the origin. Equation 6.3 corresponds to a geometrical construction, which is shown in Fig. 6.5. The apex of the reflected tip is scanned along the sample topography. The envelope of the translated tips corresponds to the SPM-image. This algorithm is called dilation.

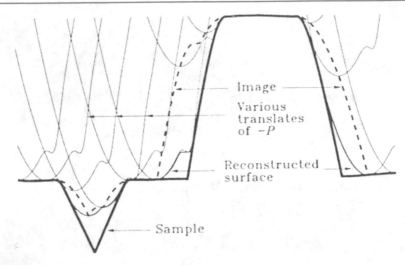

Fig. 6.6 Reconstruction of the surface topography (SPM image $i(x, y)$ and tip structure $t(u, v)$ are known). The apex of the tip is scanned along the SPM image (dashed line). The best possible reconstructed surface $s_r(x, y)$ is the envelope of the translates of the tip. Deviations between the reconstructed $s_r(x, y)$ and the real surface topography $s(x, y)$ are often found in deep valleys or at the base of hills. Reprinted from Villarrubia [670], https://pubmed.ncbi.nlm.nih.gov/27805154, Public domain

Eventhough the simulation of imaging is illustrative to visualize the tip artifact, the more practical problem is the reconstruction of the real surface from the measured SPM-topography. In case, the tip geometry $t(x, y)$ is known from imaging of needle-like structures (also called "characterizer"), the reconstruction equation is given by:

$$s_r(x, y) = min_{(u,v)}[i(x + u, y + v) - p(u, v)] \qquad (6.4)$$

where $p(u, v) = -t(-u, -v)$ is again the reflected tip and $i(x, y)$ is the topography image. $s_r(x, y)$ does not represent the real topography but it is the best possible reconstruction. It is obvious that deep valleys, which cannot be reached by the tip apex, are not accessible. The geometrical interpretation of the reconstruction equation is shown in Fig. 6.6. The tip is scanned along the SPM-profile $i(x, y)$. The reconstructed surface $s_r(x, y)$ is the minimum of the envelope of the translates of the tip. The reconstructed surface is an upper bound of the real surface topography. It is the best reconstruction because it is the surface of deepest penetration of the tip. This reconstruction is also called erosion algorithm.

In addition to the reconstruction of surfaces, it is also possible to create certainty maps $c(x, y)$, which indicate which points of the reconstructed surface coincide with the real surface and which points may or may not coincide with the surface [523]. Essentially, the algorithm distinguishes between positions with a single touch of the tip with the reconstructed surface and points with multiple touches. In the case of a single touch, the reconstructed surface coincides with the real surface, $c = 1$. In the case of multiple touches, it is unclear which point coincides with the real surface, $c = 0$.

The disadvantage of the reconstruction is the necessity to know the tip structure. Experimentally, it is not easy to determine the tip structure, which may change with time due to tip crashes. An alternative is to use blind reconstruction equations, which do not require the knowledge of the tip geometry, but reconstruct the tip geometry from the SPM image by an iterative process [668,669]. The method works reasonably, if the surface contains features that are sharper than the tip. Electronic or vibrational noise can severely disturb reconstruction algorithms. The spikes of the noise can be sharper than the tip. This problem can be overcome by using appropriate threshold parameters which suppress the influence of noise. Other limitations of reconstruction

Fig. 6.7 Characterization of tip artifacts by scanning tunneling potentiometry (STP). **a** Cross section of a rough surface and a blunt tip. **b** Actual electrical potential versus position on surface. **c** Measured SPM profile. **d** Measured profile with discontinuous jumps, which correspond to tip changes. Reprinted from [511], https://link.aps.org/doi/10.1103/PhysRevB.41.1212, with permission from AIP Publishing

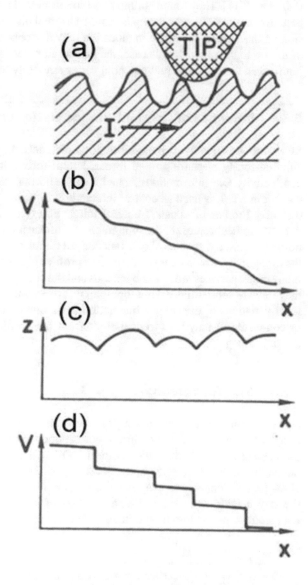

algorithms are scanner non-linearities, feedback loop overshoot resulting from fast scanning, inhomogeneous sample surfaces, and tip crashes. The above described reconstruction algorithms are commercially available [6].

Rather challenging measurements of trenches of semiconductor devices have led to the development of special tips in combination with a dedicated SFM, which are able to image sidewalls with high accuracy [227]. The dead zone (area which is not accessible by the tip) is reduced significantly by the use of 3-point tips, which are vibrated in 2 dimensions and controlled by a dual heterodyne interferometer [493].

An interesting, non-standard way to visualize tip artifacts has been proposed by Pelz et al. [511]. They found abrupt, non-monotonic behaviour in scanning tunneling potentiometry (STP).[2] As the tip is scanned a small distance across a narrow depression on the surface, the tunneling location may discontinuously jump from one side of the tip to the other. In this case, the measured potential shows a spatially abrupt jump, even though the actual potential varies smoothly with position (Fig. 6.7).

6.3   Influence of Local Inhomogenieties on Topography

SPMs are sensitive to various information channels: E.g., a force microscope can simultaneously measure normal forces, lateral forces, local compliance and local conductivity. One information channel, e.g., the normal force in SFM or the tunneling current in STM, is used as control interaction to control the probing tip to sample distance. Profiles of constant control interaction (e.g., constant force or constant current profiles) represent the topographical information. In case that the surface is homogeneous, the profiles of constant control interaction are approximately equal to the topography, which could be defined as profiles of constant total charge density. In many cases, surfaces are inhomogeneous and the constant control interaction profiles may deviate substantially from topography. An example is given in Fig. 6.8, where local variations of the control interaction (e.g., due to local variations of elasticity or conductivity) may lead to deviations from the profiles of constant total charge density (topography).

6.3.1   STM Topography

STM is sensitive to the local density of states at the Fermi energy (LDOS). Acquiring images of constant tunneling current corresponds to profiles of constant LDOS. In the case of local inhomogenieties the LDOS will change and STM profiles will deviate from the ideal topography. An illustrative example has been given by Jung et al. [322]. Copper was evaporated on a Mo(110) surface. A strong variation of the step heights was observed as a function of voltage. At some voltages, where the electrons tunnel into image states or surface states, the contrast was found to be

---

[2]In STP, a current flows within the sample. The measured electrical potential can be useful for the characterization of transport properties.

Fig. 6.8 The probing tip is scanned with constant control interaction (e.g., constant force or constant tunneling current mode). Due to inhomogenieties, such as local variations of elasticity or conductivity, the profiles of constant control interaction deviate from the profiles of constant total charge density (topography)

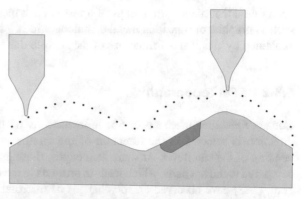

Fig. 6.9 Influence of local variations of LDOS. Top: STM-image of Cu-island deposited on Mo(110) recorded with bias voltages, where an image state leads to the increased apparent height of the Cu-islands. Bottom: STM-contrast, which is induced by a surface state. The image state produces about twice the contrast, but degrades the spatial resolution. $(90 \times 110\,nm^2)$. Reprinted from [322], https://link.aps.org/doi/10.1103/PhysRevLett.74.1641, with permission from AIP Publishing

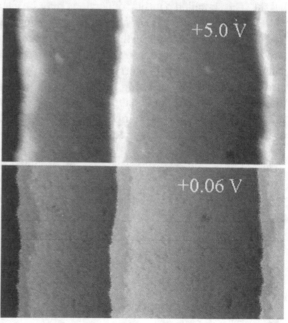

maximum. This reflects the fact that the current is proportional to the LDOS, which is drastically increased by these image or surface states. Surface states yield relatively weak contrast and are difficult to be identified, but yield high lateral resolution. Image states are in the regime between tunneling and field emission and are observed at voltages of 5–10 V.[3] Due to these large voltages, the tip is retracted and the spatial resolution is degraded (Fig. 6.9).

In some cases, STM images cannot be interpreted in terms of variations of LDOS, but are strongly influenced by forces between probing tip and sample. E.g., it has

---

[3]Image states are due to standing waves between probing tip and sample. Sometimes also called field emission resonances. They were first observed by Becker et al. [55].

been found that giant corrugations of up to 8 Å are measured on layered materials, such as graphite or transition metal dichalcogenides. These giant corrugations were explained by elastical deformations of the sample due to tip-sample forces [614].

6.3.2   SFM Topography

Contact SFM measurements can be influenced by frictional forces. If the scanning direction is selected in the direction of the cantilever, lateral forces will lead to bending of the cantilever. At sharp step edges, these lateral forces can be increased compared to the flat parts, which leads to artifacts in the topography signal. Den Boef et al. [142] have observed drastic changes of microfabricated gratings with 85 nm height. The so-called parasitic deflections may be as large as 100 nm. The influence of these lateral forces can be verified by reversing the scan direction (cf. Fig. 6.10). The effect mainly depends on the geometrical parameters of the cantilever. Long tips and small normal spring constants will lead to large topography artifacts. Therefore, triangular cantilevers with short tips are used to minimize this artifact. The effect can also be minimized by scanning in the direction perpendicular to the cantilever. The torsional spring constant is larger compared to the normal spring constant and the influence of lateral forces on topography is reduced. In tapping mode, the influence of lateral forces is drastically reduced.

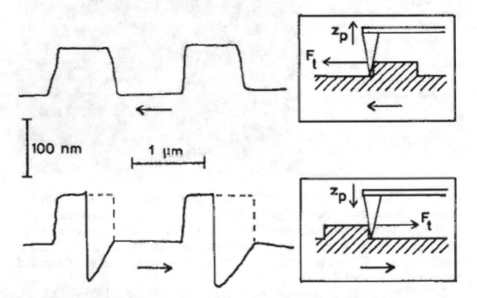

Fig. 6.10 A grating of 85 nm height is imaged by contact SFM. Depending on the scan direction, the measured profile is altered. The influence of lateral forces on topography is large, when the tip is scanned in the direction of the cantilever. Reprinted from [142], https://doi.org/10.1063/1.1142287, with permission from AIP Publishing

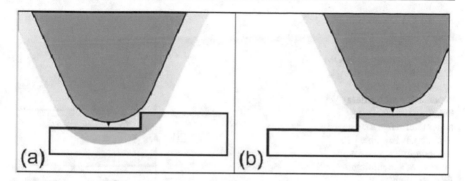

Fig. 6.11 On the lower terrace the tip effectively interacts with a large area or volume, which gives a stronger attraction. On the upper terrace the reverse happens. Reprinted from Guggisberg et al. [240], https://doi.org/10.1016/S0039-6028(00)00592-6, with permission from Elsevier

The influence of local elasticity on topography is pronounced on soft materials, such as thin organic films, where the elasticity is strongly reduced compared to the hard substrate. A film with a Youngs modulus of $E = 10$ MPa and a thickness of 20 nm will be approximately reduced in height by $\Delta z = 4.5$ nm as shown by the Hertzian penetration depth

$$\Delta z = \frac{(1 - \nu^2) F_N}{2ER} \tag{6.5}$$

where $R = 10$ nm is the tip radius, $\nu = 0.3$ is Poisson ratio and $F_N = 1$ nN is the normal force. It is assumed that the substrate and the probing tip have negligible elasticity.

In contrast to contact force microscopy, non-contact force microscopy appears to be less affected by elastic deformations due to the absence or reduction of repulsive forces. However, topography of nc-AFM may be influenced by the combination of long-range and short-range attractive forces. Guggisberg et al. could demonstrate that step heights measured in constant frequency shift mode deviate strongly from the real values, if the measurements are done close to the steps. Sometimes, even an apparent inverted topography can be observed. The explanation is illustrated in Fig. 6.11, where the interaction volume for long-range forces is increased at the lower terrace, which gives a stronger interaction. On the upper terrace the contribution of the long-range forces is reduced. Consequently, the measured step height is depending on the relative contribution of short-range and long-range forces. The effect is increased by the application of bias voltages due to the presence of the relatively large electrostatic forces.

6.4   Influence of Topography on Local Measurements

Measurements of probed interactions may be influenced by topography. The most striking examples were found in the field of SNOM. Therefore, this phenomenon is commonly also referred to as the "SNOM effect": Topography images with high

Fig. 6.12 a Distance dependence of the local property to be investigated. The light intensity of SNOM $I(z)$ is an example. Typically, a distance dependence with a decay length in the hundred nanometer range may be assumed. **b** Example of a topography scan line with maximum and minimum values z_1 and z_2. **c** Corresponding variation of local property as a function of the lateral position. The differences between the z-positions and the average plane will lead to intensity variations

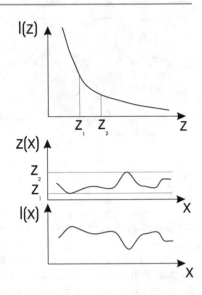

lateral resolution are acquired with the shear force sensor. Simultaneously, the light signal is collected with the SNOM-fiber. At first sight, the acquired SNOM-images reveal features with high lateral resolution. However, the SNOM-images closely resemble the topography images. An explanation is given by local variations of the probing tip to sample distance, which corresponds to variations of the evanescent field (cf. Fig. 6.12). One way to rule out this effect is to perform measurements on flat surfaces, which have only variations of the optical properties. Alternatively, constant height scans may be performed which will reveal the real resolution limits of the optical channel. Similar precautions have to be made for all other probed interactions, e.g., local barrier height measurements are often influenced by the local topography. Local damping measurements in nc-AFM are suspected to be strongly influenced by variations of the tip-sample distance. Another example are the lateral force measurements in FFM, which will be discussed below.

6.4.1   Influence of Topography on STM-Induced Photon Emission

A striking example of the influence of topography on the STM-induced photon emission has been given by Berndt et al. [74]. As described by Rohrer et al. [555], SPM can probe various interactions. One interaction, the control interaction, is used to control the the tip-sample distance. Curves of constant control interaction are acquired. Another interaction is measured during operation with constant control interaction. In the case of Berndt et al. [74], the control interaction is the tunneling current and profiles of constant tunneling current are acquired on the surface of Au(110). Atomic rows with typical corrugation heights of 0.7 Å and spacings of 8.16 Å are observed. The measured probing interaction is the STM-induced photon emission. Theoretically, it is expected that the local photon emission is laterally extended over approximately 50 Å, which is inferior to the resolution of STM. Therefore, it was surprising to observe atomic-scale variations of the photon emission. The explanation is given by the distance dependence of the photon emission, which is given by the coupling to localized plasmon modes. These plasmons are averaged over a surface area of about 50 Å and the distance dependence is also averaged over a surface area of similar lateral dimension. STM, representing the control interaction, has a higher lateral resolution, where atomic resolution is achieved. Above atomic rows the tip-sample distance averaged over the lateral extend of the plasmon is larger compared to the position in a trough. The larger distance corresponds to a weaker field strength and consequently less photon emission is excited. The photon emission is proportional to the difference between the observed STM topograph $h(x)$ and the laterally averaged surface height $\overline{h}(x)$, where

$$\overline{h}(x) = \int_{-\infty}^{\infty} h(x\prime)g(x - x\prime)dx\prime \tag{6.6}$$

is given by a convolution integral of the topography with a weight function, which represents the lateral extend of the probed interaction. The weight function may be approximated by a Gaussian function. In other words, the probed interaction decays by a certain distance dependence, where the distance is the distance between the tip apex and an average plane of the surface. Small distance variations given by the control interaction can cause variations of the probed interaction due to the distance variation between the tip and this average plane. Thus, the resolution of the probed interaction can appear better than it is in reality, because of the high lateral resolution of the control interaction. Similar effects may appear in SNOM [528], laser-driven STM [675] and plasmon-induced current experiments [463].

6.4.2 Influence of Topography on Lateral Force Measurements

Lateral force maps of surfaces that are not atomically smooth are often dominated by the topography. We distinguish two cases:

(1) The local slope of the topography, $\partial s/\partial x$ causes a lateral force $F_{topo} = \mu_{topo} \cdot F_N \cdot \cos\delta \approx \mu_{topo} \cdot F_N$, where $F_N = F_L + F_A$ is the total normal force, F_L the externally applied loading force, F_A the attractive force, $\mu_{topo} = \tan\delta = \partial s/\partial x$ and δ is indicated in Fig. 6.13. The approximation is valid for small slopes with $\delta << 1$. In principle, this kind of lateral force would even occur for frictionless sliding, originating simply from a component of the normal force. In contrast to frictional forces, the lateral force F_{topo} does not depend on scan direction and can be distinguished by examining the forward and backward scan. This effect is observed on homogeneous, rather smooth surfaces, where the radius of curvature of the probing tip is small compared with the surface roughness. Small hillocks of polycarbonate on a compact disc are shown in Fig. 6.13, where lateral forces independent of scan direction are found. The derivative image of the topography is in good agreement with the back- and forward scan. Similar examples are described in the literature [142, 193, 226, 451].

Fig. 6.13 Influence of topography on lateral forces. **a** Schematic diagram of the relevant force vectors. Due to the local slope of topography, normal forces F_N cause a component of the lateral force. **b** Topography image of a compact disc (CD), consisting of small hillocks of polycarbonate. **c** Derivative of the topography. **d** Lateral force image acquired in the forward direction. **e** Lateral force image acquired in the backward direction. The contrast in **d** and **e** is independent of scan direction and is closely related to the topography image

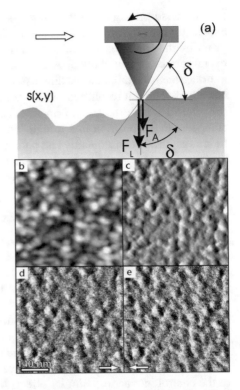

(2) The second topography effect is often dominating over the previously described topography effect and is related to local variations of the contact area. At steep slopes of the topography, where the radius of curvature becomes comparable with the curvature of the local topography, the contact area is changed and causes a direct increase or decrease of friction. In addition, the local topography can also lead to a change of attractive forces, such as van der Waals forces or capillary forces. As drawn schematically in Fig. 6.14, the van der Waals force on a summit of a hillock is weaker than in a valley, where the interaction volume is bigger. The larger attractive force has to be compensated by a larger repulsive contact force and leads to a larger contact zone and thus increases friction. Capillary forces play an analogous role: They are increased in the valley, where the liquid film thickness is increased, too. The frictional forces depend on the total normal force by $F_F = \pm\mu \cdot (F_{\text{load}} + F_a)$ and on the scan direction. This observation was first made by Mate on carbon coatings [437, 439]. An example is given in Fig. 6.14 showing a typical contrast reversal with scan direction.

In summary, the total lateral force is given by a scan direction dependent and a scan direction independent component:

$$F_{\text{Lateral}} = \pm\mu F_N + \partial s/\partial x\, F_N \tag{6.7}$$

Both components are closely related to the local topography and are superimposed on most surfaces.

(a)

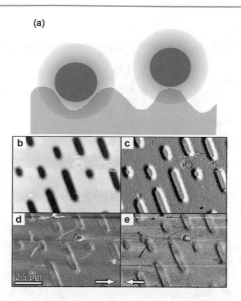

Fig. 6.14 a Influence of topography on the attractive force F_a due to van der Waals forces. On a summit the attractive force is reduced because of the smaller volume, whereas in the valley the force is increased. The larger attractive force F_a leads to an increased effective normal force F_N, which causes an increased contact area and therefore causes larger frictional forces. Analogous effects are also caused by forces such as capillary forces. **b** Topography image of a compact disc (CD) on a larger scale compared to Fig. 6.13. The bits of the CD become visible. **c** Derivative of the topography. **d** Lateral forces in the forward direction. **e** Lateral forces in the backward direction. In contrast to the example in Fig. 6.13, the lateral forces at steep slopes do depend on the scan direction and therefore are related to frictional forces

6.5   Instrumental Artifacts

6.5.1   Piezoelectric Hysteresis, Creep of Scanners, Nonlinearities and Calibration Errors

Hysteresis and creep of piezo scanners can cause severe artifacts. The basic principles were discussed in Chap. 1. In contrast to the tip artifact, piezoelectric hysteresis or creep can be removed in principle by the experimental implementation of position sensors, which measure the true motion of the tip relative to the sample. For metrology applications, hardware corrected systems are becoming common. Capacitive sensors, optical beam displacement sensors or interferometric sensors were incorporated into probe microscopes [227]. The advantage of the real-time scan correction is the easier navigation during a scanning session. The operator does not have to take into account hysteresis and creep. A common problem of these hardware corrected systems is the signal-to-noise ratio, which makes operation with high resolution difficult. Typically, the scan correction is enabled from the 100 μm down to the 10 nm scale and is disabled for high resolution imaging. An alternative way is to use software corrections, which take into account effects, such as the logarithmic time dependence of piezo motion due to creep. However, these software corrections

Fig. 6.15 Tips on tube scanners move on a nearly spherical surface. These nonlinearities can be software corrected. Different tip lengths (or sample thicknesses for samples mounted on the scanner) lead to different scan lengths, which should be taken into account for accurate calibrations

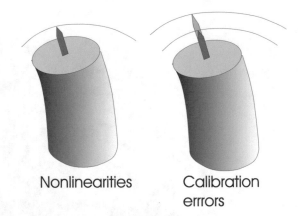

Nonlinearities   Calibration errrors

are empirical approximations and cannot correct complicated scan sequences, e.g., zoom-in may fail. Software correction of hysteresis is difficult. The usual procedure is to compare the forward- and backward scan in order to estimate the influence of hysteresis. For high resolution applications (scan range of some tens of nm) hysteresis is small, but it becomes a severe problem for large area scans (scan range of some microns). In the latter case, hardware scan corrections are useful.

Another problem is the tilting of the tube scanner or the non-orthogonal mounting of piezo bars (cf. Fig. 6.15). A x-y-motion is always accompanied by an extra motion in z-direction, Δz. These deviations from the planar geometry, also called image bow, can be accurately corrected in the software by means of polynoms. Typically, functions of the form

$$\Delta z(U_x, U_y) = a_x U_x + a_y U_y + b_x U_x^2 + b_y U_y^2 + c_x U_x^3 + c_y U_y^3 \qquad (6.8)$$

are used to linearize the scan motion, where U_x and U_y are the voltages applied to the piezos for x-y-motion. A flat surface is imaged with the probe microscope and the coefficients of equation (6.8) are determined.

Calibration errors can have many origins. It is advisable to perform calibration measurements (imaging of well-know microfabricated structures or atomic structures) regularly. The piezoelectric scanners are aging or may be partially depolarized. A common mistake is not to include the length of the probing tip in the calibration procedure. Especially, STM tips can be rather long compared to the length of the piezo (cf. Fig. 6.15). Therefore, an accurate determination of the length of the tip or the sample thickness is necessary. A non-standard way to perform calibrations was proposed by Kawakatsu et al. [340], where a dual tunneling unit STM with one xy-stage was used to perform simultaneous scanning on both the sample and the scale-reference crystal. A detailed description of calibration procedures for metrology applications are given by Edwards et al. [158, 160].

6.5.2 Tip Crashes, Feedback Oscillations, Noise, Thermal Drift

A typical source of artifacts are tip crashes. The uncontrolled contact between surface and probing tip may lead to simple effects, such as tip blunting, which then will lead to blurred images, where sharp features cannot be observed anymore. More subtle effects are the rearrangement of surface topography. Ripples can be formed in contact mode SFM on the surface by the interaction with the tip. These drastic rearrangements of the surface may lead to false interpretation of the sample topography (Fig. 6.16).

The optimization of the feedback is a rather complex procedure, which may lead to many artifacts. The optimum parameters of the feedback depend on tip conditions and may even change during scanning due to tip crashes. Therefore, experience is needed to find the optimum conditions. A typical procedure is described in Chap. 1. Unoptimized feedback parameters may lead to blurred images, because the feedback is too slow. The other extreme is the occurrence of oscillations or feedback overshoots. Since typical SPM controllers are PI-controllers, the phenomena are described in standard electronics books [647].

Noise can have various sources, such as intrinsic noise of the tunneling current (cf. Sect. 5.9) or electronic noise, e.g., Johnson noise of resistors of the preamplifier, or 50/60 Hz noise. Mode-hopping of SFM laser diodes can cause low-frequency noise, which is visible as horizontal "banding" in images. The normal procedure to exclude

Fig. 6.16 Collection of SPM artifacts. Tip crashes are uncontrolled contacts between probing tip and sample, which may lead to tip blunting or sample damage. Feedback overshoots is one possible phenomenon due incorrect feedback parameters. Noise can have various sources and various appearances, e.g., periodic noise due to 50/60 Hz. Thermal drift is characterized by the drift velocity of the probing tip relative to the sample

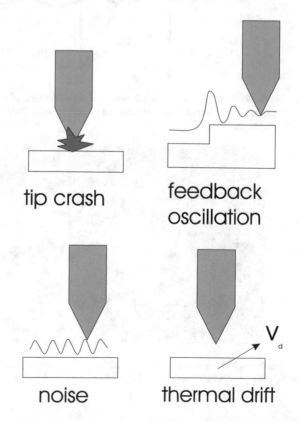

noise artifacts is to change the scan speed on the same area. If the observed features depend strongly on the scan speed, it is rather probable that noise may be the source of contrast.

Thermal drift is present in most SPMs. It becomes a real problem in high resolution applications, because the observed features eventually cannot be re-imaged due to thermal drift. Improvements can be made by faster scanning or by waiting for the thermal equilibrium. Thermal drift can also lead to wrong distance measurements. Exchange of the scan direction can clarify the situation. Ideally, the drift velocity v_d is determined by repeated imaging of a characteristic feature. Typical drift velocities at room temperature are some nm/sec. At low temperatures drift rates as small as some Ångstroms per hour are obtained.

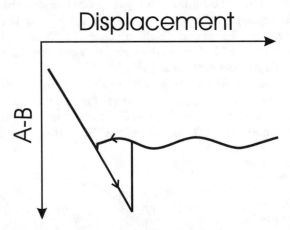

Fig. 6.17 Beam deflection signal $A-B$ versus movement of the sample with a bad focus, which leads to interference effects between the beam reflected from the sample and the beam reflected from the cantilever

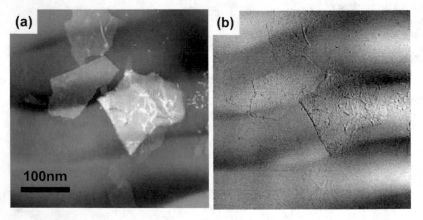

Fig. 6.18 FFM-image of MoS_2-platelets on mica, imaged with a bad focus, which leads to interference patterns (stripes across the whole image). **a** Topography **b** Lateral force map

6.5.3 Interference Patterns with Beam Deflection SFM

Figure 6.17 shows a force versus distance curve, where the focus of a laser-beam deflection SFM was not optimized. Here occurs interference between the laser beam reflected from the sample and the beam reflected from the cantilever. The distance between the interference maxima, $d_{max} = \lambda / \sin \theta$ is related to the wavelength of the laser source, λ (in this case 620 nm), where the angle of incidence of the laser beam relative to the sample surface, θ, is taken into account. These interference effects can also disturb the lateral force imaging. Figure 6.18 shows a FFM-image of MoS_2-platelets on mica. The focus was rather bad ($>30 \mu m$), which caused interference patterns, which are pronounced in the FFM-image. The use of optimized optics with small-area focus and cantilevers with gold coating ($>20 nm$) can minimize these effects. Alternatively, light emitting diodes can be used, which have a broader spectrum and small coherence length, which eliminates this problem.

Future Aspects of SPM

Abstract

Future prospects of scanning probe microscopy are discussed. Arrays of cantilevers can be used for ultrasensitive bio-sensors or be used as a mechanical
storage medium. The role of scanning probe microscopy in fields of nanoscience,
nanotechnology but also quantum information will be elaborated.

7.1 Parallel Operation of SFM Cantilever Arrays

The parallel operation of SFMs has potential in several areas of nanoscale science
and technology, such as data storage, lithography, high-speed/large-scale imaging
and molecular and atomic manipulation. Minne et al. presented the parallel operation of a 1D array of cantilevers, where the parallel acquisition of images was
demonstrated [459]. Ten cantilevers with integrated piezoresistive sensors and zinc-
oxide (ZnO) actuators were operated in parallel, where each cantilever scanned a
rectangular area of $200\,\mu m \times 2\,mm$. A total area of $2\,mm \times 2\,mm$ of a memory cell
of an integrated circuit was imaged in this way. These scan areas are adapted to
the needs of semiconductor industry, where integrated circuit chips of $100\,mm^2$ are
common. Each cantilever provided a resolution in the nm-range with a bandwidth of
$20\,kHz$. The microscope produces such a large amount of data, that it is difficult to
be handled by today's processing possibilities. These results clearly show new per-
spectives for microscopy, where large, complex structures are to be investigated on
the nanometer scale. Furthermore, new avenues were opened up in SFM-lithography
where patterns on an area of $1\,cm^2$ were written with line widths in the micron range
[459].

A 2-dimensional SFM cantilever array, called "Millipede", of the IBM Zurich
Research laboratory [666] was designed for high density data storage. The
"Millipede"-concept is illustrated in Fig. 7.1. A 32×32 cantilever array is positioned
above a polymer surface. The entire cantilever array chip is scanned in x-y-direction.
The approach in z-direction is controlled by 3 piezoresistive cantilevers and 3

© Springer Nature Switzerland AG 2021   263
E. Meyer et al., *Scanning Probe Microscopy*, Graduate Texts in Physics,
https://doi.org/10.1007/978-3-030-37089-3_7

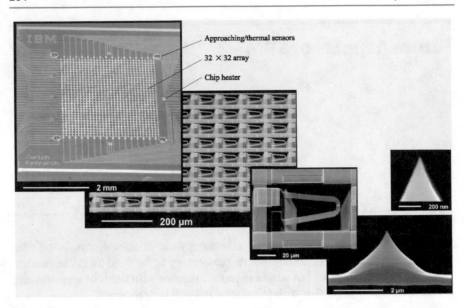

Fig. 7.1 A series of SEM-pictures, which illustrates the concept of 2D cantilever arrays for data storage called "Millipede". Reproduced with permission from Vettiger et al. [666]. Courtesy of International Business Machines Corporation, ©(2000) International Business Machines Corporation

corresponding magnetic actuators (z_1, z_2, z_3), which allow to control not only the z-distance but also the z-leveling. This concept is not based on individual z-feedbacks for each cantilever, but uses feedback control for the entire array, which simplifies the system. However, it requires accurate fabrication of the tip height and cantilever bending over the entire chip. Tip-apex height variations of less than 500 nm were demonstrated. The requirement of tip-apex uniformity is related to the uniform force needed to minimize tip and sample wear due to large force variations. During the storage operation, the chip is scanned by a magnetic x-y-actuator. Each SFM writes and reads the data in its own storage field ($\approx 100 \times 100\,\mu m^2$). Initial storage densities of 100–200 Gb/in^2 were achieved, where dots of 50 nm in diameter were written into PMMA. Assuming a storage density of 500 Gb/in^2, one storage field has a capacity of about 10 Mb and the entire 32×32-array has a capacity of 10 Gb on 3×3 mm. The Millipede is based on a thermomechanical write/read process in nanometer-thick polymer films. A similar process was demonstrated with a single cantilever by Mamin and Rugar [421], where a laser is used to heat the polymer locally. During the heat cycle the tip plastically deforms the polymers, which creates a small nanometer-sized indentation. The reading is achieved with a laser beam deflection set-up. In the case of Millipede, the tip is heated with an integrated heater and a thermal sensor is used for reading. At present, the mechanical resonance frequencies (10 kHz–1 MHz) of SFM cantilevers limit the operation speed to the microsecond time scale at best. Corresponding data rates of a single cantilever are a few Mb/s for SFM data storage. For comparison, the slow feedback speeds due to the low tunneling currents of STM-based storage application reduced the data rates even more [67]. However, conventional magnetic storage operates at best in the nanosecond time scale. Data-rates are increased due to the parallel operation of cantilevers. In the case of Millipede, a 2d-array of SFM cantilevers with integrated write/read storage functionality was implemented. Full parallel operation is performed within one row. A time-multiplexed addressing scheme is used to address the array row by row. A critical aspect is the heater time constant, which was determined to be in the microsecond regime. Therefore, multiplexing rates of 100 kHz should be possible. Applications of Millipede are to expected in areas, where small space requirements in combination with low power consumption are needed, e.g., watches, cellular phones or laptops. Another possibility are the terabyte storage systems. The implementation of very large arrays is challenged by the control of the thermal expansion of this large array. Alternatively, several medium-sized arrays may be arranged over a large media.

7.2 Novel Sensors Based on Cantilevers

The invention of force microscopy was primarily dedicated to image surfaces on the atomic scale. However, the high sensitivity of cantilevers can also be applied to sense other properties [378]. Essentially, physical or chemical properties of surfaces can be transduced into mechanical motion by the use of these micromechanical devices. Examples are the transduction of heat, surface stress, mass loading, electrical or magnetic signals into mechanical motion. Techniques from force microscopy allow to measure deflections of the order of pico-meters with a bandwidth in the kHz-regime. A typical cantilever with dimensions of about $500 \mu m$ length, $40 \mu m$ width and $0.2 \mu m$ thickness has a mass of a few nanograms. The most common modes are shown in Fig. 7.2.

7.2.1 Gravimetric Sensors

The cantilever is oscillated at its resonant frequency. A mass Δm attached to the cantilever will change (lower) the resonance frequency according to

$$\Delta m = \frac{k}{4\pi^2} \left(\frac{1}{f'^2} - \frac{1}{f^2} \right) \tag{7.1}$$

where k is the spring constant, f and f' are the resonance frequencies before and after attachment of masses. Mass changes as small as 10^{-16}g can be detected with commercial cantilevers. For example, the adsorption of water was measured [646]. Zeolite crystals were attached [574] to the apex of a cantilever, where controlled heating showed the desorption of molecules out of the zeolite pores. Recently, the fabrication of cantilevers with reduced dimensions of $10 \mu m$ length, $10 \mu m$ width and $20 nm$ thickness was demonstrated. With further optimization, the landing of single molecules on the surface of a cantilever may become detectable.

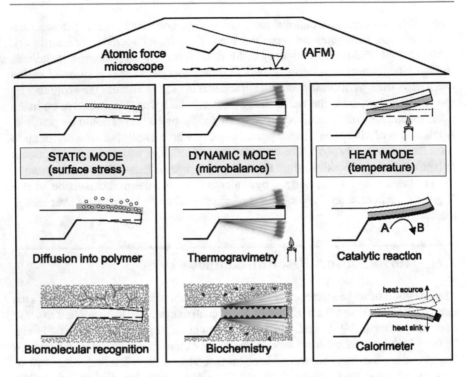

Fig. 7.2 Operation modes of cantilever-type sensors. Reproduced from [378] with permission from IOP Publishing

7.2.2 Calorimeter Sensors

If a layer of material is deposited on the cantilever, where the thermal expansion of the layer material is different from the one of the cantilever, small temperature changes will cause a deflection due to the bimetallic effect. A constant power (the unit is Watt = Joule/sec) generated on the cantilever will cause a permanent deflection. Typical values for microfabricated sensors are in the nanowatt-range and can be optimized down to the femtowatt-range. A single deposition of energy on the cantilever will cause a transient signal. Thermal response times on the millisecond scale and sensitivities in the attojoule range can be achieved by optimizing the choice of materials and geometries. This corresponds to temperature variations of the order of 10^{-5} K. Application examples are the detection of a catalytic reaction on the cantilever, where a thin Pt layer was deposited on side of the cantilever, which catalyzes the exothermic conversion of H_2 and O_2 to form water [216]. Another example is the observation of enthalpy changes associated with phase transitions of nanogram samples of alkanes [71]. Bachels and Schäfer have deposited nanometer-sized clusters on the cantilever and determined the enthalpy change due to recrystallization [39].

7.2.3 Surface Stress Sensors

An alternative way to sense the adsorption of molecules on the cantilever is to measure the deflection due to mechanical stress at the interface. Stoney's formula can be used to determine the surface stress from the cantilever curvature radius R:

$$\sigma_s = \frac{E_C t_C^2}{6R(1 - \nu_C)} \tag{7.2}$$

where ν_C is the Poisson ratio of the cantilever and

$$R^{-1} = \frac{3\Delta z}{2l^2}. \tag{7.3}$$

Typical values are $t = 100\,\text{nm}$, $l = 500\,\mu\text{m}$ and $\Delta z = 0.1\,\text{nm}$, which results in minimum detectable stresses of the order of $10^{-7}\,\text{N/m}$, corresponding to the detection of attomolar quantities. The cantilever have to be functionalized on one side in order to bind molecules. Surface stress changes were measured during the self-assembly of alkanethiols on gold [70]. Surface stress measurements are also well-suited for measurements in liquids. Fritz et al. [190] demonstrated the detection of hybridization of oligomers on the cantilever, where a single base pair mismatch could be distinguished. This detection principle has the advantage that no labeling of the nucleic acids is needed. The technique can be also extended to measure protein adsorption.

7.2.4 Cantilever Array Sensors

The parallel operation of cantilevers gives the opportunity to construct artificial electronic noses [379]. A variety of coatings is deposited on the cantilevers. Each cantilever reacts in a different way to the exposure of chemicals. These different responses are then used to discriminate or distinguish between different gases or liquids. All operation modes, described above, can be employed simultaneously with an array of cantilevers, where each cantilever is functionalized in a specific way. The goal is to achieve "orthogonal" information, which allows to distinguish between different analytes. Another major advantage of arrays is the differential measurement, where a coated cantilever is compared with an uncoated cantilever. The differential measurement is much less affected by disturbances from the environment. The sensitivity of these cantilever array sensors was improved and 1 fg of total RNA was detected, which corresponds to single bacterial cell sensitivity. The method can be applied to multidrug-resistant bacteria [292]. Single nucleotide polymorphisms (SNPs) and genes associated with antibiotic resistance were detected. This rapid and ultra-sensitive method might become a valuable diagnostic tool (Fig. 7.3).

Fig. 7.3 Example of a cantilever array sensor. 8 cantilevers with a thickness of 500 nm are operated in parallel

7.3   Molecular Electronics

The use of molecules as active elements in future devices appears promising. Molecular single electron transistors were proposed, where operation in the THz-regime appears possible [35,100]. Such a a three-terminal device incorporates source, drain and gate electrodes on a molecular level. An alternative approach is the atom relay transistor, where the mechanical motion of an atom causes conductance changes in an atom wire [679]. So far, molecular electronics is at an early stage with only a few practical examples. For example, rectification of benzene-1, 4-dithiol molecules between gold electrodes was demonstrated [548]. Fujihira demonstrated the molecular photodiode [192]. Single molecule transistors were presented, where molecules were chemically attached to electrodes of nanometer-sized break junctions [393,507]. Other examples are the transistor-like behaviour of single carbon nanotubes, which represents a molecular wire [633]. Simulations demonstrate that molecules may be used to perform arithmetic calculations [316,317]. One of the major problems is to address individual molecules. In this context, STM and SFM are used as imaging tools and as local electrodes to study the electrical properties of the molecules. Furthermore, these instruments may also be used to manipulate the molecules, e.g., move the molecules between designed electrodes. The use of parallel operation of microscopes may lead from pure laboratory experiments to practical devices. One

Fig. 7.4 Visualization of an atomic switch to control the conductance through an atomic wire. Similar concepts were proposed by Wada et al. [679]. Courtesy of H. Hidber

of the long-term goals is the molecular supercomputer with clock frequencies in the THz-regime (Fig. 7.4).

7.4 Quantum Computing and Quantum Matter

The realization of quantum computers is one of major challenges in science and technology. As proposed by Loss and Di Vicenzo [407] a promising concept is based on the spin states of coupled single-electron quantum dots. Logical operations are controlled by local gates via tunneling barriers. SPM-based lithography can be applied to fabricate nanometer-scale quantum dots and the SPM tip is used as local electrode [524]. Using electron spin resonance (ESR) techniques the spin state of quantum dots can be measured and manipulated. An impressive achievement is the writing and reading of the spin state of single atoms by Natterer et al. [479]. The spin state was controlled by STM-enabled single-atom electron spin resonance. This work shows that single-atom magnetic memory can be realized.

Coherent spin manipulation was realized by Yang et al. [712], where a magnetic STM tip was used to apply fast electrical pulses. These electrical fields lead to atom movement, which modulate the tip-atom exchange interactions, corresponding to an oscillating effective magnetic field. Spin-control protocols , such as Ramsey fringes and Hahn spin echo, demonstrated quantum dynamics of coupled titanium atoms. Therefore, SPM-based techniques have demonstrated that quantum information can be processed on the scale of single atoms.

Another approach is to use atom manipulation to create complex patterns, such as the fractal Sierpiński structure [343]. Wavefunctions delocalized over these structures show self-similar parts in the energy dispersion. Novel quantum material can be created by atomic manipulation, where effects of spin-orbit interactions and magnetic fields can be studied on electrons in non-integer dimensions. Atom manipulation can also be used to engineer topological properties of the electronic systems [344]. Topological insulators are insulating in the bulk and host protected states on the surface, where spin is locked to momentum. This robustness of spin information makes this material interesting for quantum computation as well.

7.5   Laboratory on a Tip

The idea to combine several probe methods in one instrument is obvious. Today, the combination of STM and SFM is available. The use of conductive SFM tips, such as metallic tips, doped silicon or doped diamond, gives the opportunity to measure both forces and electrical currents. Local conductivity measurements are done in the repulsive contact. Operated in the tunneling regime, frequency shifts can be detected by DFM, or operated in DFM-mode, variations of the tunneling current can be observed. In close analogy, combinations of SNOM and force detection are common, where shear force detection is used to control the probe sample distance. Other combinations are SFM with thermal probes [179] or SNOM with capacitance measurements [390]. Especially, SFM has a rich variety of possible interactions (electrostatic, magnetic and short-range interactions) which can be operated with the same instrument. The advances in this field are the implementation of microfabrication procedures, where combinations of SPM's are realized in one structure. For example, thermal sensors, which are integrated in the tip of a force sensor or microfabricated photodiodes at the apex of the probing tip were realized [136].

More complex functionality appears desirable: The combination of high lateral resolution with analysis of the local chemical composition appears to be one of the most challenging goals. In contrast to the material-specific imaging, which is observed by methods, such as friction force microscopy, phase contrast in tapping mode or local adhesion measurements, chemical analysis prerequisites that a fingerprint is created, which determines the elemental composition in a unique way. Possible realizations are the magnetic resonance force microscopy (MRFM) [562], where the local composition can be analyzed by the NMR-signal (cf. Chap. 4). Alternatively, the combination of probe microscopy with mass spectroscopy [687] or photoemission microscopy with a scanning aperture [442] appear promising (cf. Chap. 5). The use of optical spectroscopy of SNOM or the inelastic tunneling spectroscopy [621] are other methods which are candidates to perform local chemical analysis.

7.6   Local Modification Experiments

Another challenge appears to be local manipulation by probe microscopy. Previous work from Eigler and coworkers has shown that single atom manipulation by STM is possible. At present, attempts are made to perform manipulation of atoms on insulators. Sugawara et al. could demonstrate single atom manipulation by DFM on a semiconductor [487]. Gnecco et al. [220] performed wear experiments on insulators, where the debris was found to be well ordered. Thus, relatively complex structures with low defect density can be created by force microscopy. The physical and chemical properties of such artificial structures are of great scientific interest and may also lead to device fabrication. The probing tip could also be designed to perform more complex tasks. Examples are the nanotweezers [349], where clusters where picked up by the nanotweezer and electrical characterization of the nanostructures was performed. Apart from the movement of individual atoms or molecules, it appears also of interest to induce self-directed growth by the probing tip. For example, it is possible to remove single hydrogen atoms from hydrogen passivated surfaces [183]. These artificially created defects may then act as nucleation sites for the growth of molecular wires [401].

In conclusion, probe microscopes are going to overcome the limits of microscopy and will become tools for nanoscience and nanotechnology. The construction of complex structures, such as transistors, logical elements composed of atoms or molecules, or complete memory devices will be envisaged. As proposed by Feynman, small robots may be constructed by these tools, which will start to fabricate other structures. In this perspective, the current age may be called the stone-age of nanotechnology, where instruments, such as STM or AFM, were developed to perform simple modification experiments. In future, much more complex tasks will be settled on the nanometer scale, which may lead to a new era of technology.

Bibliography

1. Attocube systems ag, eglfinger weg 2, 85540 haar germany. Customized Microscopes, https://www.attocube.com/en/products/microscopes/customized-low-temperature-microscopes/milli-kelvin-challenge

2. Report Horizon 2020 MAGicSky 2016, hal-01-368830. Report on imaging of individual skyrmions of MML systems made by different techniques

3. Nt-mdt, the netherlands, hoenderparkweg 96 b, 7335 gx apeldoorn. Measuring in vacuum and controlled atmosphere, https://www.ntmdt-si.com/resources/applications/measuring-in-vacuum-and-controlled-atmosphere

4. Nanoscan ag, hermetschloostrasse 77, ch-8048 zürich, switzerland. VLS-80: High-end vacuum SPM technology, https://www.iontof.com/vls80

5. Park systems, kanc 15f, gwanggyo-ro 109, suwon 16229, korea. Park NX-Hivac, https://parksystems.com/products/small-sample-afm/park-nx-hivac

6. Some commercial software companies: Image Metrology ApS, Nils Koppels All 402,DK-2800 Lyngby, Denmark, e-mail: info@imagemet.com, http://www.imagemet.com NT-MDT, State Research Institute of Physical Problems, 103460 Moscow, Russia, e-mail: spm@ntmdt.zgrad.ru, http://www.ntmdt.ru

7. Some of the manufacturers of cantilevers (2020): Adama innovations, Dublin, Ireland, www.adama.tips Apex Probes, Bracknell, United Kingdom, www.apexprobes.uk Budget Sensors, Innovative Solutions Bulgaria, Sofia, Bulgaria, www-budgetsensors.com Bruker AFM Probes, Camarillo, USA, https://www.brukerafmprobes.com/ Mikromash AG, Sofia, Bulgaria, https://www.spmtips.com NanoAndMore GmbH, Wetzlar, Germany, https://www.nanoandmore.com/ NANOSENSORS, Neuchatel, Switzerland, https://www.nanosensors.com/ Nanonics Imaging, Jerusalem, Israel, http://www.nanonics.co.il Nanoscience, Phoenix, USA, www.nanoscience.com NanoTOOLS GmbH http://nano-tools.com Nanoworld, Neuchatel, Switzerland, www.nanoworld.com NT-MDT, Moscow, Russia, www.ntmdt-tips.ru NuNano, Bristol, United Kingdom, www.nunano.com Olympus Micro Cantilevers, Tokyo, Japan, www.probe.olympus-global.com SCL Sensortech, Vienna, Austria, www.sclsensortech.com Tipsnano,Tallinn, Estonia, www.tipsnano.com

8. Team Nanotec GmbH. Wilhelm-Schickard-Str. 10, 78052 Villingen-Schwenningen Germany, www.team-nanotec.de

9. ThermoMicroscopes, 1171 Borregas Avenue, Sunnyvale, CA 94089 Phone: (408) 747-1600 Fax: (408) 747-1601 info@thermomicro.com, www.thermomicro.com

© Springer Nature Switzerland AG 2021
E. Meyer et al., *Scanning Probe Microscopy*, Graduate Texts in Physics,
https://doi.org/10.1007/978-3-030-37089-3

10. Nanosensors, IMO Building, Im Amtmann 6, D-35578 Wetzlar, Germany, info@nanosensor.com, http://www.nanosensors.com
11. *Proceedings of the Fifth International Conference on Near Field Optics and Related Techniques (NFO-5)*, volume 194 of *Journal of Microscopy* (1999). Blackwell Science Ltd
12. A. Abdurixit, A. Baratoff, E. Meyer, Molecular dynamics simulations of dynamic force microscopy: applications to the Si (111) 7 x 7 surface. Appl. Surf. Sci. **157**(4), 355–360 (2000). ISSN 0169-4332. https://doi.org/10.1016/S0169-4332(99)00566-8, http://www.sciencedirect.com/science/article/pii/S0169433299005668
13. M. Abe, Y. Sugimoto, T. Namikawa, K. Morita, N. Oyabu, S. Morita, Drift-compensated data acquisition performed at room temperature with frequency modulation atomic force microscopy. Appl. Phys. Lett. **90**(20), 203103 (2007). ISSN 0003-6951. https://doi.org/10.1063/1.2739410. Publisher: American Institute of Physics
14. D.W. Abraham, F. Alan McDonald, Theory of magnetic force microscope images. Appl. Phys. Lett. **56**(12), 1181–1183 (1990). ISSN 0003-6951. https://doi.org/10.1063/1.102554. Publisher: American Institute of Physics
15. D.W. Abraham, C.C. Williams, H.K. Wickramasinghe, Measurement of in-plane magnetization by force microscopy. Appl. Phys. Lett. **53**(15), 1446–1448 (1988). ISSN 0003-6951. https://doi.org/10.1063/1.99964. Publisher: American Institute of Physics
16. M. Achermann, B.A. Nechay, U. Siegner, A. Hartmann, D. Oberli, E. Kapon, U. Keller, Quantization energy mapping of single v-groove gaas quantum wires by femtosecond near-field optics. Appl. Phys. Lett. **76**, 2695 (2000). https://doi.org/10.1063/1.126446
17. A.W. Adamson, *Physical Chemistry of Surfaces* (Wiley, New York, 1976). ISBN 978-0-471-00794-4
18. S.G. Addepalli, B. Ekstrom, N.P. Magtoto, J.S. Lin, J.A. Kelber, Stm atomic-scale characterization of the gamma al2o3 film on ni3al(111). Surf. Sci. **442**, 385 (1999). https://doi.org/10.1016/S0039-6028(99)00951-6
19. J.P. Aimé, R. Boisgard, L. Nony, G. Couturier, Nonlinear dynamic behavior of an oscillating tip-microlever system and contrast at the atomic scale. Phys. Rev. Lett. **82**(17), 3388 (1999). https://doi.org/10.1103/PhysRevLett.82.3388, https://link.aps.org/doi/10.1103/PhysRevLett.82.3388
20. T. Akiyama, S. Gautsch, N.F. de Roij, U. Staufer, P. Nigdermann, L. Howald, D. Muller, A. Tonin, H.-R. Hidber, W.T. Pike, M.H. Hedht, Atomic force microscope for planetary applications. Sens. Actuat. A (Phys.) **91**(3), 321 (2001). https://doi.org/10.1016/S0924-4247(01)00602-1
21. B.J. Albers, T.C. Schwendemann, M.Z. Baykara, N. Pilet, M. Liebmann, E.I. Altman, U.D. Schwarz, Three-dimensional imaging of short-range chemical forces with picometre resolution. Nature Nanotech. **4**, 307 (2009). https://doi.org/10.1038/nnano.2009.57
22. T.R. Albrecht, C.F. Quate, Atomic resolution imaging of a nonconductor by atomic force microscopy. J. Appl. Phys. **62**(7), 2599–2602 (1987). https://doi.org/10.1063/1.339435
23. T.R. Albrecht, P. Grütter, D. Horne, D. Rugar, Frequency modulation detection using high-q cantilevers for enhanced force microscope sensitivity. J. Appl. Phys. **69**, 668 (1991). https://doi.org/10.1063/1.347347
24. S. Alexander, L. Hellemans, O. Marti, J. Schneir, V. Elings, P.K. Hansma, M. Longmire, J. Gurley, An atomic resolution atomic force microscope implemented using an optical lever. J. Appl. Phys. **65**, 164 (1989). https://doi.org/10.1063/1.342563
25. B. Alldritt, P. Hapala, N. Oinonen, F. Urtev, O. Krejci, F. Federici Canova, J. Kannala, F. Schulz, P. Liljeroth, A.S. Foster, Automated structure discovery in atomic force microscopy. Sci. Adv. **6**(9) (2020). https://doi.org/10.1126/sciadv.aay6913, https://advances.sciencemag.org/content/6/9/eaay6913. Publisher: American Association for the Advancement of Science _eprint: https://advances.sciencemag.org/content/6/9/eaay6913.full.pdf
26. M.J. Allen, N.V. Hud, M. Balooch, R.J. Tench, W.J. Siekhaus, R. Balhorn, Tip-radius-induced artifacts in AFM images of protamine-complexed DNA fibers. Ultramicroscopy **42–44**, 1095–1100 (1992). ISSN 0304-3991. https://doi.org/10.1016/0304-3991(92)90408-C, http://www.sciencedirect.com/science/article/pii/030439919290408C

27. W. Allers, A. Schwarz, U.D. Schwarz, R. Wiesendanger, A scanning force microscope with atomic resolution in ultrahigh vacuum and at low temperatures. Rev. Sci. Instr. **69**, 221 (1998)
28. W. Allers, A. Schwarz, U.D. Schwarz, R. Wiesendanger, Dynamic scanning force microscopy at low temperatures on a noble-gas crystal: atomic resolution on the xenon(111) surface. Europhys. Lett. (EPL) **48**(3), 276–279 (1999). https://doi.org/10.1209/epl/i1999-00477-3
29. S.F. Alvarado, P. Renaud, Observation of spin-polarized-electron tunneling from a ferromagnet into gaas. Phys. Rev. Lett. **68**, 1387–1390 (1992). https://doi.org/10.1103/PhysRevLett.68.1387
30. N. Amos, R. Ikkawi, R. Haddon, D. Litvinov, S. Khizroev, Controlling multidomain states to enable sub-10-nm magnetic force microscopy. Appl. Phys. Lett. **93**(20), (2008). https://doi.org/10.1063/1.3036533
31. T. Ando, High speed atomic force microscopy coming of age. Nanotechnology **23**(6) (2012). https://doi.org/10.1088/0957-4484/23/6/062001
32. T. Ando, N. Kodera, E. Takai, D. Maruyama, K. Saito, A. Toda, A high-speed atomic force microscope for studying biological macromolecules. Proc. Natl. Acad. Sci. **98**(22), 12468–12472 (2001). ISSN 0027-8424. https://doi.org/10.1073/pnas.211400898, https://www.pnas.org/content/98/22/12468. Publisher: National Academy of Sciences _eprint: https://www.pnas.org/content/98/22/12468.full.pdf
33. L. Angeloni, D. Passeri, M. Reggente, D. Mantovani, M. Rossi, Removal of electrostatic artifacts in magnetic force microscopy by controlled magnetization of the tip: application to superparamagnetic nanoparticles. Sci. Rep. **6**(1), 26293 (2016). ISSN 2045-2322. https://doi.org/10.1038/srep26293
34. P. Appel, B.J. Shields, T. Kosub, N. Hedrich, R. Hübner, J. Faßbender, D. Makarov, P. Maletinsky, Nanomagnetism of magnetoelectric granular thin-film antiferromagnets. Nano Lett. **19**(3), 1682–1687 (2019). ISSN 1530-6984. https://doi.org/10.1021/acs.nanolett.8b04681. Publisher: American Chemical Society
35. A. Aviram, M.A. Ratner, Molecular rectifiers. Chem. Phys. Lett. **29**, 277 (1974). https://doi.org/10.1016/0009-2614(74)85031-1
36. R. Ph Avouris, Wolkow, Scanning tunneling microscopy of insulators: Caf2 epitaxy on si (111). Appl. Phys. Lett. **55**(11), 1074–1076 (1989). https://doi.org/10.1063/1.102457
37. I.-W. Ph Avouris, R.E. Walkup, Lyo, Real space imaging of electron scattering phenomena at metal surfaces. J. Vac. Sci. Technol. B **12**, 1447 (1994). https://doi.org/10.1116/1.587314
38. M. Baćani, M.A. Marioni, J. Schwenk, H.J. Hug, How to measure the local Dzyaloshinskii-Moriya Interaction in Skyrmion Thin-Film Multilayers. Sci. Rep. **9**(1), 3114 (2019). ISSN 2045-2322. https://doi.org/10.1038/s41598-019-39501-x
39. T. Bachels, R. Schäfer, Formation enthalpies of sn clusters: a calorimetric investigation. Chem. Phys. Lett. **300**, 177 (1999). https://doi.org/10.1016/S0009-2614(98)01376-1
40. G. Balasubramanian, P. Neumann, D. Twitchen, M. Markham, R. Kolesov, N. Mizuochi, J. Isoya, J. Achard, J. Beck, J. Tissler, V. Jacques, P.R. Hemmer, F. Jelezko, J. Wrachtrup, Ultralong spin coherence time in isotopically engineered diamond. Nat. Mater. **8**(5), 383–387 (2009). ISSN 1476-4660. https://doi.org/10.1038/nmat2420
41. M.L.M. Balistreri, J.P. Korterik, G.J. Veldhuis, L. Kuipers, N.F. van Hulst, Quantitative photon tunneling and shear-force microscopy of planar waveguide splitters and mixers. J. Appl. Phys. **89**(6), 3307 (2001). https://doi.org/10.1063/1.1347952
42. M. Bammerlin, R. Luthi, E. Meyer, A. Baratoff, J. Lu, M. Guggisberg, C. Loppacher, C. Gerber, H.J. Güntherodt, Dynamic SFM with true atomic resolution on alkali halide surfaces. Appl. Phys. A Mater. Sci. & Proc. **66**(Part 1 Suppl. S), S293–S294 (1998). ISSN 1432-0630, http://edoc.unibas.ch/dok/A5839459. Publisher: Springer-Verlag
43. G. Bar, R. Brandsch, M.H. Whangbo, Effect of tip sharpness on the relative contributions of attractive and repulsive forces in the phase imaging of tapping mode atomic force microscopy. Surf. Sci. **422**(1), L192–L199 (1999). ISSN 0039-6028. https://doi.org/10.1016/S0039-6028(98)00899-1, http://www.sciencedirect.com/science/article/pii/S0039602898008991

44. G. Bar, R. Brandsch, M.-H. Whangbo, Description of the frequency dependence of the amplitude and phase angle of a silicon cantilever tapping on a silicon substrate by the harmonic approximation. Surf. Sci. **411**(1), L802–L809 (1998). ISSN 0039-6028. https:// doi.org/10.1016/S0039-6028(98)00348-3, http://www.sciencedirect.com/science/article/pii/ S0039602898003483

45. G. Bar, R. Brandsch, M.H. Whangbo, Correlation between frequency-sweep hysteresis and phase imaging instability in tapping mode atomic force microscopy. Surf. Sci. **436**(1), L715–L723 (1999). ISSN 0039-6028. https://doi.org/10.1016/S0039-6028(99)00702-5, http:// www.sciencedirect.com/science/article/pii/S0039602899007025

46. G. Bar, R. Brandsch, M. Bruch, L. Delineau, M.-H. Whangbo, Examination of the relationship between phase shift and energy dissipation in tapping mode atomic force microscopy by frequency-sweep and force-probe measurements. Surf. Sci. **444**(1), L11–L16 (2000). ISSN 0039-6028. https://doi.org/10.1016/S0039-6028(99)00975-9, http://www.sciencedirect.com/ science/article/pii/S0039602899009759

47. A. Baratoff, Theory of scanning tunneling microscopy - methods and approximations. Phys. B **127**, 143 (1984). https://doi.org/10.1016/S0378-4363(84)80022-4

48. A.J. Bard, F.R.F. Fan, J. Kwak, O. Lev, Scanning electrochemical microscopy. Introduction and principles. Anal. Chem. **61**(2), 132–138 (1989). ISSN 0003-2700, 1520-6882. https://doi. org/10.1021/ac00177a011, https://pubs.acs.org/doi/abs/10.1021/ac00177a011

49. J. Bardeen, Tunneling from a many-body point of view. Phys. Rev. Lett. **6**, 57 (1961). https:// doi.org/10.1103/PhysRevLett.6.57

50. I. Barel, M. Urbakh, L. Jansen, A. Schirmeisen, Multibond dynamics of nanoscale friction: the role of temperature. Phys. Rev. Lett. **104** (2010). https://doi.org/10.1103/PhysRevLett. 104.066104

51. R.C. Barrett, C.F. Quate, Charge storage in a nitride oxide silicon medium by scanning capacitance microscopy. J. Appl. Phys. **70**(5), 2725–2733 (1991). https://doi.org/10.1063/1.349388

52. L. Bartels, G. Meyer, K.-H. Rieder, Basic steps of lateral manipulation of single atoms and diatomic clusters with a scanning tunneling microscope tip. Phys. Rev. Lett. **79**, 697–700 (1997). https://doi.org/10.1103/PhysRevLett.79.697

53. C. Barth, M. Reichling, Imaging the atomic arrangements on the high temperature reconstructed α-Al_2O_3-surface. Nat. Commun. **414**, 54 (2001). https://doi.org/10.1038/35102031

54. M.Z. Baykara, M. Todorović, H. Mönig, T.C. Schwendemann, Ö. Ünverdi, L. Rodrigo, E.I. Altman, R. Pérez, U.D. Schwarz, Atom-specific forces and defect identification on surface-oxidized cu(100) with combined 3d-afm and stm measurements. Phys. Rev. B **87** (2013). https://doi.org/10.1103/PhysRevB.87.155414

55. R.S. Becker, J.A. Golovchenko, B.S. Swartzentruber, Electron interferometry at crystal surfaces. Phys. Rev. Lett. **55**, 987–990 (1985). https://doi.org/10.1103/PhysRevLett.55.987

56. R.J. Behm, Scanning tunneling microscopy: metal surfaces, adsorption and surface reactions, in *Scanning Tunneling Microscopy and Related Methods*, ed. by H. Rohrer R.J. Behm, N. Garcia (Kluwer Academic Press, Dordrecht, 1990), p. 173. https://doi.org/10.1007/978-94-015-7871-4

57. O.P. Behrend, L. Odoni, J.L. Loubet, N.A. Burnham, Phase imaging: deep or superficial? Appl. Phys. Lett. **75**(17), 2551–2553 (1999). https://doi.org/10.1063/1.125074

58. L. Belliard, A. Thiaville, S. Lemerle, A. Lagrange, J. Ferré, J. Miltat, Investigation of the domain contrast in magnetic force microscopy. J. Appl. Phys. **81**(8), 3849–3851 (1997). ISSN 0021-8979. https://doi.org/10.1063/1.364730. Publisher: American Institute of Physics

59. L.M. Belova, O. Hellwig, E. Dobisz, E. Dan Dahlberg, Rapid preparation of electron beam induced deposition Co magnetic force microscopy tips with 10 nm spatial resolution. Rev. Sci. Instrum. **83**(9), 093711 (2012). ISSN 0034-6748. https://doi.org/10.1063/1.4752225. Publisher: American Institute of Physics

60. A. Benassi, M.A. Marioni, D. Passerone, H.J. Hug, Role of interface coupling inhomogeneity in domain evolution in exchange bias. Sci. Rep. **4**(1), 4508 (2014). ISSN 2045-2322. https:// doi.org/10.1038/srep04508

61. R. Bennewitz, E. Meyer, *Proceedings of the Workshop on Dynamic Force Microscopy at Ultrasonic Frequencies*, volume 27 of Surface and Interface Analysis (Wiley, Chichester, 1982). https://doi.org/10.1002/(SICI)1096-9918(199905/06)27:5/6<285::AID-SIA577>3.0. CO;2-H

62. R. Bennewitz, M. Reichling, E. Matthias, Force microscopy of cleaved and electron-irradiated CaF2(111) surfaces in ultra-high vacuum. Surf. Sci. **387**(1), 69–77 (1997). ISSN 0039-6028. https://doi.org/10.1016/S0039-6028(97)00268-9, http://www.sciencedirect.com/science/article/pii/S0039602897002689

63. R. Bennewitz, T. Gyalog, M. Guggisberg, M. Bammerlin, E. Meyer, H.-J. Güntherodt, Atomic-scale stick-slip processes on Cu(111). Phys. Rev. B **60**, R11301 (1999). https://doi.org/10.1103/PhysRevB.60.R11301

64. R. Bennewitz, A.S. Foster, L.N. Kantorovich, M. Bammerlin, Ch. Loppacher, S. Schär, M. Guggisberg, E. Meyer, A.L. Shluger, Atomically resolved edges and kinks of nacl islands on cu(111): experiment and theory. Phys. Rev. B **62**, 2074–2084 (2000). https://doi.org/10.1103/PhysRevB.62.2074

65. R. Bennewitz, C. Gerber, E. Meyer (eds.), *Proceedings of the Second International Workshop on Noncontact Atomic Force Microscopy*, volume 157 of *Applied Surface Science* (Elsevier, Amsterdam, 2000)

66. R. Bennewitz, S. Schär, V. Barwich, O. Pfeiffer, E. Meyer, F. Krok, B. Such, J. Kolodzej, M. Szymonski, Atomic-resolution images of radiation damage in KBr. Surf. Sci. **474**(1), L197–L202 (2001). ISSN 0039-6028. https://doi.org/10.1016/S0039-6028(00)01053-0, http://www.sciencedirect.com/science/article/pii/S0039602800010530

67. R. Bennewitz, J.N. Crain, A. Kirakosia, J.-L. Lin, J.L. Mc Chesney, D.Y. Petrovykh, F.J. Himpsel, Atomic scale memory at a silicon surface. Nanotechnology **13**, 495 (2002). https://doi.org/10.1088/0957-4484/13/4/312, https://iopscience.iop.org/article/10.1088/0957-4484/13/4/312

68. R. Bennewitz, O. Pfeiffer, S. Schär, V. Barwich, E. Meyer, Atomic corrugation in nc-AFM of alkali halides. Appl. Surf. Sci. **188**, 232 (2002). https://doi.org/10.1016/S0169-4332(01)00910-2

69. R. Bennewitz, S. Schär, E. Gnecco, O. Pfeiffer, M. Bammerlin, E. Meyer, Atomic structure of alkali halide surfaces. Appl. Phys. A **78**(6), 837–841 (2004). ISSN 0947-8396, 1432-0630. https://doi.org/10.1007/s00339-003-2439-3, http://link.springer.com/10.1007/s00339-003-2439-3

70. R. Berger, E. Delamarche, H.P. Lang, Ch. Gerber, J.K. Gimzewski, E. Meyer, H.-J. Güntherodt, Surface stress in the self-assembly of alkanethiols on gold probed by a force microscopy technique. Appl. Phys. A **66**, 55 (1998). https://doi.org/10.1007/s003390051099

71. R. Berger, H.P. Lang, Ch. Gerber, J.K. Gimzewski, J.H. Fabian, L. Scandella, E. Meyer, H.-J. Güntherodt, Micromechanical Thermogravimetry. Chem. Phys. Lett. **294**(4–5), 363 (1998)

72. R. Berndt, Photon emission from the scanning tunneling microscope, in *Scanning Probe Microscopy, Analytical Methods*, ed. by R. Wiesendanger (Berlin, 1998), p. 97. https://doi.org/10.1007/978-3-662-03606-8

73. R. Berndt, J.K. Gimzewski, P. Johansson, Electromagnetic interactions of metallic objects in nanometer proximity. Phys. Rev. Lett. **71**, 3493–3496 (1993)

74. R. Berndt, R. Gaisch, W.D. Schneider, J.K. Gimzewski, B. Reihl, R.R. Schlittler, M. Tschudy, Atomic resolution in photon emission induced by a scanning tunneling microscope. Phys. Rev. Lett. **74**, 102–105 (1995). https://doi.org/10.1103/PhysRevLett.74.102

75. R. Berndt, J.K. Gimzewski, P. Johansson, Inelastic tunneling excitation of tip-induced plasmon modes on noble-metal surfaces. Phys. Rev. Lett. **67**, 3796–3799 (1991). https://doi.org/10.1103/PhysRevLett.67.3796

76. H. Th Bertrams, Neddermeyer, Growth of nio(100) on ag(100): characterization by scanning tunneling microscopy. J. Vac. Sci. Technol. B **14**(2), 1141 (1996). https://doi.org/10.1116/1.588416

77. C. Beuret, T. Akiyama, U. Staufer, N.F. de Rooij, P. Niedermann, W. Hanni, Conical diamond tips realized by a double-molding process for high-resolution profilometry and atomic force. Appl. Phys. Lett. **76**(12), 1621 (2000). https://doi.org/10.1063/1.126115

78. A. Bietsch, M.A. Schneider, M.E. Welland, B. Michel, Electrical testing of gold nanostructures by conducting atomic force microscopy. J. Vac. Sci. & Technol. B: Microelectron. Nanometer Struct. Proc. Meas. Phen. **18**(3), 1160–1170 (2000). https://doi.org/10.1116/1.591353, https://avs.scitation.org/doi/abs/10.1116/1.591353

79. V.T. Binh, N. Garcia, On the electron and metallic ion emission from nanotips fabricated by field-surface-melting technique: experiments on w and au tips. Ultramicroscopy **4244**, 80 (1992). https://doi.org/10.1016/0304-3991(92)90249-J

80. G. Binnig, H. Rohrer, Scanning tunneling microscopy. Helv. Phys. Acta **55**, 726–735 (1982), https://www.e-periodica.ch/digbib/view?pid=hpa-001:1982:55#728

81. G. Binnig, H. Rohrer. Scanning tunneling microscopy. Surf. Sci. **126**(1), 236–244 (1983). ISSN 0039-6028. https://doi.org/10.1016/0039-6028(83)90716-1, http://www.sciencedirect.com/science/article/pii/0039602883907161

82. G. Binnig, H. Rohrer, Scanning tunneling microscopy - from birth to adolescence. Rev. Mod. Phys. **56**, 615 (1987). https://doi.org/10.1103/RevModPhys.59.615

83. G. Binnig, H. Rohrer, Ch. Gerber, E. Weibel, Surface studies by scanning tunneling microscopy. Phys. Rev. Lett. **49**, 57–61 (1982). https://doi.org/10.1103/PhysRevLett.49.57

84. G. Binnig, H. Rohrer, Ch. Gerber, E. Weibel, Tunneling through a controllable vacuum gap. Appl. Phys. Lett. **40**(2), 178–180 (1982). https://doi.org/10.1063/1.92999

85. G. Binnig, H. Rohrer, Ch. Gerber, E. Weibel, 7×7 reconstruction on Si(111) resolved in real space. Phys. Rev. Lett. **50**, 120–123 (1983). https://doi.org/10.1103/PhysRevLett.50.120

86. G. Binnig, C.F. Quate, Ch. Gerber, Atomic force microscope. Phys. Rev. Lett. **56**, 930–933 (1986). https://doi.org/10.1103/PhysRevLett.56.930

87. G. Binnig, Ch. Gerber, E. Stoll, T.R. Albrecht, C.F. Quate, Atomic resolution with atomic force microscope. Europhys. Lett. (EPL) **3**(12), 1281–1286 (1987). ISSN 0295-5075. https://doi.org/10.1209/0295-5075/3/12/006. Publisher: IOP Publishing

88. N. Blanc, J. Brugger, N.F. de Rooij, U. Dürig, Scanning force microscopy in the dynamic mode using microfabricated capacitive sensors. J. Vac. Sci. Technol. B **14**, 901 (1996). https://doi.org/10.1116/1.589171

89. G. Bochi, H.J. Hug, D.I. Paul, B. Stiefel, A. Moser, I. Parashikov, H.J. Güntherodt, R.C. O'Handley, Magnetic domain structure in ultrathin films. Phys. Rev. Lett. **75**, 1839–1842 (1995). https://doi.org/10.1103/PhysRevLett.75.1839

90. M. Bode, M. Getzlaff, R. Wiesendanger, Spin-polarized vacuum tunneling into the exchange-split surface state of gd(0001). Phys. Rev. Lett. **81**, 4256–4259 (1998). https://doi.org/10.1103/PhysRevLett.81.4256

91. M. Bode, S. Heinze, A. Kubetzka, O. Pietzsch, X. Nie, G. Bihlmayer, S. Blügel, R. Wiesendanger, Magnetization-direction-dependent local electronic structure probed by scanning tunneling spectroscopy. Phys. Rev. Lett. **89** (2002). https://doi.org/10.1103/PhysRevLett.89.237205

92. T. Bouhacina, J.P. Aimé, S. Gauthier, D. Michel, V. Heroguez, Tribological behavior of a polymer grafted on silanized silica probed with a nanotip. Phys. Rev. B **56**, 7694–7703 (1997)

93. A. Bouhelier, T. Huser, H. Tamaru, H.-J. Güntherodt, D.W. Pohl, F.I. Baida, D. Van Labeke, Plasmon optics of structured silver films. Phys. Rev. B **63**(15), (2001). https://doi.org/10.1103/PhysRevB.63.155404, https://link.aps.org/doi/10.1103/PhysRevB.63.155404

94. N.A. Burnham, R.J. Colton, H.M. Pollock, Work-function anisotropies as an origin of long-range surface forces. Phys. Rev. Lett. **69**(1), 144 (1992). https://doi.org/10.1103/PhysRevLett.69.144

95. C. Bustamante, S.B. Smith, J. Liphardt, D. Smith, Single molecule studies of dna mechanics. Curr. Opin. Struct. Biol. **10**, 279 (2000). https://doi.org/10.1016/s0959-440x(00)00085-3

96. H.J. Butt, M. Jaschke, Calculation of thermal noise in atomic force microscopy. Nanotechnology **6**(1), 1 (1995). https://doi.org/10.1088/0957-4484/6/1/001

97. V. Bykov, A. Gologanov, V. Shevyakov, Test structure for spm tip shape deconvolution. Appl. Phys. A **66**, 499 (1998). https://doi.org/10.1007/s003390050703

98. P.F. Carcia, Perpendicular magnetic anisotropy in Pd/Co and Pt/Co thin film layered structures. J. Appl. Phys. **63**(10), 5066–5073 (1988). ISSN 0021-8979. https://doi.org/10.1063/1.340404. Publisher: American Institute of Physics

99. R.W. Carpick, D.F. Ogletree, M. Salmeron, Lateral stiffness: a new nanomechanical measurement for the determination of shear strengths with friction force microscopy. Appl. Phys. Lett. **70**(12), 1548–1550 (1997). https://doi.org/10.1063/1.118639

100. F.L. Carter, *Molecular Electronic Devices* (Marcel Dekker, New York, 1982). ISBN 10: 0824716760 or 13: 9780824716769

101. A. Casiraghi, H. Corte-León, M. Vafaee, F. Garcia-Sanchez, G. Durin, M. Pasquale, G. Jakob, M. Kläui, O. Kazakova, Individual skyrmion manipulation by local magnetic field gradients. Commun. Phys. **2**(1), 145 (2019). ISSN 2399-3650. https://doi.org/10.1038/s42005-019-0242-5

102. J. Červenka, M.I. Katsnelson, C.F.J. Flipse, Room-temperature ferromagnetism in graphite driven by two-dimensional networks of point defects. Nat. Phys. **5**(11), 840–844 (2009). ISSN 1745-2481. https://doi.org/10.1038/nphys1399

103. A.M. Chang, H.D. Hallen, L. Harriott, H.F. Hess, H.L. Kao, J. Kwo, R.E. Miller, R. Wolfe, J. van der Ziel, T.Y. Chang, Scanning Hall probe microscopy. Appl. Phys. Lett. **61**(16), 1974–1976 (1992). ISSN 0003-6951. https://doi.org/10.1063/1.108334. Publisher: American Institute of Physics

104. T. Chang, M. Lagerquist, J. Zhu, J.H. Judy, P.B. Fischer, S.Y. Chou, Deconvolution of magnetic force images by fourier analysis. IEEE Trans. Magn. **28**(5), 3138–3140 (1992). ISSN 1941-0069. https://doi.org/10.1109/20.179737

105. C. Chappert, P. Bruno, Magnetic anisotropy in metallic ultrathin films and related experiments on cobalt films (invited). J. Appl. Phys. **64**(10), 5736–5741 (1988). ISSN 0021-8979. https://doi.org/10.1063/1.342243. Publisher: American Institute of Physics

106. S. Chatraphorn, E.F. Fleet, F.C. Wellstood, L.A. Knauss, T.M. Eiles, Scanning SQUID microscopy of integrated circuits. Appl. Phys. Lett. **76**(16), 2304–2306 (2000). ISSN 0003-6951. https://doi.org/10.1063/1.126327. Publisher: American Institute of Physics. https://doi.org/10.1103/PhysRevLett.105.217402

107. C.C. Chen, P. Chu, C.A. Bobisch, D.L. Mills, W. Ho, Viewing the interior of a single molecule: Vibronically resolved photon imaging at submolecular resolution. Phys. Rev. Lett. **105** (2010)

108. C.J. Chen, Effect of m $\neq$ 0 tip states in scanning tunneling microscopy: the explanations of corrugation reversal. Phys. Rev. Lett. **69**, 1656 (1992). https://doi.org/10.1103/PhysRevLett.69.1656

109. C.J. Chen, *Introduction to Scanning Tunneling Microscopy* (Oxford University Press, Oxford, 2007). https://doi.org/10.1093/acprof:oso/9780199211500.001.0001

110. X. Chen, M.C. Davies, C.J. Roberts, S.J.B. Tendler, P.M. Williams, N.A. Burnham, Optimizing phase imaging via dynamic force curves. Surf. Sci. **460**(1), 292–300 (2000). ISSN 0039-6028. https://doi.org/10.1016/S0039-6028(00)00574-4, http://www.sciencedirect.com/science/article/pii/S0039602800005744

111. Y.J. Chen, I.H. Wilson, C.S. Lee, J.B. Xu, M.L. Yu, Tip artifacts in atomic force microscope imaging of ion bombarded nanostructures on germanium surfaces. J. Appl. Phys. **82**(11), 5859–5861 (1997). https://doi.org/10.1063/1.366454

112. Y. Chen, Y. Li, G. Shan, Y. Zhang, Z. Wang, M. Wang, H. Li, J. Qian, Design and implementation of a novel horizontal AFM probe utilizing a quartz tuning fork. Int. J. Prec. Eng. Manuf. **19**(1), 39–46 (2018). ISSN 2005-4602. https://doi.org/10.1007/s12541-018-0005-3

113. B.M. Chernobrod, G.P. Berman, Spin microscope based on optically detected magnetic resonance. J. Appl. Phys. **97**(1), 014903 (2004). ISSN 0021-8979. https://doi.org/10.1063/1.1829373. Publisher: American Institute of Physics

114. S. Chiang, Scanning tunneling microscopy imaging of small adsorbed molecules on metal surfaces in an ultrahigh vacuum environment. Chem. Rev. **97**, 1083 (1997). https://doi.org/10.1021/cr940555a

115. S. Chiang, R.J. Wilson, C.M. Mate, H. Ohtani, Real space imaging of co-adsorbed CO and benzene molecules on Rh(111). J. Microsc. **152**(2), 567–571 (1988). ISSN 1365-2818. https://doi.org/10.1111/j.1365-2818.1988.tb01422.x, https://onlinelibrary.wiley.com/doi/abs/10.1111/j.1365-2818.1988.tb01422.x. _eprint: https://onlinelibrary.wiley.com/doi/pdf/10.1111/j.1365-2818.1988.tb01422.x

116. S. Chikazumi, E. Robert, *Physics of Magnetism* (Krieger Publishing Company, Malabar, 1964). ISBN 10: 0882756621 or 13: 9780882756622

117. W.K. Chim, K.M. Wong, Y.L. Teo, Y. Lei, Y.T. Yeow, Dopant extraction from scanning capacitance microscopy measurements of p-n junctions using combined inverse modeling and forward simulation. Appl. Phys. Lett. **80**(25), 4837–4839 (2002). ISSN 0003-6951. https://doi.org/10.1063/1.1487899. Publisher: American Institute of Physics

118. J. Chu, T. Itoh, C. Lee, T. Suga, K. Watanabe, Frequency modulation detection high vacuum scanning force microscope with a self-oscillating piezoelectric cantilever. J. Vac. Sci. & Technol. B: Microelectron. Nanometer Struct. Proc. Meas. Phen. **15**(5), 1647–1651 (1997). https://doi.org/10.1116/1.589565, https://avs.scitation.org/doi/abs/10.1116/1.589565

119. B.W. Chuia, T.W. Kenny, H.J. Mamin, B.D. Terris, D. Rugar, Independent detection of vertical and lateral forces with a sidewall-implanted dual-axis piezoresistive cantilever. Appl. Phys. Lett. **72**(11), 1388 (1998). https://doi.org/10.1063/1.121064

120. S. Ciraci, E. Tekman, A. Baratoff, I.P. Batra, Theoretical study of short- and long-range forces and atom transfer in scanning tunneling microscopy. Phys. Rev. B **46**(16), 10411 (1992). https://doi.org/10.1103/physrevb.46.10411

121. J. Clarke, Principles and applications of SQUIDs. Proc. IEEE **77**(8), 1208–1223 (1989). https://doi.org/10.1109/5.34120

122. T. Clarysse, M. Caymax, P. De Wolf, T. Trenkler, W. Vandervorost, J.S. McMurray, J. Kim, C.C. Williams, J.G. Clark, G. Neubauer, Epitaxial staircase structure for the calibration of electrical characterization techniques. J. Vac. Sci. Technol. B **16**(1), 394 (1998). https://doi.org/10.1116/1.589820

123. W. Clauss, J. Zhang, D.J. Bergeron, A.T. Johnson, Application and calibration of a quartz needle sensor for high resolution scanning force microscopy. J. Vac. Sci. Technol. B **17**, 1309 (1999). https://doi.org/10.1116/1.590751

124. J.P. Cleveland, T.E. Schäffer, and P.K. Hansma. Probing oscillatory hydration potentials using thermal-mechanical noise in an atomic force microscope. Phys. Rev. B **52**, R8692 (1995). ISSN 0163-1829. https://doi.org/10.1103/physrevb.52.r8692

125. J.P. Cleveland, B. Anczykowski, A.E. Schmid, V.B. Elings, Energy dissipation in tapping-mode atomic force microscopy. Appl. Phys. Lett. **72**, 2613 (1998). https://doi.org/10.1063/1.121434

126. J. Colchero, O. Marti, J. Mlynek, Friction on an atomic scale, vol. 286, p. 345, in *Forces in Scanning Probe Methods*, ed. by H.-J. Güntherodt, D. Anselmetti, E. Meyer, NATO ASI Series E, (Series E: Applied Sciences), vol. 286 (Springer, Dordrecht, 1995). https://doi.org/10.1007/978-94-011-0049-6_32

127. R.V. Coleman, B. Giambattista, P.K. Hansma, A. Johnson, W.W. McNairy, C.G. Slough, Scanning tunneling microscopy of charge-density waves in transition metal chalcogenides. Adv. Phys. **37**, 559 (1988). https://doi.org/10.1080/00018738800101439

128. L. Collins, A. Belianinov, R. Proksch, T. Zuo, Y. Zhang, P.K. Liaw, S.V. Kalinin, S. Jesse, G-mode magnetic force microscopy: separating magnetic and electrostatic interactions using big data analytics. Appl. Phys. Lett. **108**(19), 193103 (2016). ISSN 0003-6951. https://doi.org/10.1063/1.4948601. Publisher: American Institute of Physics

129. R.J. Colton, A. Engel, J. Frommer, H.E. Gaub, A.A. Gewirth, R. Guckenberger, W.M. Heckl, B.A. Parkinson, J.P. Rabe (eds.), *Procedures in Scanning Probe Microscopies* (Wiley, Chichester, 1998)

130. R. Colton, J. Frommer, A. Engel, H. Gaub, *Procedures in Scanning Probe Microscopies* (Wiley-VCH, Weinheim, 1998). ISBN 978-3-527-31269-6

131. M.F. Crommie, C.P. Lutz, D.M. Eigler, Imaging standing waves in a two-dimensional electron gas. Nature **363**(6429), 524–527 (1993). ISSN 1476-4687. https://doi.org/10.1038/363524a0

132. M.F. Crommie, C.P. Lutz, D.M. Eigler, Confinement of electrons to quantum corrals on a metal surface. Science **262**(5131), 218–220 (1993). ISSN 0036-8075. https://doi.org/10.1126/science.262.5131.218, https://science.sciencemag.org/content/262/5131/218

133. O.E. Dagdeviren, J. Götzen, H. Hölscher, E.I. Altman, U.D. Schwarz, Robust high-resolution imaging and quantitative force measurement with tuned-oscillator atomic force microscopy. Nanotechnology **27**(6), 065703 (2016). https://doi.org/10.1088/0957-4484/27/6/065703

134. H. Dai, J.H. Hafner, A.G. Rinzler, D.T. Colbert, R.E. Smalley, Nanotubes as nanoprobes in scanning probe microscopy. Nature **384**, 147 (1996). https://doi.org/10.1038/384147a0

135. Q. Dai, R. Vollmer, R.W. Carpick, D.F. Ogletree, M. Salmeron. A variable temperature ultra-high vacuum atomic force microscope. Rev. Sci. Instrum. **66**(11), 5266–5271 (1995). ISSN 0034-6748. https://doi.org/10.1063/1.1146097. Publisher: American Institute of Physics

136. C. Davis, C.C. Williams, P. Neuzil, A micromachined sub-micrometer photodicode for scanning probe microscopy. Appl. Phys. Lett. **66**(18), 2309 (1995). https://doi.org/10.1063/1.114223

137. L.C. Davis, M.P. Everson, R.C. Jaklevic, W. Shen, Theory of the local density of surface states on a metal: comparison with scanning tunneling spectroscopy of a au(111) surface. Phys. Rev. B **43**, 3821–3830 (1991). https://doi.org/10.1103/physrevb.43.3821

138. P.J. de Pablo, J. Colchero, M. Luna, J. Gómez-Herrero, A.M. Baró, Tip-sample interaction in tapping-mode scanning force microscopy. Phys. Rev. B **61**, 14179–14183 (2000). https://doi.org/10.1103/PhysRevB.61.14179

139. P. De Wolf, T. Clarysse, W. Vandervorst, J. Snauwaert, L. Hellemans, One and two dimensional carrier profiling in semiconductors by nanospreading resistance profiling. J. Vac. Sci. & Technol. B: Microelectron. Nanometer Struct. Proc. Meas. Phen. **14**(1), 380–385 (1996). https://doi.org/10.1116/1.588478, https://avs.scitation.org/doi/abs/10.1116/1.588478

140. E. Delamarche, B. Michel, H.A. Biebuyck, C. Gerber, Golden interfaces: the surface of self-assembled monolayers. Adv. Mater. **8**(9), 719–729 (1996). https://doi.org/10.1002/adma.19960080903

141. A.J. den Boef, Preparation of magnetic tips for a scanning force microscope. Appl. Phys. Lett. **56**(20), 2045–2047 (1990). ISSN 0003-6951. https://doi.org/10.1063/1.102991. Publisher: American Institute of Physics

142. A.J. den Boef, The influence of lateral forces in scanning force microscopy. Rev. Sci. Instrum. **62**(1), 88–92 (1991). https://doi.org/10.1063/1.1142287

143. F.J.A. den Broeder, H.C. Donkersloot, H.J.G. Draaisma, W.J.M. de Jonge, Magnetic properties and structure of Pd/Co and Pd/Fe multilayers. J. Appl. Phys. **61**(8), 4317–4319 (1987). ISSN 0021-8979. https://doi.org/10.1063/1.338459. Publisher: American Institute of Physics

144. F.J.A. den Broeder, W. Hoving, and P.J.H. Bloemen. Magnetic anisotropy of multilayers. J. Magn. Magn. Mater. **93**, 562–570 (1991). ISSN 0304-8853. https://doi.org/10.1016/0304-8853(91)90404-X, http://www.sciencedirect.com/science/article/pii/030488539190404X

145. Z. Deng, E. Yenilmez, J. Leu, J.E. Hoffman, E.W.J. Straver, H. Dai, K.A. Moler, Metal-coated carbon nanotube tips for magnetic force microscopy. Appl. Phys. Lett. **85**(25), 6263–6265 (2004). ISSN 0003-6951. https://doi.org/10.1063/1.1842374. Publisher: American Institute of Physics

146. W. Denk, D. Pohl, Local electrical dissipation imaged by scanning force microscopy. Appl. Phys. Lett. **59**(17), 2171 (1991). https://doi.org/10.1063/1.106088

147. A.C. Diebold, M.R. Kump, J.J. Kopanski, D.G. Seiler, Characterization of two-dimensional dopant profiles: Status and review. J. Vac. Sci. Technol. B **14**(1), 196 (1996). https://doi.org/10.1116/1.589028

148. B. Dieny, V.S. Speriosu, S.S.P. Parkin, B.A. Gurney, D.R. Wilhoit, D. Mauri, Giant magnetoresistive in soft ferromagnetic multilayers. Phys. Rev. B **43**(1), 1297–1300 (1991). https://doi.org/10.1103/PhysRevB.43.1297

149. C. Dietz, E.T. Herruzo, J.R. Lozano, R. Garcia, Nanomechanical coupling enables detection and imaging of 5 nm superparamagnetic particles in liquid. Nanotechnology **22**(12), 125708 (2011). https://doi.org/10.1088/0957-4484/22/12/125708

150. F. Dinelli, H.E. Assender, N. Takeda, G.A.D. Briggs, O.V. Kolosov, Elastic mapping of heterogeneous nanostructures with ultrasonic force microscopy (UFM). Surf. Interface Anal. **27**(5–6), 562–567 (1999). ISSN 0142-2421. https://doi.org/10.1002/(SICI)1096-9918(199905/06)27:5/6<562::AID-SIA538>3.0.CO;2-K. Publisher: Wiley

151. B. Doppagne, M.C. Chong, E. Lorchat, S. Berciaud, M. Romeo, H. Bulou, A. Boeglin, F. Scheurer, G. Schull, Vibronic spectroscopy with submolecular resolution from stm-induced electroluminescence. Phys. Rev. Lett. **118** (2017). https://doi.org/10.1103/PhysRevLett.118.127401

152. U. Dürig, O. Züger, D.W. Pohl, Observation of metallic adhesion using the scanning tunneling microscope. Phys. Rev. Lett. **65**(3), 349 (1990). ISSN 1079-7114. https://doi.org/10.1103/PhysRevLett.65.349

153. U. Dürig, O. Züger, L.C. Wang, H.J. Kreuzer, Adhesion in atomic-scale metal contacts. Europhys. Lett. (EPL) **23**(2), 147–152 (1993). https://doi.org/10.1209/0295-5075/23/2/012

154. Urs Dürig, Extracting interaction forces and complementary observables in dynamic probe microscopy. Appl. Phys. Lett. **76**(9), 1203 (2000). https://doi.org/10.1063/1.125983

155. Urs. Dürig, Conservative and dissipative interactions in dynamic force microscopy. Surf. Interface Anal. **27**(5–6), 467 (1999). https://doi.org/10.1002/(SICI)1096-9918(199905/06)27:5/6<467::AID-SIA519>3.0.CO;2-7

156. R. Eckert, J.M. Freyland, H. Gersen, H. Heinzelmann, G. Schürmann, W. Noell, Urs. Staufer, N.F. de Rooij, Near-field fluorescence imaging with 32 nm resolution based on microfabricated cantilevered probes. Appl. Phys. Lett. **77**(23), 3695–3697 (2000). ISSN 0003-6951. https://doi.org/10.1063/1.1330571. Publisher: American Institute of Physics

157. H. Edwards, L. Taylor, W. Duncan, A.J. Melmed, Fast, high-resolution atomic force microscopy using a quartz tuning fork as actuator and sensor. J. Appl. Phys. **82**(3), 980–984 (1997). ISSN 0021-8979. https://doi.org/10.1063/1.365936. Publisher: American Institute of Physics

158. H. Edwards, J.F. Jorgensen, J. Dagata, Y. Strausser, J. Schneir, Influence of data analysis and other factors on the short-term stability of vertical scanning-probe microscope calibration measurements. J. Vac. Sci. & Technol. B: Microelectron. Nanometer Struct. Proc. Meas. Phen. **16**(2), 633–644 (1998). ISSN 1071-1023. https://doi.org/10.1116/1.589933, https://avs.scitation.org/doi/abs/10.1116/1.589933. Publisher: American Institute of Physics

159. H. Edwards, R. McGlothlin, R. San Martin, U. Elisa, M. Gribelyuk, R. Mahaffy, C. Ken Shih, R.S. List, V.A. Ukraintsev, Scanning capacitance spectroscopy: an analytical technique for pn-junction delineation in si devices. Appl. Phys. Lett. **72**(6), 698–700 (1998). https://doi.org/10.1063/1.120849

160. H. Edwards, R. McGlothlin, U. Elisa , Vertical metrology using scanning-probe microscopes: imaging distortions and measurement repeatability. J. Appl. Phys. **83**(8), 3952–3971 (1998). ISSN 0021-8979. https://doi.org/10.1063/1.367151, https://aip.scitation.org/doi/10.1063/1.367151. Publisher: American Institute of Physics

161. D.M. Eigler, E.K. Schweizer, Positioning single atoms with a scanning tunnelling microscope. Nature **344**(6266), 524–526 (1990). ISSN 1476-4687. https://doi.org/10.1038/344524a0

162. D.M. Eigler, C.P. Lutz, W.E. Rudge, An atomic switch realized with the scanning tunnelling microscope. Nature **352**(6336), 600–603 (1991). ISSN 1476-4687. https://doi.org/10.1038/352600a0

163. M. Ellner, N. Pavlicek, P. Pou, B. Schuler, N. Moll, G. Meyer, L. Gross, R. Perez, The electric field of co tips and its relevance for atomic force microscopy. Nano Lett. **16**, 1924 (2016). https://doi.org/10.1021/acs.nanolett.5b05251

164. L.M. Eng, H.J. Güntherodt, G. Rosenman, A. Skliar, M. Oron, M. Katz, D. Eger, Nondestructive imaging and characterization of ferroelectric domains in periodically poled crystals. J. Appl. Phys. **83**(11), 5973 (1998). https://doi.org/10.1063/1.367462

165. A. Engel, H.E. Gaub, D.J. Müller, Atomic force microscopy: a forceful way with single molecules. Curr. Biol. **9**(4), R133–R136 (1999). ISSN 0960-9822. https://doi.org/10.1016/S0960-9822(99)80081-5, http://www.sciencedirect.com/science/article/pii/S0960982299800815

166. H.A. Engel, D. Loss, Detection of single spin decoherence in a quantum dot via charge currents. Phys. Rev. Lett. **86**, 4648 (2001). https://doi.org/10.1103/PhysRevLett.86.4648

167. R. Erlandsson, L. Olsson, Force interaction in low-amplitude ac mode atomic force microscopy: cantilever simulations and comparison with data from si (111) 7 x 7. Appl. Phys. A **66**, S879 (1998). https://doi.org/10.1103/PhysRevB.62.13680

168. R. Erlandsson, V. Yakimov, Force interaction between a w tip and si(111) investigated under ultrahigh vacuum conditions. Phys. Rev. B **62**, 13680 (2000), https://link.aps.org/doi/10.1103/PhysRevB.62.13680

169. R. Erlandsson, L. Olsson, P. Martensson, Inequivalent atoms and imaging mechanisms in ac-mode atomic force microscopy of Si (111). Phys. Rev. B **54**(12), R8309–8312 (1996). https://doi.org/10.1103/physrevb.54.r8309

170. Evan Evans, Probing the relation between force-lifetime-and chemistry in single molecular bonds. Ann. Rev. Biophys. Biomol. Struct. **30**(1), 105–128 (2001). https://doi.org/10.1146/annurev.biophys.30.1.105

171. R.F.L. Evans, R. Yanes, O. Mryasov, R.W. Chantrell, O. Chubykalo-Fesenko, On beating the superparamagnetic limit with exchange bias. EPL (Europhys. Lett.) **88**(5), 57004 (2009). ISSN 0295-5075. https://doi.org/10.1209/0295-5075/88/57004. Publisher: IOP Publishing

172. M.P. Everson, R.C. Jaklevic, W. Shen, Measurement of the local density of states on a metal surface: Scanning tunneling spectroscopic imaging of au(111). J. Vac. Sci. & Technol. A **8**(5), 3662–3665 (1990). https://doi.org/10.1116/1.576476

173. P. Eyben, M. Xu, N. Duhayon, T. Clarysse, S. Callewaert, W. Vandervorst, Scanning spreading resistance microscopy and spectroscopy for routine and quantitative two dimensional carrier profiling. J. Vac. Sci. Technol. B **20**(1), 471 (2002). https://doi.org/10.1116/1.1424280

174. G.E. Fantner, R.J. Barbero, D.S. Gray, A.M. Belcher, Kinetics of antimicrobial peptide activity measured on individual bacterial cells using high-speedatomic force microscopy. Nature Nanotechnol. **5**, 280 (2010). https://doi.org/10.1038/nnano.2010.29

175. G.E. Fantner, G. Schitter, J.H. Kindt, T. Ivanov, K. Ivanova, R. Patel, N. Holten-Andersen, J. Adams, P.J. Thurner, I.W. Rangelow, P.K. Hansma, Components for high speed atomic force microscopy, in *Proceedings of the Seventh International Conference on Scanning Probe Microscopy, Sensors and Nanostructures*, vol. 106 no. 8 (2006), pp. 881–887. ISSN 0304-3991. https://doi.org/10.1016/j.ultramic.2006.01.015, http://www.sciencedirect.com/science/article/pii/S0304399106000593

176. R.M. Feenstra, J.A. Stroscio, A.P. Fein, Tunneling spectroscopy of the Si(111)2 × 1 surface. Surf. Sci. **181**(1), 295–306 (1987). ISSN 0039-6028. https://doi.org/10.1016/0039-6028(87)90170-1, http://www.sciencedirect.com/science/article/pii/0039602887901701

177. R.M. Feenstra, J. Tersoff, A.P. Fein, Atom-selective imaging of the GaAs(110) surface. Phys. Rev. Lett. **58**, 1192 (1987). https://doi.org/10.1103/PhysRevLett.58.1192

178. R. Feynman, "there's plenty of room at the bottom", 1959. "This talk given at the annual meeting of the American Physical Society at the California Institute of Technology was first published in the February 1960 issue of Caltech's *Engineering and Science*, which owns the copyright. It has been made available on the web at http://www.zyvex.com/nanotech/feynman.html with their kind permission"

179. G.B.M. Fiege, V. Feige, J.C.H. Phang, M. Maywald, S. Gorlich, L.J. Balk, Failure analysis of integrated devices by scanning thermal microscopy (sthm). Microelectron. Reliab. **38**(6–8), 957 (1998). https://doi.org/10.1016/S0026-2714(98)00086-9

180. A. Finkler, Y. Segev, Y. Myasoedov, M.L. Rappaport, L. Ne'eman, D. Vasyukov, E. Zeldov, M.E. Huber, J. Martin, A. Yacoby, Self-aligned nanoscale SQUID on a tip. Nano Lett. **10**(3), 1046–1049 (2010). ISSN 1530-6984. https://doi.org/10.1021/nl100009r. Publisher: American Chemical Society

181. P.B. Fischer, M.S. Wei, S.Y. Chou, Ultrahigh resolution magnetic force microscope tip fabricated using electron beam lithography. J. Vac. Sci.& Technol. B: Microelectron. Nanometer Struct. Proc. Meas. Phen. **11**(6), 2570–2573 (1993). ISSN 1071-1023. https://doi.org/10.1116/1.586626, https://avs.scitation.org/doi/abs/10.1116/1.586626. Publisher: American Institute of Physics

182. C. Fokas, V. Deckert, Towards in situ raman microscopy of single catalytic sites. Appl. Spectrosc. **56**(2), 192 (2002). https://doi.org/10.1366/0003702021954665

183. E.T. Foley, A.F. Kam, J.W. Lyding, Ph Avouris, Cryogenic uhv-stm study of hydrogen and deuterium desorption from si(100). Phys. Rev. Lett. **80**, 1336–1339 (1998). https://doi.org/10.1103/PhysRevLett.80.1336

184. L. Folks, M.E. Best, P.M. Rice, B.D. Terris, D. Weller, J.N. Chapman, Perforated tips for high-resolution in-plane magnetic force microscopy. Appl. Phys. Lett. **76**(7), 909–911 (2000). ISSN 0003-6951. https://doi.org/10.1063/1.125626. Publisher: American Institute of Physics

185. R.E. Fontana Jr, S.S.P. Parkin, Magnetic tunnel junction device with longitudinal biasing, March 17 1998. US Patent 5,729,410

186. D. Forchheimer, D. Platz, E.A. Tholén, D.B. Haviland, Simultaneous imaging of surface and magnetic forces. Appl. Phys. Lett. **103**(1), 013114 (2013). ISSN 0003-6951. https://doi.org/10.1063/1.4812979. Publisher: American Institute of Physics

187. S. Foss, R. Proksch, E.D. Dahlberg, B. Moskowitz, B. Walsh, Localized micromagnetic perturbation of domain walls in magnetite using a magnetic force microscope. Appl. Phys. Lett. **69**(22), 3426–3428 (1996). https://doi.org/10.1063/1.117281

188. J.S. Foster, J.E. Frommer, P.C. Arnett, Molecular manipulation using a tunneling microscope. Nature Commun. **331**, 324 (1988). https://doi.org/10.1038/331324a0

189. J. Frenkel, On the electrical resistance of contacts between solid conductors. Phys. Rev. **36**, 1604–1618 (1930). https://doi.org/10.1103/PhysRev.36.1604

190. J. Fritz, M.K. Baller, H.P. Lang, H. Rothuizen, P. Vettiger, E. Meyer, H.J. Güntherodt, Ch. Gerber, J.K. Gimzewski, Translating biomolecular recognition into nanomechanics. Science **288**(5464), 316–318 (2000). ISSN 0036-8075. https://doi.org/10.1126/science.288.5464.316, https://science.sciencemag.org/content/288/5464/316. Publisher: American Association for the Advancement of Science _eprint: https://science.sciencemag.org/content/288/5464/316.full.pdf

191. J. Frommer, Scanning tunneling microscopy and atomic force microscopy in organic chemistry. Angewandte Chemie International Edition in English **31**(10), 1298–1328 (1992). ISSN 0570-0833. https://doi.org/10.1002/anie.199212981. Publisher: Wiley

192. M. Fujihira, N. Ohishi, T. Osa, Photocell using covalently-bound dyes on semiconductor surfaces. Nature **268**(5617), 226–228 (1977). ISSN 1476-4687. https://doi.org/10.1038/268226a0

193. S. Fujisawa, E. Kishi, Y. Sugawara, S. Morita, Two-dimensionally quantized friction observed with two-dimensional frictional force microscope. Tribol. Lett. **1**(2), 121–127 (1995). ISSN 1573-2711. https://doi.org/10.1007/BF00209767

194. R. Garcia, E.T. Herruzo, The emergence of multifrequency force microscopy. Nat. Nanotechnol. **7**, 217 (2012). https://doi.org/10.1038/nnano.2012.38

195. R. Garcia, R. Perez, Dynamic atomic force microscopy methods. Surf. Sci. Rep. **47**(6), 197–301 (2002). ISSN 0167-5729. https://doi.org/10.1016/S0167-5729(02)00077-8, http://www.sciencedirect.com/science/article/pii/S0167572902000778

196. R. Garcia, A.S. Paulo, Attractive and repulsive tip-sample interaction regimes in tapping-mode atomic force microscopy. Phys. Rev. B **60**, 4961–4967 (1999). https://doi.org/10.1103/PhysRevB.60.4961

197. R. Garcia, A.S. Paulo, Dynamics of a vibrating tip near or in intermittent contact with a surface. Phys. Rev. B **61**, R13381–R13384 (2000). https://doi.org/10.1103/PhysRevB.61.R13381

198. M. Gauthier, M. Tsukada, Theory of noncontact dissipation force microscopy. Phys. Rev. B **60**(16), 11716 (1999). https://doi.org/10.1103/PhysRevB.60.11716

199. M. Gauthier, M. Tsukada, Damping mechanism in dynamic force microscopy. Phys. Rev. Lett. **85**(25), 5348 (2000). https://doi.org/10.1103/PhysRevLett.85.5348

200. M. Gauthier, N. Sasaki, M. Tsukada, Dynamics of the cantilever in noncontact dynamic force microscopy: the steady-state approximation and beyond. Phys. Rev. B **64** (2001). https://doi.org/10.1103/PhysRevB.64.085409

201. M. Gelbert, A. Roters, M. Schimmel, J. Rühe, D. Johannsmann, Viscoelastic spectra of soft polymer interfaces obtained by noise analysis of afm cantilevers. Surf. Interface Anal. **27**, 572 (1999). ISSN 0142-2421. https://doi.org/10.1002/(SICI)1096-9918(199905/06)27:5/6<572::AID-SIA487>3.0.CO;2-K

202. M. Gerken, A. Solignac, D. Momeni Pakdehi, A. Manzin, T. Weimann, K. Pierz, S. Sievers, H.W. Schumacher, Traceably calibrated scanning Hall probe microscopy at room temperature. J. Sensors Sensor Syst. **9**(2), 391–399 (2020). https://doi.org/10.5194/jsss-9-391-2020, https://jsss.copernicus.org/articles/9/391/2020/

203. F.J. Giessibl, Atomic resolution of silicon(111)7x7 by atomic force microscopy through repulsive and attractive forces. Science **267**(68), 1451–1455 (1995). ISSN 00368075, 10959203, http://www.jstor.org/stable/2886041

204. F.J. Giessibl, Forces and frequency shifts in atomic resolution dynamic-force microscopy. Phys. Rev. B **56**(24), 16010–16015 (1997). https://doi.org/10.1103/PhysRevB.56.16010

205. F.J. Giessibl, High-speed force sensor for force microscopy and profilometry utilizing quartz tuning fork. Appl. Phys. Lett. **73**, 3956 (1998). https://doi.org/10.1063/1.122948

206. F.J. Giessibl, A direct method to calculate tip-sample forces from frequency shifts in frequency-modulation atomic force microscopy. Appl. Phys. Lett. **78**(1), 123 (2001). https://doi.org/10.1063/1.1335546

207. F.J. Giessibl, Noncontact atomic force microscopy, in ed. by S. Morita, R. Wiesendanger, E. Meyer (Springer, Berlin, 2002), p. 11. https://doi.org/10.1007/9783319155883

208. F.J. Giessibl, H. Bielefeldt, Physical interpretation of frequency-modulation atomic force microscopy. Phys. Rev. B **61**(15), 9968 (2000). https://doi.org/10.1103/PhysRevB.61.9968

209. F.J. Giessibl, Atomic resolution on Si(111)-(7×7) by noncontact atomic force microscopy with a force sensor based on a quartz tuning fork. Appl. Phys. Lett. **76**(11), 1470–1472 (2000). ISSN 0003-6951. https://doi.org/10.1063/1.126067. Publisher: American Institute of Physics

210. F.J. Giessibl, Advances in atomic force microscopy. Rev. Mod. Phys. **75**, 949–983 (2003). https://doi.org/10.1103/RevModPhys.75.949

211. F.J. Giessibl, H. Bielefeldt, Physical interpretation of frequency modulation atomic force microscopy. Phys. Rev. B **61**, 9968–9971 (2000). https://doi.org/10.1103/PhysRevB.61.9968

212. R. Giles, J.P. Cleveland, S. Manne, P.K. Hansma, B. Drake, P. Maivald, C. Boles, J. Gurley, V. Elings, Noncontact force microscopy in liquids. Appl. Phys. Lett. **63**(5), 617–618 (1993). ISSN 0003-6951. https://doi.org/10.1063/1.109967. Publisher: American Institute of Physics

213. J.K. Gimzewski, A. Humbert, J.G. Bednorz, B. Reihl, Silver films condensed at 300 and 90 k: Scanning tunneling microscopy of their surface topography. Phys. Rev. Lett. **55**, 951–954 (1985). https://doi.org/10.1103/PhysRevLett.55.951

214. J.K. Gimzewski, B. Reihl, J.H. Coombs, R.R. Schlittler, Photon emission with the scanning tunneling microscope. Zeitschrift für Physik B Condensed Matter **72**(4), 497–501 (1988). ISSN 1431-584X. https://doi.org/10.1007/BF01314531

215. J.K. Gimzewski, C. Joachim, Nanoscale science of single molecules using local probes. Science **283**, 1683 (1999), https://science.sciencemag.org/content/283/5408/1683

216. J.K. Gimzewski, C. Gerber, E. Meyer, R.R. Schlittler, Observation of a chemical reaction using a micromechanical sensor. Chem. Phys. Lett. **217**, 589 (1993). https://doi.org/10.1016/0009-2614(93)E1419-H

217. J.K. Gimzewski, T.A. Jung, M.T. Cuberes, R.R. Schlittler, Scanning tunneling microscopy of individual molecules: beyond imaging. Surf. Sci. **386**, 101–114 (1997), https://www.sciencedirect.com/science/article/pii/S0039602897003014

218. J.K. Gimzewski, C. Joachim, R.R. Schlittler, H. Tang, I. Johannsen, Rotation of a single molecule within a supramolecular bearing. Science **281**, 531 (1998), https://science.sciencemag.org/content/281/5376/531

219. E. Gnecco, R. Bennewitz, T. Gyalog, Ch. Loppacher, M. Bammerlin, E. Meyer, H.J. Güntherodt, Velocity dependence of atomic friction. Phys. Rev. Lett. **84**, 1172 (2000). https://doi.org/10.1103/PhysRevLett.84.1172

220. E. Gnecco, R. Bennewitz, E. Meyer, Abrasive wear on the atomic scale. Phys. Rev. Lett. **88** (2002). https://doi.org/10.1103/PhysRevLett.88.215501

221. T. Göddenhenrich, U. Hartmann, M. Anders, C. Heiden, Investigation of bloch wall fine structures by magnetic force microscopy. J. Microsc. **152**(2), 527–536 (1988)

222. R.D. Gomez, E.R. Burke, A.A. Adly, I.D. Mayergoyz, Magnetic field imaging by using magnetic force scanning tunneling microscopy. Appl. Phys. Lett. **60**(7), 906–908 (1992). ISSN 0003-6951. https://doi.org/10.1063/1.107442. Publisher: American Institute of Physics

223. K. Goto, K. Hane, Application of a semiconductor tip to capacitance microscopy. Appl. Phys. Lett. **73**(4), 544 (1998). https://doi.org/10.1063/1.121927

224. B. Gotsmann, C. Seidel, B. Anczykowski, H. Fuchs, Conservative and dissipative tip-sample interaction forces probed with dynamic afm. Phys. Rev. B **60**(15), 11051 (1999). https://doi.org/10.1103/PhysRevB.60.11051

225. T. Gotszalk, P. Grabiec, I.W. Rangelow, Piezoresistive sensors for scanning probe microscopy. Ultramicroscopy **82**, 39 (2000). https://doi.org/10.1016/S0304-3991(99)00171-0

226. S. Grafström, M. Neitzert, T. Hagen, J. Ackermann, R. Neumann, O. Probst, M. Wörtge, The role of topography and friction for the image contrast in lateral force microscopy. Nanotechnology **4**, 143 (1993). https://doi.org/10.1088/0957-4484/4/3/003

227. J.E. Griffith, D.A. Grigg, Dimensional metrology with scanning probe microscopes. J. Appl. Phys. **74**(9), R83 (1993). https://doi.org/10.1063/1.354175

228. M.S. Grinolds, S. Hong, P. Maletinsky, L. Luan, M.D. Lukin, R.L. Walsworth, A. Yacoby, Nanoscale magnetic imaging of a single electron spin under ambient conditions. Nat. Phys. **9**(4), 215–219 (2013). ISSN 1745-2481. https://doi.org/10.1038/nphys2543

229. R.D. Grober, J. Acimovic, J. Schuck, D. Hessman, P.J. Kindlemann, J. Hespanha, A.S. Morse, K. Karrai, I. Tiemann, S. Manus, Fundamental limits to force detection using quartz tuning forks. Rev. Sci. Instr. **71**(7), 2776 (2000). https://doi.org/10.1063/1.1150691

230. I. Gross, W. Akhtar, A. Hrabec, J. Sampaio, L.J. Martinez, S. Chouaieb, B.J. Shields, P. Maletinsky, A. Thiaville, S. Rohart, V. Jacques, Skyrmion morphology in ultrathin magnetic films. Phys. Rev. Mater. **2** (2018). https://doi.org/10.1103/PhysRevMaterials.2.024406

231. L. Gross, F. Mohn, N. Moll, P. Liljeroth, G. Meyer, The chemical structure of a molecule resolved by atomic force microscopy. Science **325**, 1110 (2009). https://doi.org/10.1126/science.1176210

232. L. Gross, F. Mohn, N. Moll, B. Schuler, A. Criado, E. Guitian, D. Pena, A. Gourdon, G. Meyer, Bond-order discrimination by atomic force microscopy. Science **337**, 1326 (2012). https://doi.org/10.1126/science.1225621

233. L. Gross, Recent advances in submolecular resolution with scanning probe microscopy. Nat. Chem. **3**(4), 273–278 (2011). ISSN 1755-4349. https://doi.org/10.1038/nchem.1008

234. A. Gruber, A. Dräbenstedt, C. Tietz, L. Fleury, J. Wrachtrup, C. von Borczyskowski, Scanning confocal optical microscopy and magnetic resonance on single defect centers. Science **276**(5321), 2012–2014 (1997). ISSN 0036-8075. https://doi.org/10.1126/science.276.5321.2012, https://science.sciencemag.org/content/276/5321/2012. Publisher: American Association for the Advancement of Science _eprint: https://science.sciencemag.org/content/276/5321/2012.full.pdf

235. P. Grütter, E. Meyer, H. Heinzelmann, L. Rosenthaler, H.-R. Hidber, H.-J. Güntherodt, Application of atomic force microscopy to magnetic materials. J. Vacuum Sci. Technol. A **6**(2), 279–282 (1988). ISSN 0734-2101. https://doi.org/10.1116/1.575425. Publisher: American Vacuum Society

236. P. Grütter, D. Rugar, H.J. Mamin, G. Castillo, S.E. Lambert, C.J. Lin, R.M. Valletta, O. Wolter, T. Bayer, J. Greschner, Batch fabricated sensors for magnetic force microscopy. Appl. Phys. Lett. **57**(17), 1820–1822 (1990). https://doi.org/10.1063/1.104030

237. P. Grütter, H.J. Mamin, D. Rugar, *Magnetic Force Microscopy (MFM)* (Springer, Berlin, 1992), pp. 151–207. ISBN 978-3-642-97363-5. https://doi.org/10.1007/978-3-642-97363-5_5

238. P. Grütter, W. Zimmermann-Edling, D. Brodbeck, Tip artifacts of microfabricated force sensors for atomic force microscopy. Appl. Phys. Lett. **60**(22), 2741–2743 (1992). https://doi.org/10.1063/1.106862

239. P. Grütter, Y. Liu, P. LeBlanc, Magnetic dissipation force microscopy. Appl. Phys. Lett. **71**(2), 279–281 (1997). https://doi.org/10.1063/1.119519

240. M. Guggisberg, M. Bammerlin, A. Baratoff, R. Lüthi, Ch. Loppacher, F.M. Battiston, J. Lü, R. Bennewitz, E. Meyer, H.-J. Güntherodt, Dynamic force microscopy across steps on the si(111) (7x7) surface. Surf. Sci. **461**(1), 255–265 (2000). ISSN 0039-6028. https://doi.org/10.1016/S0039-6028(00)00592-6, http://www.sciencedirect.com/science/article/pii/S0039602800005926

241. M. Guggisberg, M. Bammerlin, Ch. Loppacher, O. Pfeiffer, A. Abdurixit, V. Barwich, R. Bennewitz, A. Baratoff, E. Meyer, H.-J. Güntherodt, Separation of interactions by noncontact force microscopy. Phys. Rev. B **61**, 11151–11155 (2000). https://doi.org/10.1103/PhysRevB.61.11151

242. P. Günther, U.Ch. Fischer, K. Dransfeld, Scanning near-field acoustic microscopy. Appl. Phys. B **48**(1), 89–92 (1989). ISSN 1432-0649. https://doi.org/10.1007/BF00694423

243. H.-J. Güntherodt, R. Wiesendanger, *Scanning Tunneling Microscopy I*, volume 20 of Springer Series in Surface Science (Springer, Berlin, 1994). https://doi.org/10.1007/978-3-642-97343-7

244. P. Güthner, Untersuchung der lokalen piezoelektrischen Eigenschaften ferroelektrischer Polymerfilme. Dissertation, Hartung Gorre Verlag, Konstanz (1992)

245. T. Gyalog, M. Bammerlin, R. Lüthi, E. Meyer, H. Thomas, Mechanism of atomic friction. Europhys. Lett. **31**, 269 (1995). https://doi.org/10.1209/0295-5075/31/5-6/004

246. H. Fink, Mono-atomic tips for scanning tunneling microscopy. IBM J. Res. Dev. **30**(5), 460–465 (1986). ISSN 0018-8646. https://doi.org/10.1147/rd.305.0460

247. J.H. Hafner, C.L. Cheung, C.M. Lieber, Direct growth of single-walled nanotube scanning probe microscopy tips. J. Amer. Chem. Soc. **121**, 9750 (1999). https://doi.org/10.1021/ja992761b

248. V.M. Hallmark, S. Chiang, J.F. Rabolt, J.D. Swalen, R.J. Wilson, Observation of atomic corrugation on au(111) by scanning tunneling microscopy. Phys. Rev. Lett. **59**(25), 2879 (1987). https://doi.org/10.1103/PhysRevLett.59.2879

249. R.J. Hamers, D.G. Cahill, Ultrafast time resolution in scanned probe microscopies. Appl. Phys. Lett. **57**(19), 2031 (1990). https://doi.org/10.1063/1.103997

250. A. Hammiche, D.J. Hourston, H.M. Pollock, M. Reading, M. Song, Scanning thermal microscopy: subsurface imaging, thermal mapping of polymer blends, and localized calorimetry. J. Vac. Sci. Technol. B **14**(2), 1486 (1996). https://doi.org/10.1103/PhysRevLett.59.2879

251. C. Hanneken, F. Otte, A. Kubetzka, B. Dupé, N. Romming, K. von Bergmann, R. Wiesendanger, S. Heinze, Electrical detection of magnetic skyrmions by tunnelling non-collinear magnetoresistance. Nat. Nanotechnol. **10**(12), 1039–1042 (2015). ISSN 1748-3395. https://doi.org/10.1038/nnano.2015.218

252. K.H. Hansen, T. Worren, S. Stempel, E. Laegsgaard, M. Bäumer, H.J. Freund, F. Besenbacher, I. Stensgaard, Palladium nanocrystals on al2o3: structure and adhesion energy. *prl*, **83**(20), 4120 (1999). https://doi.org/10.1103/PhysRevLett.83.4120, https://link.aps.org/doi/10.1103/PhysRevLett.83.4120

253. P.K. Hansma, J.P. Cleveland, M. Radmacher, D.A. Walters, P.E. Hillner, M. Bezanilla, M. Fritz, D. Vie, H.G. Hansma, C.B. Prater, J. Massie, L. Fukunaga, J. Gurley, V. Elings, Tapping mode atomic force microscopy in liquids. Appl. Phys. Lett. **64**(13), 1738–1740 (1994). https://doi.org/10.1063/1.111795

254. P.K. Hansma, *Tunneling Spectroscopy: Capabilities, Applications, and New Techniques* (Plenum Press, New York, 1982). https://doi.org/10.1007/978-1-4684-1152-2

255. P.K. Hansma, B. Drake, O. Marti, S.A.C. Gould, C.B. Prater, The scanning ion conductance microscope. Science **243**, 641 (1989). https://doi.org/10.1126/science.2464851

256. L. Hao, J.C. Macfarlane, J.C. Gallop, D. Cox, J. Beyer, D. Drung, T. Schurig, Measurement and noise performance of nano-superconducting-quantum-interference devices fabricated by focused ion beam. Appl. Phys. Lett. **92**(19), 192507 (2008). ISSN 0003-6951. https://doi.org/10.1063/1.2917580. Publisher: American Institute of Physics

257. P. Hapala, G. Kichin, C. Wagner, F.S. Tautz, R. Temirov, P. Jelínek, Mechanism of high-resolution stm/afm imaging with functionalized tips. Phys. Rev. B **90**, 085421 (2014). https://doi.org/10.1103/PhysRevB.90.085421

258. Y. Hasegawa, Ph Avouris, Direct observation of standing wave formation at surface steps using scanning tunneling. Phys. Rev. Lett. **71**, 1071 (1993). https://doi.org/10.1103/PhysRevLett. 71.1071

259. N. Hatter, B.W. Heinrich, M. Ruby, J.I. Pascual, K.J. Franke, Magnetic anisotropy in Shiba bound states across a quantum phase transition. Nat. Commun. **6**(1), 8988 (2015). ISSN 2041-1723. https://doi.org/10.1038/ncomms9988

260. N. Hauptmann, J.W. Gerritsen, D. Wegner, A.A. Khajetoorians, Sensing noncollinear magnetism at the atomic scale combining magnetic exchange and spin-polarized imaging. Nano Lett. **17**(9), 5660–5665 (2017). ISSN 1530-6984. https://doi.org/10.1021/acs.nanolett. 7b02538. Publisher: American Chemical Society

261. N. Hauptmann, M. Dupé, T.-C. Hung, A.K. Lemmens, D. Wegner, B. Dupé, A.A. Khajetoorians, Revealing the correlation between real-space structure and chiral magnetic order at the atomic scale. Phys. Rev. B **97** (2018). https://doi.org/10.1103/PhysRevB.97.100401

262. N. Hauptmann, S. Haldar, T.-C. Hung, W. Jolie, M. Gutzeit, D. Wegner, S. Heinze, A.A. Khajetoorians, Quantifying exchange forces of a spin spiral on the atomic scale. Nat. Commun. **11**(1), 1197 (2020). ISSN 2041-1723. https://doi.org/10.1038/s41467-020-15024-2

263. S. Hearman, Elastic constants of anisotropic materials. Adv. Phys. **5**, 323 (1956). https://doi. org/10.1080/00018732.1956.tADP0323

264. W. Hebenstreit, J. Redinger, Z. Horozova, M. Schmid, R. Podloucky, P. Varga, Atomic resolution by STM on ultra-thin films of alkali halides: experiment and local density calculations. Surf. Sci. **424**, L321 (1999).

265. B. Hecht, H. Bielefeldt, Y. Inouye, D.W. Pohl, Facts and artifacts in near-field optical microscopy. J. Appl. Phys. **81**, 2492 (1997). ISSN 0021-8979. https://doi.org/10.1063/1. 363956

266. M. Heigl, C. Vogler, A.-O. Mandru, X. Zhao, H.J. Hug, D. Suess, M. Albrecht, Microscopic origin of magnetization reversal in nanoscale exchange-coupled Ferri/Ferromagnetic bilayers: implications for high energy density permanent magnets and spintronic devices. ACS Appl. Nano Mater. **3**(9), 9218–9225 (2020). https://doi.org/10.1021/acsanm.0c01835. Publisher: American Chemical Society

267. M. Heim, R. Steigerwald, R. Guckenberger, Hydration scanning tunneling microscopy of dna and a bacterial surface protein. J. Struct. Biol. **119**, 212 (1997). ISSN 1047-8477. https://doi.org/10.1006/jsbi.1997.3874, http://www.sciencedirect.com/science/article/ pii/S1047847797938740

268. A.J. Heinrich, J.A. Gupta, C.P. Lutz, D.M. Eigler, Single-atom spin-flip spectroscopy. Science **306**(5695), 466–469 (2004). ISSN 0036-8075. https://doi.org/10.1126/science.1101077, https://science.sciencemag.org/content/306/5695/466. Publisher: American Association for the Advancement of Science _eprint: https://science.sciencemag.org/content/306/5695/466. full.pdf

269. S. Heinze, M. Bode, A. Kubetzka, O. Pietzsch, Y. Nie, S. Blügel, R. Wiesendanger, Real-space imaging of two-dimensional antiferromagnetism on the atomic scale. Science **288**(9 June), 1805 (2000). ISSN 0036-8075. https://doi.org/10.1126/science.288.5472.1805, https:// science.sciencemag.org/content/288/5472/1805

270. S. Heisig, O. Rudow, E. Oesterschulze, Scanning near-field optical microscopy in the near-infrared region using light emitting cantilever probes. Appl. Phys. Lett. **77**(8), 1071 (2000). ISSN 0003-6951. https://doi.org/10.1063/1.1289261

271. L. Hellemans, K. Waeyaert, F. Hennau, L. Stockman, I. Heyvaert, C. Van Haesendonck, Can atomic force microscopy tips be inspected by atomic force microscopy? J. Vac. Sci. & Technol. B: Microelectron. Nanometer Struct. Proc. Meas. Phen. **9**(2), 1309–1312 (1991). https://doi. org/10.1116/1.585185

272. E.J. Heller, M.F. Crommie, C.P. Lutz, D.M. Eigler, Scattering and absorption of surface electron waves in quantum corrals. Nature **369**(6480), 464–466 (1994). ISSN 1476-4687. https:// doi.org/10.1038/369464a0

273. S. Hembacher, F.J. Giessibl, J. Mannhart, Force microscopy with light-atom probes. Science **305**(5682), 380–383 (2004). ISSN 0036-8075. https://doi.org/10.1126/science.1099730, https://science.sciencemag.org/content/305/5682/380

274. M.C. Hersam, A.C.F. Hoole, S.J.O Shea, M.E. Welland, Potentiometry and repair of electrically stressed nanowires using atomic force microscopy. Appl. Phys. Lett. **72**(8), 915 (1998). ISSN 0003-6951. https://doi.org/10.1063/1.120872

275. H.F. Hess, R.B. Robinson, R.C. Dynes, J.M. Valles Jr., J.V. Waszczak, Scanning-tunneling-microscope observation of the abrikosov flux lattice and the density of states near and inside a fluxoid. Phys. Rev. Lett. **62**(2), 214 (1989). https://doi.org/10.1103/PhysRevLett.62.214

276. H.F. Hess, R.B. Robinson, J.V. Waszczak, Vortex-core structure observed with a scanning tunneling microscope. Phys. Rev. Lett. **64**(22), 2711 (1990)

277. M. Hipp, H. Bielefeldt, J. Colchero, O. Marti, J. Mlynek, A stand-alone scanning force and friction microscope. Ultramicroscopy **42–44**, 1498 (1992). ISSN 0304-3991. https://doi.org/10.1016/0304-3991(92)90472-V, http://www.sciencedirect.com/science/article/pii/030439919290472V

278. C.F. Hirjibehedin, C.-Y. Lin, A.F. Otte, M. Ternes, C.P. Lutz, B.A. Jones, A.J. Heinrich, Large magnetic anisotropy of a single atomic spin embedded in a surface molecular network. Science **317**(5842), 1199–1203 (2007). ISSN 0036-8075. https://doi.org/10.1126/science.1146110, https://science.sciencemag.org/content/317/5842/1199. Publisher: American Association for the Advancement of Science _eprint: https://science.sciencemag.org/content/317/5842/1199.full.pdf

279. A. Baratoff, H.J. Hug, Innovation and intellectual property rights, in *Noncontact Atomic Force Microscopy* ed. by S. Morita, R. Wiesendanger, E. Meyer (Chap. 20) (Springer Science & Business Media, Oxford, 2002), pp. 359–431. https://doi.org/10.1007/9783319155883

280. T. Hochwitz, A.K. Henning, C. Levey, C. Daghlian, J. Slinkman, Capacitive effects on quantitative dopant profiling with scanned electrostatic force microscopes. J. Vac. Sci. & Technol. B: Microelectron. Nanometer Struct. Proc. Meas. Phen. **14**(1), 457–462 (1996). https://doi.org/10.1116/1.588494

281. P. Hoffmann, B. Dutoit, R.-P. Salathe, Comparison of mechanically drawn and protection layer chemically etched optical fiber tips. Ultramicroscopy **61**, 165 (1995). ISSN 0304-3991. https://doi.org/10.1016/0304-3991(95)00122-0, http://www.sciencedirect.com/science/article/pii/0304399195001220

282. P.M. Hoffmann, A. Oral, R.A. Grimble, H.O. Oezer, S. Jeffery, J.B. Pethica, Direct measurement of interatomic force gradients using an ultra-low-amplitude atomic force microscope. Proc. R. Soc. Lond. Ser. A: Math. Phys. Eng. Sci. **457**(2009), 1161–1174 (2001). https://doi.org/10.1098/rspa.2000.0713. Publisher: Royal Society

283. M.S. Hoogeman, D. Glastra van Loon, R.W.M. Loos, H.G. Ficke, E. de Haas, J.J. van der Linden, H. Zeijlemaker, L. Kuipers, M.F. Chang, M.A.J. Klik, J.W.M. Frenken, Design and performance of a programmable-temperature scanning tunneling microscope. Rev. Sci. Instr. **69**, 2072 (1998). ISSN 0034-6748. https://doi.org/10.1063/1.1148901

284. B.W. Hoogenboom, P.L.T.M. Frederix, J.L. Yang, S. Martin, Y. Pellmont, M. Steinacher, S. Zäch, E. Langenbach, H.-J. Heimbeck, A. Engel, H.J. Hug, A Fabry–Perot interferometer for micrometer-sized cantilevers. Appl. Phys. Lett. **86**(7), 074101 (2005). ISSN 0003-6951. https://doi.org/10.1063/1.1866229, https://aip.scitation.org/doi/abs/10.1063/1.1866229. Publisher: American Institute of Physics

285. S. Hosaka, A. Kikukawa, Y. Honda, H. Koyanagi, Study of magnetic stray field measurement on surface using new force microscope. Jpn. J. Appl. Phys. **31**(Part 2, No. 7A), L908–L911 (1992). https://doi.org/10.1143/jjap.31.l908

286. H. Hosoi, K. Sueoka, K. Hayakawa, K. Mukasa, Atomic resolved imaging of cleaved NiO(100) surfaces by NC-AFM. Appl. Surf. Sci. **157**(4), 218–221 (2000). ISSN 0169-4332. https://doi.org/10.1016/S0169-4332(99)00529-2, http://www.sciencedirect.com/science/article/pii/S0169433299005292

287. L. Howald, E. Meyer, R. Lüthi, H. Haefke, R. Overney, H. Rudin, H.J. Guentherodt, Multi-functional probe microscope for facile operation in ultrahigh vacuum. Appl. Phys. Lett. **63**(1), 117–119 (1993). ISSN 0003-6951. https://doi.org/10.1063/1.109732. Publisher: American Institute of Physics

288. L. Howald, R. Lüthi, E. Meyer, P. Güthner, H.J. Güntherodt, Scanning force microscopy on the si (111)surface reconstruction. Zeitschrift für Physik B Condensed Matter **93**(3), 267–268 (1994). ISSN 1431-584X. https://doi.org/10.1007/BF01312696

289. A. Hrabec, J. Sampaio, M. Belmeguenai, I. Gross, R. Weil, S.M. Chérif, A. Stashkevich, V. Jacques, A. Thiaville, S. Rohart, Current-induced skyrmion generation and dynamics in symmetric bilayers. Nat. Commun. **8**(1), 15765 (2017). ISSN 2041-1723. https://doi.org/10. 1038/ncomms15765

290. X. Hu, G. Dai, S. Sievers, A. Fernández-Scarioni, H. Corte-León, R. Puttock, C. Barton, O. Kazakova, M. Ulvr, P. Klapetek, M. Havlíček, D. Nečas, Y. Tang, V. Neu, H.W. Schumacher, Round robin comparison on quantitative nanometer scale magnetic field measurements by magnetic force microscopy. J. Magn. Magn. Mater. **511**, 166947 (2020). ISSN 0304-8853. https://doi.org/10.1016/j.jmmm.2020.166947, http://www.sciencedirect.com/science/article/pii/S0304885320300366

291. Y. Huang, C.C. Williams, J. Slinkman, Quantitative two dimensional dopant profile measurement and inverse modeling by scanning capacitance microscopy. Appl. Phys. Lett. **66**(3), 344–346 (1995). ISSN 0003-6951. https://doi.org/10.1063/1.114207, https://aip.scitation.org/doi/10.1063/1.114207. Publisher: American Institute of Physics

292. F. Huber, H.P. Lang, D. Lang, D. Wüthrich, V. Hinić, C. Gerber, A. Egli, E. Meyer, Rapid and ultrasensitive detection of mutations and genes relevant to antimicrobial resistance in bacteria. Glob. Chall. **n/a**(n/a), 2000066 (2020). ISSN 2056-6646. https://doi.org/10.1002/gch2.202000066. Publisher: John Wiley & Sons, Ltd

293. A. Hubert, W. Rave, S.L. Tomlinson, Imaging magnetic charges with magnetic force microscopy. Phys. Status Solidi (b) **204**(2), 817–828 (1997). https://doi.org/10.1002/1521-3951(199712)204:2<817::AID-PSSB817>3.0.CO;2-D

294. H.J. Hug, B. Stiefel, A. Moser, I. Parashikov, A. Klicznik, D. Lipp, H.J. Güntherodt, G. Bochi, D.I. Paul, R.C.O. Handley, Magnetic domain structure in ultrathin Cu/Ni/Cu/Si(001) films (invited). J. Appl. Phys. **79**(8), 5609–5614 (1996). ISSN 0021-8979. https://doi.org/10.1063/1.362258, https://aip.scitation.org/doi/abs/10.1063/1.362258. Publisher: American Institute of Physics

295. H.J. Hug, B. Stiefel, P.J.A. van Schendel, A.Moser, R. Hofer, S. Martin, H.J. Güntherodt, S. Porthun, L. Abelmann, J.C. Lodder, G. Bochi, R.C.O. Handley, Quantitative magnetic force microscopy on perpendicularly magnetized samples. J. Appl. Phys. **83**(11), 5609–5620 (1998). ISSN 0021-8979. https://doi.org/10.1063/1.367412. Publisher: American Institute of Physics

296. H.J. Hug, B. Stiefel, P.J.A. van Schendel, A. Moser, R. Hofer, S. Martin, H.J. Güntherodt, S. Porthun, L. Abelmann, J.C. Lodder, G. Bochi, R.C.O. Handley, Quantitative magnetic force microscopy on perpendicularly magnetized samples. J. Appl. Phys. **83**(11), 5609–5620 (1998). ISSN 0021-8979. https://doi.org/10.1063/1.367412. Publisher: American Institute of Physics

297. H.J. Hug, B. Stiefel, P.J.A. van Schendel, A. Moser, S. Martin, H.-J. Güntherodt, A low temperature ultrahigh vaccum scanning force microscope. Rev. Sci. Instr. **70**(9), 3625–3640 (1999). ISSN 0034-6748. https://doi.org/10.1063/1.1149970. Publisher: American Institute of Physics

298. H.J. Hug, Low Temperature Magnetic Force Microscopy. Application on Magnetic Materials and Superconductors. Ph.D. thesis, Department of Physics, University of Basel, CH-4056 Basel, Switzerland (1993)

299. H.J. Hug, A. Moser, I. Parashikov, B. Stiefel, O. Fritz, H.-J. Güntherodt, H. Thomas, Observation and manipulation of vortices in a YBa2Cu3O7 thin film with a low temperature magnetic force microscope. Phys. C: Supercond. **235–240**, 2695–2696 (1994). ISSN 0921-4534. https://doi.org/10.1016/0921-4534(94)92568-2, http://www.sciencedirect.com/science/article/pii/0921453494925682

300. L. Th Huser, N. Lacoste Th., R. Eckert, H. Heinzelmann, Observation and analysis of near-field optical diffraction. J. Opt. Soc. Amer. A **16**(1), 141–148 (1999). https://doi.org/10.1364/JOSAA.16.000141

301. C. Hutter, D. Platz, E.A. Tholen, T.H. Hansson, D.B. Haviland, Reconstructing nonlinearities with intermodulation spectroscopy. Phys. Rev. Lett. **104**(5), 050801 (2010). https://doi.org/10.1103/PhysRevLett.104.050801

302. J.L. Hutter, J. Bechhoefer, Manipulation of van der Waals forces to improve image resolution in atomic force microscopy. J. Appl. Phys. **73**(9), 4123–4129 (1993). ISSN 0021-8979. https://doi.org/10.1063/1.352845, https://aip.scitation.org/doi/10.1063/1.352845. Publisher: American Institute of Physics

303. J.P. Ibe, P.P. Bey, S.L. Brandow, R.A. Brizzolara, N.A. Burnham, D.P. DiLella, K.P. Lee, C.R.K. Marrian, R.J. Colton, On the electrochemical etching of tips for scanning tunneling microscopy. J. Vac. Sci. & Technol. A **8**(4), 3570–3575 (1990). ISSN 0734-2101. https://doi.org/10.1116/1.576509, https://avs.scitation.org/doi/10.1116/1.576509. Publisher: American Vacuum Society

304. S. Kitamura, M. Iwatsuki, Observation of 7 x 7 reconstructed structure on the silicon (111) surface using ultrahigh vacuum noncontact atomic force microscopy. Jpn. J. Appl. Phys. **34**(Part 2, No. 1B), L145–L148 (1995). https://doi.org/10.1143/jjap.34.l145, https://doi.org/10.1143/l145

305. Ó. Iglesias-Freire, J.R. Bates, Y. Miyahara, A. Asenjo, P.H. Grütter, Tip-induced artifacts in magnetic force microscopy images. Appl. Phys. Lett. **102**(2), 022417 (2013). ISSN 0003-6951. https://doi.org/10.1063/1.4776669. Publisher: American Institute of Physics

306. J.N. Israelachvili, *Intermolecular and Surface Forces* (Academic, London, 2011). https://doi.org/10.1016/C2009-0-21560-1

307. M. Jaafar, O. Iglesias-Freire, L.S. Ramon, M.R. Ibarra, J.M. de Teresa, A. Asenjo, Distinguishing magnetic and electrostatic interactions by a kelvin probe force microscopy magnetic force microscopy combination. Beilstein J. Nanotechnol. **2**, 552–560 (2011). ISSN 2190-4286. https://doi.org/10.3762/bjnano.2.59

308. J.D. Jackson, *Classical Electrodynamics*, 3rd edn. (Wiley, New York, 1998). ISBN 978-0-471-30932-1

309. L. Jansen, H. Hölscher, H. Fuchs, A. Schirmeisen, Temperature dependence of atomic scale stick-slip friction. Phys. Rev. Lett. **104** (2010). https://doi.org/10.1103/PhysRevLett.104.256101

310. S.P. Jarvis, H. Yamada, S.I. Yamamoto, H. Tokumoto, A new force controlled atomic force microscope for use in ultrahigh vacuum. Rev. Sci. Instr. **67**(6), 2281–2285 (1996). ISSN 0034-6748. https://doi.org/10.1063/1.1147047, https://aip.scitation.org/doi/10.1063/1.1147047. Publisher: American Institute of Physics

311. S.P. Jarvis, H. Yamada, K. Kobayashi, A. Toda, H. Tokumoto, Normal and lateral force investigation using magnetically activated force sensors. Appl. Surf. Sci. **157**(4), 314–319 (2000). ISSN 0169-4332. https://doi.org/10.1016/S0169-4332(99)00545-0, http://www.sciencedirect.com/science/article/pii/S0169433299005450

312. S.P. Jarvis, T. Uchihashi, T. Ishida, H. Tokumoto, Y. Nakayama, Local solvation shell measurement in water using a carbon nanotube probe. J. Phys. Chem. B **104**(26), 6091–6094 (2000). ISSN 1520-6106. https://doi.org/10.1021/jp001616d. Publisher: American Chemical Society

313. F. Jelezko, J. Wrachtrup, Single defect centres in diamond: a review. Phys. Status Solidi (a) **203**(13), 3207–3225 (2006). ISSN 1862-6300. https://doi.org/10.1002/pssa.200671403. Publisher: John Wiley & Sons, Ltd

314. P. Jelínek, High resolution SPM imaging of organic molecules with functionalized tips. J. Phys.: Conden. Matter **29**(34), 343002 (2017). ISSN 0953-8984. https://doi.org/10.1088/1361-648x/aa76c7. Publisher: IOP Publishing

315. S. Jesse, S.V. Kalinin, R. Proksch, A.P. Baddorf, B.J. Rodriguez, The band excitation method in scanning probe microscopy for rapid mapping of energy dissipation on the nanoscale. Nanotechnology **18**(43), 435503 (2007). ISSN 0957-4484. https://doi.org/10.1088/0957-4484/18/43/435503. Publisher: IOP Publishing

316. C. Joachim, Molecular and intramolecular electronics. Superlattices and Microstruct. **28**(4), 305–315 (2000). ISSN 0749-6036. https://doi.org/10.1006/spmi.2000.0918, http://www.sciencedirect.com/science/article/pii/S0749603600909182

317. C. Joachim, J.K. Gimzewski, A. Aviram, Electronics using hybrid-molecular and mono-molecular devices. Nature **408**(6812), 541–548 (2000). ISSN 1476-4687. https://doi.org/10.1038/35046000, https://www.nature.com/articles/35046000

318. K.L. Johnson, *Contact Mechanics* (Cambridge University Press, Cambridge, 1985). https://doi.org/10.1017/CBO9781139171731

319. M. Johnson, J. Clarke, Spin polarized scanning tunneling microscope: concept, design, and preliminary results from a prototype operated in air. J. Appl. Phys. **67**(10), 6141–6152 (1990). ISSN 0021-8979. https://doi.org/10.1063/1.345176, https://aip.scitation.org/doi/10.1063/1.345176. Publisher: American Institute of Physics

320. N.R. Joshi, S. Özer, T.V. Ashworth, P.G. Stickar, S. Romer, M.A. Marioni, H.J. Hug, Engineering the ferromagnetic domain size for optimized imaging of the pinned uncompensated spins in exchange-biased samples by magnetic force microscopy. Appl. Phys. Lett. **98**(8), 082502 (2011). ISSN 0003-6951. https://doi.org/10.1063/1.3559228. Publisher: American Institute of Physics

321. M. Julliere, Tunneling between ferromagnetic films. Phys. Lett. A **54**(3), 225–226 (1975). ISSN 0375-9601. https://doi.org/10.1016/0375-9601(75)90174-7, http://www.sciencedirect.com/science/article/pii/0375960175901747

322. T. Jung, Y.W. Mo, F.J. Himpsel, Identification of metals in scanning tunneling microscopy via image states. Phys. Rev. Lett. **74**, 1641–1644 (1995)

323. U. Kaiser, A. Schwarz, R. Wiesendanger, Magnetic exchange force microscopy with atomic resolution. Nature **446**(7135), 522–525 (2007). ISSN 1476-4687. https://doi.org/10.1038/nature05617

324. S.V. Kalinin, E. Strelcov, A. Belianinov, S. Somnath, R.K. Vasudevan, E.J. Lingerfelt, R.K. Archibald, C. Chen, R. Proksch, N. Laanait, S. Jesse, Big, Deep, and Smart Data in Scanning Probe Microscopy. ACS Nano **10**(10), 9068–9086 (2016). ISSN 1936-0851. https://doi.org/10.1021/acsnano.6b04212. Publisher: American Chemical Society

325. C.J. Kang, C.K. Kim, J.D. Lera, Y. Kuk, K.M. Mang, J.G. Lee, K.S. Suh, C.C. Williams, Depth dependent carrier density profile by scanning capacitance microscopy. Appl. Phys. Lett. **71**(11), 1546–1548 (1997). ISSN 0003-6951. https://doi.org/10.1063/1.119961. Publisher: American Institute of Physics

326. L.N. Kantorovich, A.S. Foster, A.L. Shluger, A.M. Stoneham, Role of image forces in non-contact scanning force microscope images of ionic surfaces. Surf. Sci. **445**(2), 283–299 (2000). ISSN 0039-6028. https://doi.org/10.1016/S0039-6028(99)01086-9, http://www.sciencedirect.com/science/article/pii/S0039602899010869

327. P. Kappenberger, S. Martin, Y. Pellmont, H.J. Hug, J.B. Kortright, O. Hellwig, E.E. Fullerton, Direct imaging and determination of the uncompensated spin density in exchange-biased CoO/(CoPt) multilayers. Phys. Rev. Lett. **91**, 267202 (2003). https://doi.org/10.1103/PhysRevLett.91.267202

328. P. Kappenberger, I. Schmid, H.J. Hug, Investigation of the exchange bias effect by quantitative magnetic force microscopy. Adv. Eng. Mater. **7**(5), 332–338 (2005). ISSN 1438-1656. https://doi.org/10.1002/adem.200500088. Publisher: John Wiley & Sons, Ltd

329. R.R. Katti, P. Rice, J.C. Wu, H.L. Stadler, Domain imaging in magnetic garnets using tunneling-stabilized magnetic force microscopy. IEEE Trans. Magn. **28**(5), 2913–2915 (1992). https://doi.org/10.1109/20.179670

330. S. Kawai, T. Glatzel, S. Koch, B. Such, A. Baratoff, E. Meyer, Time-averaged cantilever deflection in dynamic force spectroscopy. Phys. Rev. B **80** (2009). https://doi.org/10.1103/PhysRevB.80.085422

331. S. Kawai, T. Glatzel, S. Koch, B. Such, A. Baratoff, E. Meyer, Systematic achievement of improved atomic-scale contrast via bimodal dynamic force microscopy. Phys. Rev. Lett. **103** (2009). https://doi.org/10.1103/PhysRevLett.103.220801

332. S. Kawai, T. Glatzel, S. Koch, B. Such, A. Baratoff, E. Meyer, Ultrasensitive detection of lateral atomic-scale interactions on graphite (0001) via bimodal dynamic force measurements. Phys. Rev. B **81**(8), 085420 (2010). ISSN 1098-0121, 1550-235X. https://doi.org/10.1103/PhysRevB.81.085420, https://link.aps.org/doi/10.1103/PhysRevB.81.085420

333. S. Kawai, F.F. Canova, T. Glatzel, A.S. Foster, E. Meyer, Atomic-scale dissipation processes in dynamic force spectroscopy. Phys. Rev. B **84**(11), 115415 (2011). ISSN 1098-0121, 1550-235X. https://doi.org/10.1103/PhysRevB.84.115415, https://link.aps.org/doi/10.1103/PhysRevB.84.115415

334. S. Kawai, T. Glatzel, S. Koch, A. Baratoff, E. Meyer, Interaction-induced atomic displacements revealed by drift-corrected dynamic force spectroscopy. Phys. Rev. B **83** (2011). https://doi.org/10.1103/PhysRevB.83.035421

335. S. Kawai, S. Hafizovic, T. Glatzel, A. Baratoff, E. Meyer, Rapid reconstruction of a strong nonlinear property by a multiple lock-in technique. Phys. Rev. B **85** (2012). https://doi.org/10.1103/PhysRevB.85.165426

336. S. Kawai, A.S. Foster, F.F. Canova, H. Onodera, S. Kitamura, E. Meyer, Atom manipulation on an insulating surface at room temperature. Nat. Commun. **5**(1), 4403 (2014). ISSN 2041-1723. https://doi.org/10.1038/ncomms5403

337. S. Kawai, M. Koch, E. Gnecco, A. Sadeghi, R. Pawlak, T. Glatzel, J. Schwarz, S. Goedecker, S. Hecht, A. Baratoff, L. Grill, E. Meyer, Quantifying the atomic-level mechanics of single long physisorbed molecular chains. Proc. Natl. Acad. Sci. **111**(11), 3968–3972 (2014). ISSN 0027-8424. https://doi.org/10.1073/pnas.1319938111, https://www.pnas.org/content/111/11/3968

338. S. Kawai, A. Benassi, E. Gnecco, H. Söde, R. Pawlak, X. Feng, K. Müllen, D. Passerone, C.A. Pignedoli, P. Ruffieux, R. Fasel, E. Meyer, Superlubricity of graphene nanoribbons on gold surfaces. Science **351**(6276), 957–961 (2016). ISSN 0036-8075. https://doi.org/10.1126/science.aad3569, https://science.sciencemag.org/content/351/6276/957

339. S. Kawai, A.S. Foster, T. Björkman, S. Nowakowska, J. Bjoerk, F.F. Canova, L.H. Gade, T.A. Jung, E. Meyer, Van der Waals interactions and the limits of isolated atom models at interfaces. Nat. Commun. **7**(1), 11559 (2016). ISSN 2041-1723. https://doi.org/10.1038/ncomms11559, http://www.nature.com/articles/ncomms11559

340. H. Kawakatsu, T. Higuchi, H. Kougami, M. Kawai, M. Watanabe, Y . Hoshi, N. Nishioki, Comparison measurement in the hundred nanometer range with a crystalline lattice using a dual tunneling unit scanning tunneling microscope. J. Vac. Sci. & Technol. B: Microelectron. Nanometer Struct. Proc. Meas. Phen. **12**(3), 1681–1685 (1994). https://doi.org/10.1116/1.587262

341. O. Kazakova, R. Puttock, C. Barton, H. Corte-León, M. Jaafar, V. Neu, A. Asenjo, Frontiers of magnetic force microscopy. J. Appl. Phys. **125**(6), 060901 (2019). ISSN 0021-8979. https://doi.org/10.1063/1.5050712. Publisher: American Institute of Physics. https://doi.org/10.1103/PhysRevB.59.13267

342. S.H. Ke, T. Uda, K. Terakura, Quantity measured in frequency-shift-mode atomic force microscopy: an analysis with a numerical model. Phys. Rev. B **59**, 13267–13272 (1999)

343. S.N. Kempkes, M.R. Slot, S.E. Freeney, S.J.M. Zevenhuizen, D. Vanmaekelbergh, I. Swart, C. Morais Smith, Design and characterization of electrons in a fractal geometry. Nat. Phys. **15**(2), 127–131 (2019). ISSN 1745-2481. https://doi.org/10.1038/s41567-018-0328-0

344. S.N. Kempkes, M.R. Slot, J.J. van den Broeke, P. Capiod, W.A. Benalcazar, D. Vanmaekelbergh, D. Bercioux, I. Swart, C. Morais Smith, Robust zero-energy modes in an electronic higher-order topological insulator. Nat. Mater. **18**(12), 1292–1297 (2019). ISSN 1476-4660. https://doi.org/10.1038/s41563-019-0483-4

345. S.D. Kevan, R.H. Gaylord, High-resolution photoemission study of the electronic structure of the noble-metal (111) surfaces. Phys. Rev. B **36**, 5809–5818 (1987). https://doi.org/10.1103/physrevb.36.5809

346. A.A. Khajetoorians, B. Chilian, J. Wiebe, S. Schuwalow, F. Lechermann, R. Wiesendanger, Detecting excitation and magnetization of individual dopants in a semiconductor. Nature **467**(7319), 1084–1087 (2010). ISSN 1476-4687. https://doi.org/10.1038/nature09519

347. G. Kichin, C. Weiss, C. Wagner, F.S. Tautz, R. Temirov, Single molecule and single atom sensors for atomic resolution imaging of chemically complex surfaces. J. Amer. Chem. Soc. **133**(42), 16847–16851 (2011). ISSN 1520-5126. https://doi.org/10.1021/ja204624g

348. A. Kikukawa, S. Hosaka, R. Imura, Silicon pn junction imaging and characterizations using sensitivity enhanced kelvin probe force microscopy. Appl. Phys. Lett. **66**(25), 3510–3512 (1995). https://doi.org/10.1063/1.113780

349. Y.J. Kim, C.M. Lieber, Nanotube nanotweezers. Science **286**, 2148 (1999), https://science.sciencemag.org/content/286/5447/2148

350. Y.J. Kim, C. Westphal, R.X. Ynzunza, Z. Wang, H.C. Galloway, M. Salmeron, M.A. Van Hove, C.S. Fadley, The growth of iron oxide films on Pt(111): a combined XPD, STM, and LEED study. Surf. Sci. **416**(1), 68–111 (1998). ISSN 0039-6028. https://doi.org/10.1016/S0039-6028(98)00506-8, http://www.sciencedirect.com/science/article/pii/S0039602898005068

351. J.R. Kirtley, Fundamental studies of superconductors using scanning magnetic imaging. Rep. Prog. Phys. **73**(12), 126501 (2010). ISSN 0034-4885. https://doi.org/10.1088/0034-4885/73/12/126501. Publisher: IOP Publishing

352. J.R. Kirtley, M.B. Ketchen, K.G. Stawiasz, J.Z. Sun, W.J. Gallagher, S.H. Blanton, S.J. Wind, High-resolution scanning SQUID microscope. Appl. Phys. Lett. **66**(9), 1138–1140 (1995). ISSN 0003-6951. https://doi.org/10.1063/1.113838. Publisher: American Institute of Physics

353. J.R. Kirtley, C.C. Tsuei, J.Z. Sun, C.C. Chi, L.S. Yu-Jahnes, A. Gupta, M. Rupp, M.B. Ketchen, Symmetry of the order parameter in the high-Tc superconductor YBa2Cu3O7- δ. Nature **373**(6511), 225–228 (1995). ISSN 1476-4687. https://doi.org/10.1038/373225a0

354. M. Kisiel, E. Gnecco, U. Gysin, L. Marot, S. Rast, E. Meyer, Suppression of electronic friction on Nb films in the superconducting state. Nat. Mater. **10**(2), 119–122 (2011). ISSN 1476-1122, 1476-4660. https://doi.org/10.1038/nmat2936, http://www.nature.com/articles/nmat2936

355. S. Kitamura, K. Suzuki, M. Iwatsuki, C.B. Mooney, Atomic-scale variations in contact potential difference on au/si(111) 7 x 7surface in ultrahigh vacuum. Appl. Surf. Sci. **157**(4), 222–227 (2000). ISSN 0169-4332. https://doi.org/10.1016/S0169-4332(99)00530-9, http://www.sciencedirect.com/science/article/pii/S0169433299005309

356. K. Kobayashi, H. Yamada, K. Matsushige, Frequency noise in frequency modulation atomic force microscopy. Rev. Sci. Instr. **80**(4), 043708–9 (2009). https://doi.org/10.1063/1.3120913

357. G.P. Kochanski, Nonlinear alternating-current tunneling microscopy. Phys. Rev. Lett. **62**, 2285–2288 (1989). https://doi.org/10.1103/PhysRevLett.62.2285

358. J. Koglin, U.C. Fischer, H. Fuchs, Material contrast in scanning near-field optical microscopy at 1–10 nm resolution. Phys. Rev. B **55**(12), 7977–7984 (1997). https://doi.org/10.1103/PhysRevB.55.7977

359. D.D. Koleske, W.R. Barger, G.U. Lee, R.J. Colton, Scanning probe microscope study of mixed chain-length phase-segregated Langmuir-Blodgett monolayers. MRS Online Proceedings Library Archive, p. 464 (1996). ISSN 0272–9172, 1946–4274

360. L. Kong, S.Y. Chou, Study of magnetic properties of magnetic force microscopy probes using micronscale current rings. J. Appl. Phys. **81**(8), 5026–5028 (1997). ISSN 0021-8979. https://doi.org/10.1063/1.364499. Publisher: American Institute of Physics

361. L. Kong, S.Y. Chou, Quantification of magnetic force microscopy using a micronscale current ring. Appl. Phys. Lett. **70**(15), 2043–2045 (1997). ISSN 0003-6951. https://doi.org/10.1063/1.118808. Publisher: American Institute of Physics

362. L. Kong, R.C. Shi, P.R. Krauss, S.Y. Chou, Writing bits of longitudinal quantized magnetic disk using magnetic force microscope tip. Jpn. J. Appl. Phys. **36**(Part 1, No. 8), 5109–5111 (1997). https://doi.org/10.1143/jjap.36.5109

363. D.O. Koralek, W.F. Heinz, M.D. Antonik, A. Baik, J.H. Hoh, Probing deep interaction potentials with white-noise-driven atomic force microscope cantilevers. Appl. Phys. Lett. **76**(20), 2952–2954 (2000). ISSN 0003-6951. https://doi.org/10.1063/1.126527, https://aip.scitation.org/doi/10.1063/1.126527. Publisher: American Institute of Physics

364. Y.E. Korchev, C.L. Bashford, M. Milovanovic, I. Vodyanoy, M.J. Lab, Scanning ion conductance microscopy of living cells. Biophys. J. **73**(2), 653–658 (1997). ISSN 0006-3495, https://www.ncbi.nlm.nih.gov/pmc/articles/PMC1180964/

365. Y.E. Korchev, J. Gorelik, M.J. Lab, E.V. Sviderskaya, C.L. Johnston, C.R. Coombes, I. Vodyanoy, C.R. Edwards, Cell volume measurement using scanning ion conductance microscopy. Biophys. J. **78**(1), 451–457 (2000). ISSN 0006-3495. https://doi.org/10.1016/S0006-3495(00)76607-0

366. T. Kosub, M. Kopte, R. Hühne, P. Appel, B. Shields, P. Maletinsky, R. Hübner, M.O. Liedke, J. Fassbender, O.G. Schmidt, D. Makarov, Purely antiferromagnetic magnetoelectric random access memory. Nat. Commun. **8**(1), 13985 (2017). ISSN 2041-1723. https://doi.org/10.1038/ncomms13985

367. N. Kresz, J. Kokavecz, T. Smausz, B. Hopp, M. Csete, S. Hild, O. Marti, Investigation of pulsed laser deposited crystalline ptfe thin layer with pulsed force mode afm. Thin Solid Films **453–454**, 239–244 (2004). ISSN 0040-6090. https://doi.org/10.1016/j.tsf.2003.11.254, http://www.sciencedirect.com/science/article/pii/S0040609003017292. Proceedings of Symposium H on Photonic Processing of Surfaces, Thin Films and Devices, of the E-MRS 2003 Spring Conference

368. H.-U. Krotil, E. Weilandt, Th Stifter, O. Marti, S. Hild, Dynamic friction force measurement with the scanning force microscope. Surf. Int. Anal. **27**(5–6), 341–347 (1999). ISSN 1096-9918. https://doi.org/10.1002/(SICI)1096-9918(199905/06)27:5/6<341::AID-SIA513>3.0.CO;2-B, https://onlinelibrary.wiley.com/doi/abs/10.1002/(SICI)1096-9918(199905/06)27:5/6<341::AID-SIA513>3.0.CO;2-B

369. H.-U. Krotil, Th. Stifter, O. Marti, Concurrent measurement of adhesive and elastic surface properties with a new modulation technique for scanning force microscopy. Rev. Sci. Instr. **71**(7), 2765–2771 (2000). ISSN 0034-6748. https://doi.org/10.1063/1.1150689, https://aip.scitation.org/doi/10.1063/1.1150689. Publisher: American Institute of Physics

370. H.-U. Krotil, T. Stifter, H. Waschipky, K. Weishaupt, S. Hild, O. Marti, Pulsed force mode: a new method for the investigation of surface properties. Surf. Int. Anal. **27**(5–6), 336–340 (1999). ISSN 0142-2421. https://doi.org/10.1002/(SICI)1096-9918(199905/06)27:5/6<336::AID-SIA512>3.0.CO;2-0. Publisher: John Wiley and Sons, Ltd

371. A. Kubetzka, C. Hanneken, R. Wiesendanger, K. von Bergmann, Impact of the skyrmion spin texture on magnetoresistance. Phys. Rev. B **95** (2017)

372. A. Kühle, A.H. Sorensen, J.B. Zandbergen, J. Bohr, Contrast artifacts in tapping tip atomic force microscopy. Appl. Phys. A **66**(1), S329–S332 (1998). ISSN 1432-0630. https://doi.org/10.1007/s003390051156

373. K.H. Kuit, J.R. Kirtley, W. van der Veur, C.G. Molenaar, F.J.G. Roesthuis, A.G.P. Troeman, J.R. Clem, H. Hilgenkamp, H. Rogalla, J. Flokstra, Vortex trapping and expulsion in thin-film $yba_2cu_3O_{7-\delta}$ strips. Phys. Rev. B **77** (2008). https://doi.org/10.1103/PhysRevB.77.134504

374. M. Kulawik, M. Nowicki, G. Thielsch, L. Cramer, H.-P. Rust, H.-J. Freund, T.P. Pearl, P.S. Weiss, A double lamellae dropoff etching procedure for tungsten tips attached to tuning fork atomic force microscopy/scanning tunneling microscopy sensors. Rev. Sci. Instr. **74**(2), 1027–1030 (2003). ISSN 0034-6748. https://doi.org/10.1063/1.1532833. Publisher: American Institute of Physics

375. H. Kuramochi, T. Uzumaki, M. Yasutake, A. Tanaka, H. Akinaga, H. Yokoyama, A magnetic force microscope using CoFe-coated carbon nanotube probes. Nanotechnology **16**(1), 24–27 (2004). https://doi.org/10.1088/0957-4484/16/1/006

376. A. Labuda, K. Kobayashi, Y. Miyahara, P. Grütter, Retrofitting an atomic force microscope with photothermal excitation for a clean cantilever response in low Q environments. Rev. Sci. Instr. **83**(5), 053703 (2012). ISSN 0034-6748. https://doi.org/10.1063/1.4712286, https://aip.scitation.org/doi/10.1063/1.4712286. Publisher: American Institute of Physics

377. Th. Lacoste, Th. Huser, R. Prioli, H. Heinzelmann, Contrast enhancement using polarization-modulation scanning near-field optical microscopy (PM-SNOM). Ultramicroscopy **71**(1), 333–340 (1998). ISSN 0304-3991. https://doi.org/10.1016/S0304-3991(97)00093-4, http://www.sciencedirect.com/science/article/pii/S0304399197000934

378. H.P. Lang, M. Hegner, E. Meyer, Ch. Gerber, Nanomechanics from atomic resolution to molecular recognition based on atomic force microscopy technology. Nanotechnology **13**(5), R29–R36 (2002). https://doi.org/10.1088/0957-4484/13/5/202

379. H.P. Lang, R. Berger, F. Battiston, J.-P. Ramseyer, E. Meyer, C. Andreoli, J. Brugger, P. Vettiger, M. Despont, T. Mezzacasa, L. Scandella, H.-J. Güntherodt, C. Gerber, J.K. Gimzewski, A chemical sensor based on a micromechanical cantilever array for the identification of gases and vapors. Appl. Phys. A **66**(1), S61–S64 (1998). ISSN 1432-0630. https://doi.org/10.1007/s003390051100

380. M. Langer, M. Kisiel, R. Pawlak, F. Pellegrini, G.E. Santoro, R. Buzio, A. Gerbi, G. Balakrishnan, A. Baratoff, E. Tosatti, E. Meyer, Giant frictional dissipation peaks and charge-density-wave slips at the NbSe2 surface. Nat. Mater. **13**(2), 173–177 (2014). ISSN 1476-4660. https://doi.org/10.1038/nmat3836

381. M.A. Lantz, S.J.O. Shea, A.C.F. Hoole, M.E. Welland, Lateral stiffness of the tip and tip-sample contact in frictional force microscopy. Appl. Phys. Lett. **70**(8), 970–972 (1997). ISSN 0003-6951, 1077-3118. https://doi.org/10.1063/1.118476, http://aip.scitation.org/doi/10.1063/1.118476

382. M.A. Lantz, S.J.O. Shea, M.E. Welland, Simultaneous force and conduction measurements in atomic force microscopy. Phys. Rev. B **56**(23), 15345–15352 (1997). https://doi.org/10.1103/PhysRevB.56.15345

383. M.A. Lantz, S.J.O. Shea, M.E. Welland, Characterization of tips for conducting atomic force microscopy in ultrahigh vacuum. Rev. Sci. Instr. **69**(4), 1757–1764 (1998). ISSN 0034-6748. https://doi.org/10.1063/1.1148838, https://aip.scitation.org/doi/10.1063/1.1148838. Publisher: American Institute of Physics

384. M.A. Lantz, H.J. Hug, P.J.A. van Schendel, R. Hoffmann, S. Martin, A. Baratoff, A. Abdurixit, H.-J. Güntherodt, Ch. Gerber, Low temperature scanning force microscopy of the Si(111) − (7 × 7) surface. Phys. Rev. Lett. **84**, 2642–2645 (2000). https://doi.org/10.1103/PhysRevLett.84.2642

385. M.A. Lantz, H.J. Hug, R. Hoffmann, P.J.A. van Schendel, P. Kappenberger, S. Martin, A. Baratoff, H.-J. Güntherodt, Quantitative Measurement of Short-Range Chemical Bonding Forces. Science **291**(5513), 2580–2583 (2001). ISSN 0036-8075, 1095-9203. https://doi.org/10.1126/science.1057824, https://science.sciencemag.org/content/291/5513/2580. Publisher: American Association for the Advancement of Science Section: Report

386. B.M. Law, F. Rieutord, Electrostatic forces in atomic force microscopy. Phys. Rev. B **66** (2002). https://doi.org/10.1063/1.363884

387. C. Lazo, V. Caciuc, H. Hölscher, S. Heinze, Role of tip size, orientation, and structural relaxations in first-principles studies of magnetic exchange force microscopy and spin-polarized scanning tunneling microscopy. Phys. Rev. B **78** (2008). https://doi.org/10.1103/PhysRevB.78.214416

388. C. Lazo, S. Heinze, First-principles study of magnetic exchange force microscopy with ferromagnetic and antiferromagnetic tips. Phys. Rev. B **84** (2011). https://doi.org/10.1103/PhysRevB.84.144428

389. T. Leinhos, M. Stopka, E. Oesterschulze, Micromachined fabrication of Si cantilevers with Schottky diodes integrated in the tip. Appl. Phys. A **66**(1), S65–S69 (1998). ISSN 1432-0630. https://doi.org/10.1007/s003390051101

390. J.K. Leong and C. C. Williams. Shear force microscopy with capacitance detection for near field scanning optical microscopy. Appl. Phys. Lett. **66**(11), 1432–1434 (1995). ISSN 0003-6951. https://doi.org/10.1063/1.113269. Publisher: American Institute of Physics

391. J.W. Li, J.P. Cleveland, R. Proksch, Bimodal magnetic force microscopy: Separation of short and long range forces. Appl. Phys. Lett. **94**(16), 163118 (2009). ISSN 0003-6951. https://doi.org/10.1063/1.3126521. Publisher: American Institute of Physics

392. J. Li, W.-D. Schneider, R. Berndt, S. Crampin, Electron confinement to nanoscale ag islands on ag(111): a quantitative study. Phys. Rev. Lett. **80**, 3332–3335 (1998). https://doi.org/10.1103/PhysRevLett.80.3332

393. W. Liang, M.P. Shores, M. Bockrath, J.R. Long, H. Park, Kondo resonance in a single-molecule transistor. Nature **417**(6890), 725–729 (2002). ISSN 1476-4687. https://doi.org/10.1038/nature00790

394. T.R. Linderoth, S. Horch, L. Petersen, S. Helveg, E. Lægsgaard, I. Stensgaard, F. Besenbacher, Novel mechanism for diffusion of one-dimensional clusters: Pt/Pt(110) − (1 × 2). Phys. Rev. Lett. **82**, 1494–1497 (1999). https://doi.org/10.1103/PhysRevLett.82.1494

395. X.X. Liu, R. Heiderhoff, H.P. Abicht, L.J. Balk, Scanning near-field acoustic study of ferroelectric BaTiO3ceramics. J. Phys. D: Appl. Phys. **35**(1), 74–87 (2001). https://doi.org/10.1088/0022-3727/35/1/313

396. Y. Liu, P. Grütter, Magnetic dissipation force microscopy studies of magnetic materials (invited). J. Appl. Phys. **83**(11), 7333–7338 (1998). ISSN 0021-8979. https://doi.org/10.1063/1.367825. Publisher: American Institute of Physics

397. Y. Liu, B. Ellman, P. Grütter, Theory of magnetic dissipation imaging. Appl. Phys. Lett. **71**(10), 1418–1420 (1997). ISSN 0003-6951. https://doi.org/10.1063/1.119911. Publisher: American Institute of Physics

398. A.I. Livshits, A.L. Shluger, Self-lubrication in scanning-force-microscope image formation on ionic surfaces. Phys. Rev. B **56**, 12482–12489 (1997). https://doi.org/10.1103/PhysRevB.56.12482

399. A.I. Livshits, A.L. Shluger, A.L. Rohl, A.S. Foster, Model of noncontact scanning force microscopy on ionic surfaces. Phys. Rev. B **59**, 2436–2448 (1999). https://doi.org/10.1103/PhysRevB.59.2436

400. J. Lohau, S. Kirsch, A. Carl, G. Dumpich, E.F. Wassermann, Quantitative determination of effective dipole and monopole moments of magnetic force microscopy tips. J. Appl. Phys. **86**(6), 3410–3417 (1999). ISSN 0021-8979. https://doi.org/10.1063/1.371222. Publisher: American Institute of Physics

401. G.P. Lopinski, D.D.M. Wayner, R.A. Wolkow, Self-directed growth of molecular nanostructures on silicon. Nature **406**(6791), 48–51 (2000). ISSN 1476-4687. https://doi.org/10.1038/35017519

402. C. Loppacher, R. Bennewitz, O. Pfeiffer, M. Guggisberg, M. Bammerlin, S. Schär, V. Barwich, A. Baratoff, E. Meyer, Experimental aspects of dissipation force microscopy. Phys. Rev. B **62**(20), 13674–13679 (2000). https://doi.org/10.1103/PhysRevB.62.13674

403. Ch. Loppacher, M. Bammerlin, M. Guggisberg, F. Battiston, R. Bennewitz, S. Rast, A. Baratoff, E. Meyer, H.J. Güntherodt, Phase variation experiments in non-contact dynamic force microscopy using phase locked loop techniques. Appl. Surf. Sci. **140**(3), 287–292 (1999). ISSN 0169-4332. https://doi.org/10.1016/S0169-4332(98)00542-X, http://www.sciencedirect.com/science/article/pii/S016943329800542X

404. Ch. Loppacher, M. Bammerlin, M. Guggisberg, S. Schär, R. Bennewitz, A. Baratoff, E. Meyer, H.J. Güntherodt, Dynamic force microscopy of copper surfaces: atomic resolution and distance dependence of tip-sample interaction and tunneling current. Phys. Rev. B **62**, 16944–16949 (2000). https://doi.org/10.1103/PhysRevB.62.16944

405. Ch. Loppacher, M. Bammerlin, M. Guggisberg, E. Meyer, H.-J. Güntherodt, R. Lüthi, R. Schlittler, J.K. Gimzewski, Forces with submolecular resolution between the probing tip and Cu-TBPP molecules on Cu(100) observed with a combined AFM/STM. Appl. Phys. A **72**(1), S105–S108 (2001). ISSN 1432-0630. https://doi.org/10.1007/s003390100643

406. W.J. Lorenz, W. Plieth (eds.), *Electrochemical Nanotechnology* (Wiley-VCH, Weinheim, 1998). ISBN 978-3-527-61214-7

407. D. Loss, D.P. DiVincenzo, Quantum computation with quantum dots. Phys. Rev. A **57**, 120–126 (1998). https://doi.org/10.1103/PhysRevA.57.120

408. J. Lübbe, L. Tröger, S. Torbrügge, R. Bechstein, C. Richter, A. Kühnle, M. Reichling, Achieving high effectiveQ-factors in ultra-high vacuum dynamic force microscopy. Meas. Sci. Technol. **21**(12), 125501 (2010). ISSN 0957-0233. https://doi.org/10.1088/0957-0233/21/12/125501. Publisher: IOP Publishing

409. K. Luo, Z. Shi, J. Varesi, A. Majumdar, Sensor nanofabrication, performance, and conduc-
     tion mechanisms in scanning thermal microscopy. J. Vac. Sci. & Technol. B: Microelectron.
     Nanometer Struct. Proc. Meas. Phen. **15**(2), 349–360 (1997). ISSN 1071-1023. https://doi.org/
     10.1116/1.589319, https://avs.scitation.org/doi/abs/10.1116/1.589319. Publisher: American
     Institute of Physics
410. R. Lüthi, E. Meyer, M. Bammerlin, A. Baratoff, T. Lehmann, L. Howald, Ch. Gerber, H.J.
     Güntherodt, Atomic Resolution in Dynamic Force Microscopy Across Steps on Si (111)7 × 7.
     Z. Phys. B **100**, 165 (1996), http://edoc.unibas.ch/dok/A5839473
411. R. Lüthi, E. Meyer, M. Bammerlin, L. Howald, H. Haefke, T. Lehmann, C. Loppacher, H.J.
     Güntherodt, T. Gyalog, H. Thomas, Friction on the atomic scale: An ultrahigh vacuum atomic
     force microscopy study on ionic crystals. J. Vac. Sci. & Technol. B: Microelectron. Nanometer
     Struct. Proc. Meas. Phen. **14**(2), 1280–1284 (1996). ISSN 1071-1023. https://doi.org/10.1116/
     1.589081, https://avs.scitation.org/doi/abs/10.1116/1.589081. Publisher: American Institute
     of Physics
412. R. Lüthi, E. Meyer, M. Bammerlin, A. Baratoff, L. Howald, Ch. Gerber, H.J. Güntherodt,
     Ultrahigh vacuum atomic force microscopy: true atomic resolution. Surf. Rev. Lett. **04**(05),
     1025–1029 (1997). ISSN 0218-625X. https://doi.org/10.1142/S0218625X9700122X. Pub-
     lisher: World Scientific Publishing Co
413. J.W. Lyding, T.C. Shen, J.S. Hubacek, J.R. Tucker, G.C. Abeln, Nanoscale patterning and
     oxidation of h passivated si (100) 2 x 1 surfaces with an ultrahigh vacuum scanning tunneling
     microscope. Appl. Phys. Lett. **64**(15), 2010–2012 (1994). https://doi.org/10.1063/1.111722
414. V. Madhavan, W. Chen, T. Jamneala, M.F. Crommie, N.S. Wingreen, Tunneling into a single
     magnetic atom: spectroscopic evidence of the kondo resonance. Science **280**(5363), 567–
     569 (1998). ISSN 0036-8075. https://doi.org/10.1126/science.280.5363.567, https://science.
     sciencemag.org/content/280/5363/567. Publisher: American Association for the Advance-
     ment of Science _eprint: https://science.sciencemag.org/content/280/5363/567.full.pdf
415. Y. Maeda, T. Matsumoto, T. Kawai, Observation of single- and double-stranded DNA
     using non-contact atomic force microscopy. Appl. Surf. Sci. **140**(3), 400–405 (1999). ISSN
     0169-4332. https://doi.org/10.1016/S0169-4332(98)00562-5, http://www.sciencedirect.com/
     science/article/pii/S0169433298005625
416. I. Maggio-Aprile, Ch. Renner, A. Erb, E. Walker, Ø. Fischer, Direct vortex lattice imaging and
     tunneling spectroscopy of flux lines on yba$_2$cu$_3$O$_{7-\delta}$. Phys. Rev. Lett. **75**, 2754–2757 (1995).
     https://doi.org/10.1103/PhysRevLett.75.2754
417. S.N. Magonov, V. Elings, M.-H. Whangbo, Phase imaging and stiffness in tapping-mode
     atomic force microscopy. Surf. Sci. **375**(2), L385–L391 (1997). ISSN 0039-6028. https://
     doi.org/10.1016/S0039-6028(96)01591-9, http://www.sciencedirect.com/science/article/pii/
     S0039602896015919
418. A. Majumdar, J.P. Carrejo, J. Lai, Thermal imaging using the atomic force microscope. Appl.
     Phys. Lett. **62**(20), 2501–2503 (1993). ISSN 0003-6951. https://doi.org/10.1063/1.109335.
     Publisher: American Institute of Physics
419. A. Majumdar, K. Luo, J. Lai, Z. Shi, Failure analysis of sub-micrometer devices and struc-
     tures using scanning thermal microscopy, in *Proceedings of International Reliability Physics
     Symposium* (1996). https://doi.org/10.1109/RELPHY.1996.492140
420. P. Maletinsky, S. Hong, M.S. Grinolds, B. Hausmann, M.D. Lukin, R.L. Walsworth, M. Loncar,
     A. Yacoby, A robust scanning diamond sensor for nanoscale imaging with single nitrogen-
     vacancy centres. Nat. Nanotechnol. **7**(5), 320–324 (2012). ISSN 1748-3395. https://doi.org/
     10.1038/nnano.2012.50
421. H.J. Mamin, D. Rugar, Thermomechanical writing with an atomic force microscope tip. Appl.
     Phys. Lett. **61**(8), 1003–1005 (1992). ISSN 0003-6951. https://doi.org/10.1063/1.108460,
     https://aip.scitation.org/doi/10.1063/1.108460. Publisher: American Institute of Physics
422. H.J. Mamin, D. Rugar, Sub-attonewton force detection at millikelvin temperatures. Appl.
     Phys. Lett. **79**(20), 3358–3360 (2001). ISSN 0003-6951. https://doi.org/10.1063/1.1418256.
     Publisher: American Institute of Physics

423. H.J. Mamin, D. Rugar, J.E. Stern, B.D. Terris, S.E. Lambert, Force microscopy of magnetization patterns in longitudinal recording media. Appl. Phys. Lett. **53**(16), 1563–1565 (1988). ISSN 0003-6951. https://doi.org/10.1063/1.99952, https://aip.scitation.org/doi/10.1063/1.99952. Publisher: American Institute of Physics

424. H.J. Mamin, D. Rugar, J.E. Stern, R.E. Fontana, P. Kasiraj, Magnetic force microscopy of thin permalloy films. Appl. Phys. Lett. **55**(3), 318–320 (1989). https://doi.org/10.1063/1.101898

425. S. Manalis, K. Babcock, J. Massie, V. Elings, M. Dugas, Submicron studies of recording media using thin-film magnetic scanning probes. Appl. Phys. Lett. **66**(19), 2585–2587 (1995). ISSN 0003-6951. https://doi.org/10.1063/1.113509. Publisher: American Institute of Physics

426. M.H. Jericho, P.J. Mulhern, W. Xu, B.L. Blackford, L. Fritz, Scanning force microscopy and STM imaging of B-chitin. **1855** (1993). https://doi.org/10.1117/12.146379

427. H.C. Manoharan, C.P. Lutz, D.M. Eigler, Quantum mirages formed by coherent projection of electronic structure. Nature **403**(6769), 512–515 (2000). ISSN 1476-4687. https://doi.org/10.1038/35000508

428. M. Mansuripur, R.C. Giles, Demagnetizing field computation for dynamic simulation of the magnetization reversal process (1988). https://doi.org/10.1109/20.92100

429. M.A. Marioni, N. Pilet, T.V. Ashworth, R.C. O'Handley, H.J. Hug, Remanence due to wall magnetization and counterintuitive magnetometry data in 200-nm films of ni. Phys. Rev. Lett. **97** (2006). https://doi.org/10.1103/PhysRevLett.97.027201

430. M.A. Marioni, M. Penedo, M. Baćani, J. Schwenk, H.J. Hug, Halbach Effect at the Nanoscale from Chiral Spin Textures. Nano Lett. **18**(4), 2263–2267 (2018). ISSN 1530-6984. https://doi.org/10.1021/acs.nanolett.7b04802. Publisher: American Chemical Society

431. M. Marth, D. Maier, J. Honerkamp, R. Brandsch, G. Bar, A unifying view on some experimental effects in tapping-mode atomic force microscopy. J. Appl. Phys. **85**(10), 7030–7036 (1999). https://doi.org/10.1063/1.370508

432. O. Marti, V. Elings, M. Haugan, C.E. Bracker, J. Schneir, B. Drake, S.A.C. Gould, J. Gurley, L. Hellemans, K. Shaw, A.L. Weisenhorn, J. Zasadzinski, P.K. Hansma, Scanning probe microscopy of biological samples and other surfaces. J. Microsc. **152**(3), 803–809 (1988). ISSN 0022-2720. https://doi.org/10.1111/j.1365-2818.1988.tb01452.x. Publisher: John Wiley & Sons, Ltd

433. Y. Martin, H.K. Wickramasinghe, Magnetic imaging by force microscopy with 1000 å resolution. Appl. Phys. Lett. **50**(20), 1455–1457 (1987). ISSN 0003-6951. https://doi.org/10.1063/1.97800. Publisher: American Institute of Physics

434. Y. Martin, C.C. Williams, H.K. Wickramasinghe, Atomic force microscope force mapping and profiling on a sub 100 Å scale. J. Appl. Phys. **61**(10), 4723–4729 (1987). ISSN 0021-8979. https://doi.org/10.1063/1.338807. Publisher: American Institute of Physics

435. Y. Martin, D. Rugar, H.K. Wickramasinghe, High-resolution magnetic imaging of domains in TbFe by force microscopy. Appl. Phys. Lett. **52**(3), 244–246 (1988). ISSN 0003-6951. https://doi.org/10.1063/1.99482. Publisher: American Institute of Physics

436. D. Martínez-Martín, M. Jaafar, R. Pérez, J. Gómez-Herrero, A. Asenjo, Upper bound for the magnetic force gradient in graphite. Phys. Rev. Lett. **105** (2010)

437. C.M. Mate, Nanotribology of lubricated and unlubricated carbon overcoats on magnetic disks studied by friction force microscopy. Surf. Coat. Technol. **62**(1), 373–379 (1993). ISSN 0257-8972. https://doi.org/10.1016/0257-8972(93)90270-X, http://www.sciencedirect.com/science/article/pii/025789729390270X

438. C.M. Mate, G.M. McClelland, R. Erlandsson, S. Chiang, Atomic-scale friction of a tungsten tip on a graphite surface. Phys. Rev. Lett. **59**(17), 1942–1945 (1987). https://doi.org/10.1103/PhysRevLett.59.1942

439. C.M. Mate, Nanotribology studies of carbon surfaces by force microscopy. Wear **168**(1), 17–20 (1993). ISSN 0043-1648. https://doi.org/10.1016/0043-1648(93)90192-O, http://www.sciencedirect.com/science/article/pii/004316489390192O

440. J.R. Matey, J. Blanc, Scanning capacitance microscopy. J. Appl. Phys. **57**(5), 1437–1444 (1985). https://doi.org/10.1063/1.334506

441. G. Matteucci, M. Muccini, U. Hartmann, Flux measurements on ferromagnetic microprobes by electron holography. Phys. Rev. B **50**, 6823–6828 (1994). https://doi.org/10.1103/PhysRevB. 50.6823

442. G.M. McClelland, C.T. Rettner, Scanning aperture photoemission microscopy for magnetic imaging. Appl. Phys. Lett. **77**(10), 1511–1513 (2000). https://doi.org/10.1063/1.1290721

443. G.M. McClelland, J.N. Glosli, Friction at the atomic scale, in *Fundamentals of Friction: Macroscopic and Microscopic Processes, NATO-ASI E*, ed. by I. Singer, H. Pollock, vol. 220 (Kluwer Academic, Dordrecht, 1992), p. 427. https://doi.org/10.1007/978-94-011-2811-7

444. T.W. McDaniel, Ultimate limits to thermally assisted magnetic recording. J. Phys.: Conden. Matter **17**(7), R315–R332 (2005). ISSN 0953-8984. https://doi.org/10.1088/0953-8984/17/ 7/r01, http://dx.doi.org/10.1088/0953-8984/17/7/R01. Publisher: IOP Publishing

445. F. Meier, L. Zhou, J. Wiebe, R. Wiesendanger, Revealing Magnetic Interactions from Single-Atom Magnetization Curves. Science **320**(5872), 82 (2008). https://doi.org/10.1126/science. 1154415

446. K.-Y. Meng, A.S. Ahmed, M. Baćani, A.-O. Mandru, X. Zhao, N. Bagués, B.D. Esser, J. Flores, D.W. McComb, H.J. Hug, F. Yang, Observation of Nanoscale Skyrmions in SrIrO3/SrRuO3 Bilayers. Nano Lett. **19**(5), 3169–3175 (2019). ISSN 1530-6984. https://doi.org/10.1021/acs. nanolett.9b00596. Publisher: American Chemical Society

447. F. Menges, P. Mensch, H. Schmid, H. Riel, A. Stemmer, B. Gotsmann, Temperature mapping of operating nanoscale devices by scanning probe thermometry. Nat. Commun. **7**(1), 10874 (2016). ISSN 2041-1723. https://doi.org/10.1038/ncomms10874

448. F. Menges, H. Riel, A. Stemmer, B. Gotsmann, Nanoscale thermometry by scanning thermal microscopy. Rev. Sci. Instr. **87**(7), 074902 (2016). ISSN 0034-6748. https://doi.org/10.1063/1. 4955449, https://aip.scitation.org/doi/abs/10.1063/1.4955449. Publisher: American Institute of Physics

449. E. Meyer, H. Heinzelmann, H. Rudin, H.J. Güntherodt, Atomic resolution on LiF (001) by atomic force microscopy. Zeitschrift für Physik B Condensed Matter **79**(1), 3–4 (1990). ISSN 0722-3277, 1434-6036. https://doi.org/10.1007/BF01387818, http://link.springer.com/ 10.1007/BF01387818

450. E. Meyer, H.-J. Güntherodt, H. Haefke, G. Gerth, M. Krohn, Atomic resolution on the AgBr(001) surface by atomic force microscopy. Europhys. Lett. (EPL) **15**(3), 319–323 (1991). https://doi.org/10.1209/0295-5075/15/3/015

451. E. Meyer, R. Overney, D. Brodbeck, L. Howald, R. Lüthi, J. Frommer, H.-J. Güntherodt, Friction force microscopy of Langmuir-Blodgett films, in *Fundamentals of Friction, NATO-ASI E*, ed. by I. Singer, H. Pollock, vol. 220 (1992), p. 427. https://doi.org/10.1007/978-94-011-2811-7

452. E. Meyer, T. Gyalog, R.M. Overney, K. Dransfeld, *Nanoscience: Friction and Rheology on the Nanometer Scale* (World Scientific, Singapore, 1998). https://doi.org/10.1142/3026

453. G. Meyer, N.M. Amer, Novel optical approach to atomic force microscopy. Appl. Phys. Lett. **53**(12), 1045–1047 (1988). ISSN 0003-6951. https://doi.org/10.1063/1.100061. Publisher: American Institute of Physics

454. G. Meyer, N.M. Amer, Simultaneous measurement of lateral and normal forces with an optical beam deflection atomic force microscope. Appl. Phys. Lett. **57**(20), 2089–2091 (1990). ISSN 0003-6951. https://doi.org/10.1063/1.103950. Publisher: American Institute of Physics

455. J. Michaelis, C. Hettich, J. Mlynek, V. Sandoghdar, Optical microscopy using a single-molecule light source. Nature **405**(6784), 325–328 (2000). ISSN 1476-4687. https://doi.org/ 10.1038/35012545

456. A. Michels, F. Meinen, T. Murdfield, W. Göhde, U.C. Fischer, E. Beckmann, H. Fuchs, 1 mhz quartz length extension resonator as a probe for scanning near-field acoustic microscopy. Thin Solid Films **264**(2), 172–175 (1995). ISSN 0040-6090. https://doi.org/10.1016/0040-6090(95)05853-2, http://www.sciencedirect.com/science/article/pii/0040609095058532

457. P. Milde, D. Köhler, J. Seidel, L.M. Eng, A. Bauer, A. Chacon, J. Kindervater, S. Mühlbauer, C. Pfleiderer, S. Buhrandt, C. Schütte, A. Rosch, Unwinding of a skyrmion lattice by magnetic monopoles. Science **340**(6136), 1076 (2013). https://doi.org/10.1126/science.1234657

458. G. Mills, H. Zhou, A. Midha, L. Donaldson, J.M.R. Weaver, Scanning thermal microscopy using batch fabricated thermocouple probes. Appl. Phys. Lett. **72**(22), 2900–2902 (1998). ISSN 0003-6951. https://doi.org/10.1063/1.121453. Publisher: American Institute of Physics
459. S.C. Minne, J.D. Adams, G. Yaralioglu, S.R. Manalis, A. Atalar, C.F. Quate, Centimeter scale atomic force microscope imaging and lithography. Appl. Phys. Lett. **73**(12), 1742–1744 (1998). ISSN 0003-6951. https://doi.org/10.1063/1.122263. Publisher: American Institute of Physics
460. C. Möller, M. Allen, V. Elings, A. Engel, D.J. Müller, Tapping-mode atomic force microscopy produces faithful high-resolution images of protein surfaces. Biophys. J. **77**(2), 1150–1158 (1999). ISSN 0006-3495. https://doi.org/10.1016/S0006-3495(99)76966-3, http://www.sciencedirect.com/science/article/pii/S0006349599769663
461. R. Möller, Thermovoltages in scanning tunneling microscopy, in *Scanning Probe Microscopy: Analytical Methods*, ed. by R. Wiesendanger. Nanoscience and Technology (Springer, Heidelberg, 1998), p. 49. https://doi.org/10.1007/978-3-662-03606-8
462. R. Möller, A. Esslinger, B. Koslowski, Noise in vacuum tunneling: application for a novel scanning microscope. Appl. Phys. Lett. **55**(22), 2360–2362 (1989). ISSN 0003-6951. https://doi.org/10.1063/1.102018. Publisher: American Institute of Physics
463. R. Möller, U. Albrecht, J. Boneberg, B. Koslowski, P. Leiderer, K. Dransfeld, Detection of surface plasmons by scanning tunneling microscopy. J. Vac. Sci. & Technol. B: Microelectron. Nanometer Struct. Proc. Meas. Phen. **9**(2), 506–509 (1991). https://doi.org/10.1116/1.585557
464. H. Mönig, D.R. Hermoso, O. Díaz Arado, M. Todorović, A. Timmer, S. Schüer, G. Langewisch, R. Pérez, H. Fuchs, Submolecular imaging by noncontact atomic force microscopy with an oxygen atom rigidly connected to a metallic probe. ACS Nano **10**(1), 1201–1209 (2016). ISSN 1936-0851. https://doi.org/10.1021/acsnano.5b06513. Publisher: American Chemical Society
465. J.S. Moodera, L.R. Kinder, T.M. Wong, R. Meservey, Large magnetoresistance at room temperature in ferromagnetic thin film tunnel junctions. Phys. Rev. Lett. **74**, 3273–3276 (1995). https://doi.org/10.1103/PhysRevLett.74.3273
466. C. Moreau-Luchaire, C. Moutafis, N. Reyren, J. Sampaio, C.A.F. Vaz, N. Van Horne, K. Bouzehouane, K. Garcia, C. Deranlot, P. Warnicke, P. Wohlhüter, J.-M. George, M. Weigand, J. Raabe, V. Cros, A. Fert, Additive interfacial chiral interaction in multilayers for stabilization of small individual skyrmions at room temperature. Nat. Nanotechnol. **11**(5), 444–448 (2016). ISSN 1748-3395. https://doi.org/10.1038/nnano.2015.313
467. J. Moreland, P. Rice, Tunneling stabilized magnetic force microscopy: prospects for low temperature applications to superconductors. IEEE Transactions on Magnetics **27**(2), 1198–1201 (1991). https://doi.org/10.1109/20.133399
468. J. Moreland, P. Rice, Tunneling stabilized, magnetic force microscopy with a gold-coated, nickel-film tipa). J. Appl. Phys. **70**(1), 520–522 (1991). ISSN 0021-8979. https://doi.org/10.1063/1.350266. Publisher: American Institute of Physics
469. S. Morita (Ed.), *Proceedings of the International Workshop on Noncontact Atomic Force Microscopy*, volume 140 of Applied Surface Science (Elsevier, Amsterdam, 1999), https://www.sciencedirect.com/journal/applied-surface-science/vol/188/issue/3
470. S. Morita, F.J. Giessibl, E. Meyer, R. Wiesendanger, *Noncontact Atomic Force Microscopy*. NanoScience and Technology (Springer, Berlin, 2015). https://doi.org/10.1007/978-3-319-15588-3
471. A. Moser, H.J. Hug, B. Stiefel, H.J. Güntherodt, Low temperature magnetic force microscopy on YBa2Cu3O7-δ thin films. J. Magn. Magn. Mater. **190**(1), 114–123 (1998). ISSN 0304-8853. https://doi.org/10.1016/S0304-8853(98)00273-X, http://www.sciencedirect.com/science/article/pii/S030488539800273X
472. A. Moser, M. Xiao, P. Kappenberger, K. Takano, W. Weresin, Y. Ikeda, H. Do, H.J. Hug, High-resolution magnetic force microscopy study of high-density transitions in perpendicular recording media. J. Magn. Magn. Mater. **287**, 298–302 (2005). ISSN 0304-8853. https://doi.org/10.1016/j.jmmm.2004.10.048, http://www.sciencedirect.com/science/article/pii/S0304885304011266

473. D. Mulin, D. Courjon, J.-P. Malugani, B. Gauthier-Manuel, Use of solid electrolytic erosion for generating nano-aperture near-field collectors. Appl. Phys. Lett. **71**(4), 437–439 (1997). ISSN 0003-6951. https://doi.org/10.1063/1.120439, https://aip.scitation.org/doi/10.1063/1.120439. Publisher: American Institute of Physics

474. A.D. Müller, F. Müller, M. Hietschold, Localized electrochemical deposition of metals using micropipettes. Thin Solid Films **366**(1), 32–36 (2000). ISSN 0040-6090. https://doi.org/10.1016/S0040-6090(99)01117-7, http://www.sciencedirect.com/science/article/pii/S0040609099011177

475. P. Muralt, D.W. Pohl, W. Denk, Wide-range, low-operating-voltage, bimorph Stm - application as potentiometer (1986), https://infoscience.epfl.ch/record/88737. Issue: 5 Library Catalog: infoscience.epfl.ch Number: ARTICLE Pages: 443 Volume: 30

476. M.P. Murrell, M.E. Welland, S.J. O'Shea, T.M.H. Wong, J.R. Barnes, A.W. McKinnon, M. Heyns, S. Verhaverbeke, Spatially resolved electrical measurements of SiO2 gate oxides using atomic force microscopy. Appl. Phys. Lett. **62**(7), 786–788 (1993). ISSN 0003-6951. https://doi.org/10.1063/1.108579. Publisher: American Institute of Physics

477. S. Nadj-Perge, I.K. Drozdov, J. Li, H. Chen, S. Jeon, J. Seo, A.H. MacDonald, B. Andrei Bernevig, A. Yazdani, Observation of Majorana fermions in ferromagnetic atomic chains on a superconductor. Science **346**(6209), 602–607 (2014). ISSN 0036-8075. https://doi.org/10.1126/science.1259327, https://science.sciencemag.org/content/346/6209/602. Publisher: American Association for the Advancement of Science _eprint: https://science.sciencemag.org/content/346/6209/602.full.pdf

478. O. Nakabeppu, M. Chandrachood, Y. Wu, J. Lai, A. Majumdar, Scanning thermal imaging microscopy using composite cantilever probes. Appl. Phys. Lett. **66**(6), 694–696 (1995). ISSN 0003-6951. https://doi.org/10.1063/1.114102, https://aip.scitation.org/doi/10.1063/1.114102. Publisher: American Institute of Physics

479. F.D. Natterer, K. Yang, W. Paul, P. Willke, T. Choi, T. Greber, A.J. Heinrich, C.P. Lutz, Reading and writing single-atom magnets. Nature **543**(7644), 226–228 (2017). ISSN 0028-0836, 1476-4687. https://doi.org/10.1038/nature21371, http://www.nature.com/articles/nature21371

480. B. Naydenov, F. Reinhard, A. Lämmle, V. Richter, R. Kalish, U.F.S. D'Haenens-Johansson, M. Newton, F. Jelezko, J. Wrachtrup, Increasing the coherence time of single electron spins in diamond by high temperature annealing. Appl. Phys. Lett. **97**(24), 242511 (2010). ISSN 0003-6951. https://doi.org/10.1063/1.3527975. Publisher: American Institute of Physics

481. D. Nečas, P. Klapetek, V. Neu, M. Havlíček, R. Puttock, O. Kazakova, X. Hu, L. Zajíčková, Determination of tip transfer function for quantitative MFM using frequency domain filtering and least squares method. Sci. Rep. **9**(1), 3880 (2019). ISSN 2045-2322. https://doi.org/10.1038/s41598-019-40477-x

482. N. Néel, S. Schröder, N. Ruppelt, P. Ferriani, J. Kröger, R. Berndt, S. Heinze, Tunneling anisotropic magnetoresistance at the single-atom limit. Phys. Rev. Lett. **110** (2013). https://doi.org/10.1103/PhysRevLett.110.037202

483. V. Neu, S. Vock, T. Sturm, L. Schultz, Epitaxial hard magnetic SmCo5 MFM tips - a new approach to advanced magnetic force microscopy imaging. Nanoscale **10**(35), 16881–16886 (2018). https://doi.org/10.1039/C8NR03997F. Publisher: The Royal Society of Chemistry

484. Ph. Niedermann, W. Hänni, N. Blanc, R. Christoph, J. Burger, Chemical vapor deposition diamond for tips in nanoprobe experiments. J. Vac. Sci. & Technol. A **14**(3), 1233–1236 (1996). ISSN 0734-2101. https://doi.org/10.1116/1.580273, https://avs.scitation.org/doi/abs/10.1116/1.580273. Publisher: American Vacuum Society

485. A.P. Nievergelt, J.D. Adams, P.D. Odermatt, G.E. Fantner, High-frequency multimodal atomic force microscopy. Beilstein J. Nanotechnol. **5**(1), 2459–2467 (2014). https://doi.org/10.3762/bjnano.5.255

486. A.P. Nievergelt, B.W. Erickson, N. Hosseini, J.D. Adams, G.E. Fantner, Studying biological membranes with extended range high-speed atomic force microscopy. Sci. Rep. **5**(1), 11987 (2015). ISSN 2045-2322. https://doi.org/10.1038/srep11987

487. R. Nishi, S. Araragi, K. Shirai, Y. Sugawara, S. Morita, Atom selective imaging by NC-AFM: case of oxygen adsorbed on a si (111) 7 x 7 surface. Appl. Surf. Sci. **210**(1), 90–92 (2003). ISSN 0169-4332. https://doi.org/10.1016/S0169-4332(02)01485-X, http://www.sciencedirect.com/science/article/pii/S016943320201485X

488. M. Nonnenmacher, H.K. Wickramasinghe, Scanning probe microscopy of thermal conductivity and subsurface properties. Appl. Phys. Lett. **61**(2), 168–170 (1992). ISSN 0003-6951. https://doi.org/10.1063/1.108207. Publisher: American Institute of Physics

489. M. Nonnenmacher, M.P. O'Boyle, H.K. Wickramasinghe, Kelvin probe force microscopy. Appl. Phys. Lett. **58**(25), 2921–2923 (1991). ISSN 0003-6951. https://doi.org/10.1063/1.105227. Publisher: American Institute of Physics

490. L. Nony, R. Boisgard, J.P. Aim, Nonlinear dynamical properties of an oscillating tip cantilever system in the tapping mode. J. Chem. Phys. **111**(4), 1615–1627 (1999). https://doi.org/10.1063/1.479422

491. D. Nowak, W. Morrison, H.K. Wickramasinghe, J. Jahng, E. Potma, L. Wan, R. Ruiz, T.R. Albrecht, K. Schmidt, J. Frommer, D.P. Sanders, S. Park, Nanoscale chemical imaging by photoinduced force microscopy. Sci. Adv. **2**(3) (2016). https://doi.org/10.1126/sciadv.1501571, https://advances.sciencemag.org/content/2/3/e1501571. Publisher: American Association for the Advancement of Science eprint: https://advances.sciencemag.org/content/2/3/e1501571.full.pdf

492. J.N. Nxumalo, D.T. Shimizu, D.J. Thomson, Cross sectional imaging of semiconductor device structures by scanning resistance microscopy. J. Vac. Sci. & Technol. B: Microelectron. Nanometer Struct. Proc. Meas. Phen. **14**(1), 386–389 (1996). ISSN 1071-1023. https://doi.org/10.1116/1.588479, https://avs.scitation.org/doi/10.1116/1.588479. Publisher: American Institute of Physics

493. D. Nyyssonen, L. Landstein, E. Coombs, Two dimensional atomic force microprobe trench metrology system. J. Vac. Sci. & Technol. B: Microelectron. Nanometer Struct. Proc. Meas. Phen. **9**(6), 3612–3616 (1991). ISSN 1071-1023. https://doi.org/10.1116/1.585855, https://avs.scitation.org/doi/10.1116/1.585855. Publisher: American Institute of Physics

494. M.L.O. Malley, G.L. Timp, S.V. Moccio, J.P. Garno, R.N. Kleiman, Quantification of scanning capacitance microscopy imaging of the pn junction through electrical simulation. Appl. Phys. Lett. **74**(2), 272–274 (1999). ISSN 0003-6951. https://doi.org/10.1063/1.123278, https://aip.scitation.org/doi/10.1063/1.123278. Publisher: American Institute of Physics

495. S.J.O. Shea, R.M. Atta, M.E. Welland, Characterization of tips for conducting atomic force microscopy. Rev. Sci. Instr. **66**(3), 2508–2512 (1995). ISSN 0034-6748. https://doi.org/10.1063/1.1145649, https://aip.scitation.org/doi/10.1063/1.1145649. Publisher: American Institute of Physics

496. E. Oesterschulze, M. Stopka, L. Ackermann, W. Scholz, S. Werner, Thermal imaging of thin films by scanning thermal microscope. J. Vac. Sci. & Technol. B: Microelectron. Nanometer Struct. Proc. Meas. Phen. **14**(2), 832–837 (1996). ISSN 1071-1023. https://doi.org/10.1116/1.588724, https://avs.scitation.org/doi/10.1116/1.588724. Publisher: American Institute of Physics

497. D.F. Ogletree, R.W. Carpick, M. Salmeron, Calibration of frictional forces in atomic force microscopy. Rev. Sci. Instr. **67**(9), 3298–3306 (1996). ISSN 0034-6748. https://doi.org/10.1063/1.1147411, https://aip.scitation.org/doi/10.1063/1.1147411. Publisher: American Institute of Physics

498. H. Ohldag, A. Scholl, F. Nolting, E. Arenholz, S. Maat, A.T. Young, M. Carey, J. Stöhr, Correlation between exchange bias and pinned interfacial spins. Phys. Rev. Lett. **91** (2003). https://doi.org/10.1103/PhysRevLett.91.017203

499. F. Ohnesorge, G. Binnig, True atomic resolution by atomic force microscopy through repulsive and attractive forces. Science **260**(5113), 1451–1456 (1993). ISSN 0036-8075. https://doi.org/10.1126/science.260.5113.1451, https://science.sciencemag.org/content/260/5113/1451

500. K. Ohno, F. Joseph Heremans, L.C. Bassett, B.A. Myers, D.M. Toyli, A.C. Bleszynski Jayich, C.J. Palmstrøm, D.D. Awschalom, Engineering shallow spins in diamond with nitrogen delta-doping. Appl. Phys. Lett. **101**(8), 082413 (2012). ISSN 0003-6951. https://doi.org/10.1063/1.4748280, https://doi.org/10.1063/1.4748280. Publisher: American Institute of Physics

501. A. Oral, S.J. Bending, M. Henini, Real-time scanning Hall probe microscopy. Appl. Phys. Lett. **69**(9), 1324–1326 (1996). ISSN 0003-6951. https://doi.org/10.1063/1.117582. Publisher: American Institute of Physics

502. S. Orisaka, T. Minobe, T. Uchihashi, Y. Sugawara, S. Morita, The atomic resolution imaging of metallic Ag(111) surface by noncontact atomic force microscope. Appl. Surf. Sci. **140**(3), 243–246 (1999). ISSN 0169-4332. https://doi.org/10.1016/S0169-4332(98)00534-0, http://www.sciencedirect.com/science/article/pii/S0169433298005340

503. J.O. Oti, P. Rice, S.E. Russek, Proposed antiferromagnetically coupled dual-layer magnetic force microscope tips. J. Appl. Phys. **75**(10), 6881–6883 (1994). ISSN 0021-8979. https://doi.org/10.1063/1.356815. Publisher: American Institute of Physics

504. R.C. Palmer, E.J. Denlinger, H. Kawamoto, Capacitive-pickup circuitry videodiscs. RCA Rev. **43**, 194 (1982)

505. L. Pan, D.B. Bogy, Heat-assisted magnetic recording. Nat. Photonics **3**(4), 189–190 (2009). ISSN 1749-4893. https://doi.org/10.1038/nphoton.2009.40

506. V. Panchal, H. Corte-León, B. Gribkov, L.A. Rodriguez, E. Snoeck, A. Manzin, E. Simonetto, S. Vock, V. Neu, O. Kazakova, Calibration of multi-layered probes with low/high magnetic moments. Sci. Rep. **7**(1), 7224 (2017). ISSN 2045-2322. https://doi.org/10.1038/s41598-017-07327-0

507. J. Park, A.N. Pasupathy, J.I. Goldsmith, C. Chang, Y. Yaish, J.R. Petta, M. Rinkoski, J.P. Sethna, H.D. Abruña, P.L. McEuen, D.C. Ralph, Coulomb blockade and the Kondo effect in single-atom transistors. Nature **417**(6890), 722–725 (2002). ISSN 1476-4687. https://doi.org/10.1038/nature00791

508. R. Pawlak, M. Kisiel, J. Klinovaja, T. Meier, S. Kawai, T. Glatzel, D. Loss, E. Meyer, Probing atomic structure and Majorana wavefunctions in mono-atomic Fe chains on superconducting Pb surface. NPJ Q. Inf. **2**(1), 16035 (2016). ISSN 2056-6387. https://doi.org/10.1038/npjqi.2016.35

509. R. Pawlak, T. Meier, N. Renaud, M. Kisiel, A. Hinaut, T. Glatzel, D. Sordes, C. Durand, W.-H. Soe, A. Baratoff, C. Joachim, C.E. Housecroft, E.C. Constable, E. Meyer, Design and characterization of an electrically powered single molecule on gold. ACS Nano **11**(10), 9930–9940 (2017). ISSN 1936-0851. https://doi.org/10.1021/acsnano.7b03955. Publisher: American Chemical Society

510. R. Pawlak, J.G. Vilhena, P. D'Astolfo, X. Liu, G. Prampolini, T. Meier, T. Glatzel, J.A. Lemkul, R. Häner, S. Decurtins, A. Baratoff, R. Pérez, S.-X. Liu, E. Meyer, Sequential bending and twisting around C-C single bonds by mechanical lifting of a pre-adsorbed polymer. Nano Lett. **20**(1), 652–657 (2020). ISSN 1530-6984. https://doi.org/10.1021/acs.nanolett.9b04418, https://doi.org/10.1021/acs.nanolett.9b04418. Publisher: American Chemical Society

511. J.P. Pelz, R.H. Koch, Tip-related artifacts in scanning tunneling potentiometry. Phys. Rev. B **41**, 1212–1215 (1990). https://doi.org/10.1103/physrevb.41.1212

512. M. Penedo, H.J. Hug, Off-resonance intermittent contact mode multi-harmonic scanning force microscopy. Appl. Phys. Lett. **113**(2), 023103 (2018). ISSN 0003-6951. https://doi.org/10.1063/1.5026657. Publisher: American Institute of Physics

513. R. Pérez, M.C. Payne, I. Štich, K. Terakura, Role of covalent tip-surface interactions in non-contact atomic force microscopy on reactive surfaces. Phys. Rev. Lett. **78**, 678–681 (1997). https://doi.org/10.1103/PhysRevLett.78.678

514. R. Perez, I. Stich, M.C. Payne, K. Terakura, Surface-tip interactions in noncontact atomic force microscopy on reactive surfaces: Si (111). Phys. Rev. B **58**, 10835–10849 (1998). https://doi.org/10.1103/PhysRevB.58.10835

515. B.N.J. Persson, A. Baratoff, Theory of photon emission in electron tunneling to metallic particles. Phys. Rev. Lett. **68**, 3224–3227 (1992). https://doi.org/10.1103/PhysRevLett.68.3224

516. C.L. Petersen, T.M. Hansen, P. Bøggild, A. Boisen, O. Hansen, T. Hassenkam, F. Grey, Scanning microscopic four-point conductivity probes. Sens. Actuators A: Phys. **96**(1), 53–58 (2002). ISSN 0924-4247. https://doi.org/10.1016/S0924-4247(01)00765-8, http://www.sciencedirect.com/science/article/pii/S0924424701007658

517. J.B. Pethica, Comment on "interatomic forces in scanning tunneling microscopy: Giant corrugations of the graphite surface". Phys. Rev. Lett. **57**, 3235 (1986). https://doi.org/10.1103/PhysRevLett.57.3235

518. O. Pfeiffer, C. Loppacher, C. Wattinger, M. Bammerlin, U. Gysin, M. Guggisberg, S. Rast, R. Bennewitz, E. Meyer, H.J. Güntherodt, Using higher flexural modes in non-contact force microscopy. Appl. Surf. Sci. **157**(4), 337–342 (2000). ISSN 0169-4332. https://doi.org/10.1016/S0169-4332(99)00548-6, http://www.sciencedirect.com/science/article/pii/S0169433299005486

519. S.-H. Phark, D. Sander, Spin-polarized scanning tunneling microscopy with quantitative insights into magnetic probes. Nano Converg. **4**(1), 8 (2017). ISSN 2196-5404. https://doi.org/10.1186/s40580-017-0102-5

520. G.N. Phillips, M. Siekman, L. Abelmann, J.C. Lodder, High resolution magnetic force microscopy using focused ion beam modified tips. Appl. Phys. Lett. **81**(5), 865–867 (2002). ISSN 0003-6951. https://doi.org/10.1063/1.1497434. Publisher: American Institute of Physics

521. L.M. Picco, L. Bozec, A. Ulcinas, D.J. Engledew, M. Antognozzi, M.A. Horton, M.J. Miles, Breaking the speed limit with atomic force microscopy. Nanotechnology **18**(4), 044030 (2006). ISSN 0957-4484. https://doi.org/10.1088/0957-4484/18/4/044030. Publisher: IOP Publishing

522. F. Pielmeier, F.J. Giessibl, Spin resolution and evidence for superexchange on nio(001) observed by force microscopy. Phys. Rev. Lett. **110** (2013). https://doi.org/10.1103/PhysRevLett.110.266101

523. G.S. Pingali, R.C. Jain, Restoration of scanning probe microscope images, in *[1992] Proceedings IEEE Workshop on Applications of Computer Vision* (1992). https://doi.org/10.1109/ACV.1992.240301

524. A. Pioda, S. Kičin, T. Ihn, M. Sigrist, A. Fuhrer, K. Ensslin, A. Weichselbaum, S.E. Ulloa, M. Reinwald, W. Wegscheider, Spatially resolved manipulation of single electrons in quantum dots using a scanned probe. Phys. Rev. Lett. **93** (2004). https://doi.org/10.1063/1.1994339

525. D. Platz, E.A. Tholn, D. Pesen, D.B. Haviland, Intermodulation atomic force microscopy. Appl. Phys. Lett. **92**(15), 153106 (2008). ISSN 0003-6951. https://doi.org/10.1063/1.2909569, https://aip.scitation.org/doi/full/10.1063/1.2909569. Publisher: American Institute of Physics

526. D. Platz, E.A. Tholn, D. Pesen, D.B. Haviland, Intermodulation atomic force microscopy. Appl. Phys. Lett. **92**(15), 153106 (2008). ISSN 0003-6951. https://doi.org/10.1063/1.2909569, https://aip.scitation.org/doi/full/10.1063/1.2909569. Publisher: American Institute of Physics

527. I.V. Pobelov, C. Li, T. Wandlowski, *Electrochemical Scanning Tunneling Microscopy* (Springer Netherlands, Dordrecht, 2012), pp. 688–702. ISBN 978-90-481-9751-4. https://doi.org/10.1007/978-90-481-9751-4_46

528. D.W. Pohl, D. Courjon, *Near Field Optics*, volume 241 of NATO Advanced Studies Institutes Series E (Kluwer Academic Publishers, Dordrecht, 1993). https://doi.org/10.1007/978-94-011-1978-8

529. G.E. Poirier, Characterization of organosulfur molecular monolayers on Au(111) using scanning tunneling microscopy. Chem. Rev. **97**(4), 1117–1128 (1997). ISSN 1520-6890. https://doi.org/10.1021/cr960074m

530. S. Porthun, L. Abelmann, S.J.L. Vellekoop, J.C. Lodder, H.J. Hug, Optimization of lateral resolution in magnetic force microscopy. Appl. Phys. A **66**(1), 1185–1189 (1998). ISSN 1432-0630. https://doi.org/10.1007/s003390051323

531. S. Porthun, L. Abelmann, C. Lodder, Magnetic force microscopy of thin film media for high density magnetic recording. J. Magn. Magn. Mater. **182**(1), 238–273

(1998). ISSN 0304-8853. https://doi.org/10.1016/S0304-8853(97)01010-X, http://www.sciencedirect.com/science/article/pii/S030488539701010X

532. C.B. Prater, P.K. Hansma, M. Tortonese, C.F. Quate, Improved scanning ion conductance microscope using microfabricated probes. Rev. Sci. Instr. **62**(11), 2634–2638 (1991). ISSN 0034-6748. https://doi.org/10.1063/1.1142244, https://aip.scitation.org/doi/10.1063/1.1142244. Publisher: American Institute of Physics

533. I.L. Prejbeanu, S. Bandiera, J. Alvarez-Hérault, R.C. Sousa, B. Dieny, J.-P. Nozières, Thermally assisted MRAMs: ultimate scalability and logic functionalities. J. Phys. D: Appl. Phys. **46**(7), 074002 (2013). ISSN 0022-3727. https://doi.org/10.1088/0022-3727/46/7/074002. Publisher: IOP Publishing

534. M.W.J. Prins, R. Jansen, H. van Kempen, Spin-polarized tunneling with GaAs tips in scanning tunneling microscopy. Phys. Rev. B **53**, 8105–8113 (1996). https://doi.org/10.1103/PhysRevB.53.8105

535. R. Proksch, R. Lal, P.K. Hansma, D. Morse, G. Stucky, Imaging the internal and external pore structure of membranes in fluid: tappingMode scanning ion conductance microscopy. Biophys. J. **71**(4), 2155–2157 (1996). ISSN 0006-3495. https://doi.org/10.1016/S0006-3495(96)79416-X, https://pubmed.ncbi.nlm.nih.gov/8889191

536. R. Proksch, Recent advances in magnetic force microscopy. Curr. Opin. Solid State Mater. Sci. **4**(2), 231–236 (1999). ISSN 1359-0286. https://doi.org/10.1016/S1359-0286(99)00002-9, http://www.sciencedirect.com/science/article/pii/S1359028699000029

537. C.A.J. Putman, K.O. Van der Werf, B.G. De Grooth, N.F. Van Hulst, J. Greve, Tapping mode atomic force microscopy in liquid. Appl. Phys. Lett. **64**(18), 2454–2456 (1994). ISSN 0003-6951. https://doi.org/10.1063/1.111597, https://aip.scitation.org/doi/10.1063/1.111597. Publisher: American Institute of Physics

538. X.H. Qiu, G.V. Nazin, W. Ho, Vibrationally resolved fluorescence excited with submolecular precision. Science **299**(5606), 542–546 (2003). ISSN 0036-8075. https://doi.org/10.1126/science.1078675, https://science.sciencemag.org/content/299/5606/542. Publisher: American Association for the Advancement of Science _eprint: https://science.sciencemag.org/content/299/5606/542.full.pdf

539. C.F. Quate, The acoustic microscope. Sci. Amer. **241**(October):58 (1979), https://www.scientificamerican.com/article/the-acoustic-microscope/

540. C.F. Quate, Vacuum tunneling: a new technique for microscopy. Phys. Today August, 26 (1986). https://doi.org/10.1063/1.881071

541. U. Rabe, K. Janser, W. Arnold, Vibrations of free and surface coupled atomic force microscope cantilevers: theory and experiment. Rev. Sci. Instr. **67**(9), 3281–3293 (1996). ISSN 0034-6748. https://doi.org/10.1063/1.1147409, https://aip.scitation.org/doi/10.1063/1.1147409. Publisher: American Institute of Physics

542. E. Rabinowicz, *Friction and Wear of Materials* (Wiley, New York, 1995). ISBN 978-0-471-83084-9

543. F. Radu, R. Abrudan, I. Radu, D. Schmitz, H. Zabel, Perpendicular exchange bias in ferrimagnetic spin valves. Nat. Commun. **3**(1), 715 (2012). ISSN 2041-1723. https://doi.org/10.1038/ncomms1728

544. I. Rajapaksa, K. Uenal, H.K. Wickramasinghe, Image force microscopy of molecular resonance: a microscope principle. Appl. Phys. Lett. **97**(7), 073121 (2010). ISSN 0003-6951. https://doi.org/10.1063/1.3480608. Publisher: American Institute of Physics

545. S. Rast, C. Wattinger, U. Gysin, E. Meyer, Dynamics of damped cantilevers. Rev. Sci. Instr. **71**(7), 2772–2775 (2000). https://doi.org/10.1063/1.1150690

546. S. Rast, C. Wattinger, U. Gysin, E. Meyer, The noise of cantilevers. Nanotechnology **11**(3), 169–172 (2000). https://doi.org/10.1088/0957-4484/11/3/306

547. G. Reecht, F. Scheurer, V. Speisser, Y.J. Dappe, F. Mathevet, G. Schull, Electroluminescence of a polythiophene molecular wire suspended between a metallic surface and the tip of a scanning tunneling microscope. Phys. Rev. Lett. **112** (2014). https://doi.org/10.1103/PhysRevLett.112.047403

548. M.A. Reed, C. Zhou, C.J. Muller, T.P. Burgin, J.M. Tour, Conductance of a molecular junction. Science **278**(5336), 252–254 (1997). ISSN 0036-8075. https://doi.org/10.1126/science.278. 5336.252, https://science.sciencemag.org/content/278/5336/252

549. M. Reichling, M. Huisinga, S. Gogoll, C. Barth, Degradation of the CaF2(111) surface by air exposure. Surf. Sci. **439**(1), 181–190 (1999). ISSN 0039-6028. https://doi.org/10.1016/S0039-6028(99)00760-8, http://www.sciencedirect.com/science/article/pii/S0039602899007608

550. Ch. Renner, B. Revaz, J.-Y. Genoud, K. Kadowaki, Ø. Fischer, Pseudogap precursor of the superconducting gap in under- and overdoped $bi_2sr_2cacu_2O_{8+\delta}$. Phys. Rev. Lett. **80**, 149–152 (1998). https://doi.org/10.1103/PhysRevLett.80.149

551. J. Repp, G. Meyer, S.M. Stojković, A. Gourdon, C. Joachim, Molecules on insulating films: scanning-tunneling microscopy imaging of individual molecular orbitals. Phys. Rev. Lett. **94** (2005). https://doi.org/10.1103/PhysRevLett.94.026803

552. E. Riedo, F. Lévy, H. Brune, Kinetics of capillary condensation in nanoscopic sliding friction. Phys. Rev. Lett. **88** (2002). https://doi.org/10.1103/PhysRevLett.88.185505

553. T.R. Rodriguez, R. Garcia, Compositional mapping of surfaces in atomic force microscopy by excitation of the second normal mode of the microcantilever. Appl. Phys. Lett. **84**(3), 449–451 (2004). https://doi.org/10.1063/1.1642273

554. F.P. Rogers, A device for experimental observation of flux vortices trapped in superconducting thin film. Master's thesis, Massachusetts Institute of Technology, Cambridge, MA (1983), http://hdl.handle.net/1721.1/98942

555. H. Rohrer, in *Scanning Tunneling Microscopy and Related Methods*, vol. 184, ed. by R.J. Behm, N. Garcia, H. Rohrer. NATO Advanced Studies Institutes Series E (Academic Publisher, Dordrecht, 1990), p. 1. ISBN 9789401578714. https://doi.org/10.1007/9789401578714

556. N. Romming, A. Kubetzka, C. Hanneken, K. von Bergmann, R. Wiesendanger, Field-dependent size and shape of single magnetic skyrmions. Phys. Rev. Lett. **114** (2015). https://doi.org/10.1103/PhysRevLett.114.177203

557. L. Rondin, J.-P. Tetienne, T. Hingant, J.-F. Roch, P. Maletinsky, V. Jacques, Magnetometry with nitrogen-vacancy defects in diamond. Rep. Prog. Phys. **77**(5), 056503 (2014). ISSN 0034-4885. https://doi.org/10.1088/0034-4885/77/5/056503. Publisher: IOP Publishing

558. M. Roseman, P. Grütter, Magnetic imaging and dissipation force microscopy of vortices on superconducting Nb films. Appl. Surf. Sci. **188**(3), 416–420 (2002). ISSN 0169-4332. https://doi.org/10.1016/S0169-4332(01)00960-6, http://www.sciencedirect.com/science/article/pii/S0169433201009606

559. M. Roseman, P. Grütter, Determination of Tc, vortex creation and vortex imaging of a superconducting Nb film using low-temperature magnetic force microscopy. J. Appl. Phys. **91**(10), 8840–8842 (2002). ISSN 0021-8979. https://doi.org/10.1063/1.1456055, https://aip.scitation.org/doi/abs/10.1063/1.1456055. Publisher: American Institute of Physics

560. D. Rugar, H.J. Mamin, P. Guethner, Improved fiber optic interferometer for atomic force microscopy. Appl. Phys. Lett. **55**(25), 2588–2590 (1989). ISSN 0003-6951, 1077-3118. https://doi.org/10.1063/1.101987, http://aip.scitation.org/doi/10.1063/1.101987

561. D. Rugar, H.J. Mamin, P. Guethner, S.E. Lambert, J.E. Stern, I. McFadyen, T. Yogi, Magnetic force microscopy: general principles and application to longitudinal recording media. J. Appl. Phys. **68**(3), 1169–1183 (1990). https://doi.org/10.1063/1.346713

562. D. Rugar, O. Züger, S. Hoen, C.S. Yannoni, H.-M. Vieth, R.D. Kendrick, Force detection of nuclear magnetic resonance. Science **264**(5165), 1560–1563 (1994). ISSN 0036-8075. https://doi.org/10.1126/science.264.5165.1560, https://science.sciencemag.org/content/264/5165/1560. Publisher: American Association for the Advancement of Science _eprint: https://science.sciencemag.org/content/264/5165/1560.full.pdf

563. D. Rugar, B.C. Stipe, H.J. Mamin, C.S. Yannoni, T.D. Stowe, K.Y. Yasumura, T.W. Kenny, Adventures in attonewton force detection. Appl. Phys. A **72**(1), S3–S10 (2001). ISSN 1432-0630. https://doi.org/10.1007/s003390100729

564. M. Rührig, S. Porthun, J.C. Lodder, Magnetic force microscopy using electron-beam fabricated tips. Rev. Sci. Instr. **65**(10), 3224–3228 (1994). ISSN 0034-6748. https://doi.org/10.1063/1.1144554. Publisher: American Institute of Physics

565. J.E. Sader, S.P. Jarvis, Accurate formulas for interaction force and energy in frequency modulation force spectroscopy. Appl. Phys. Lett. **84**(10), 1801–1803 (2004). ISSN 0003-6951. https://doi.org/10.1063/1.1667267. Publisher: American Institute of Physics

566. J.E. Sader, J.W.M. Chon, P. Mulvaney, Calibration of rectangular atomic force microscope cantilevers. Rev. Sci. Instr. **70**(10), 3967–3969 (1999). ISSN 0034-6748. https://doi.org/10.1063/1.1150021, https://aip.scitation.org/doi/10.1063/1.1150021. Publisher: American Institute of Physics

567. J.E. Sader, Parallel beam approximation for v shaped atomic force microscope cantilevers. Rev. Sci. Instr. **66**(9), 4583–4587 (1995). https://doi.org/10.1063/1.1145292

568. J.J. Sáenz, N. García, P. Grütter, E. Meyer, H. Heinzelmann, R. Wiesendanger, L. Rosenthaler, H.R. Hidber, H.-J. Güntherodt, Observation of magnetic forces by the atomic force microscope. J. Appl. Phys. **62**(10), 4293–4295 (1987). ISSN 0021-8979. https://doi.org/10.1063/1.339105. Publisher: American Institute of Physics

569. O. Sahin, S. Magonov, C. Su, C.F. Quate, O. Solgaard, An atomic force microscope tip designed to measure time-varying nanomechanical forces. Nat. Nanotechnol. **2**(8), 507–514 (2007). ISSN 1748-3395. https://doi.org/10.1038/nnano.2007.226

570. M.S. Jean, S. Hudlet, C. Guthmann, J. Berger. Van der Waals and capacitive forces in atomic force microscopies. J. Appl. Phys. **86**(9), 5245–5248 (1999). ISSN 0021-8979. https://doi.org/10.1063/1.371506. Publisher: American Institute of Physics

571. A. Sandhu, K. Kurosawa, M. Dede, A. Oral, 50 nm hall sensors for room temperature scanning hall probe microscopy. Jpn. J. Appl. Phys. **43**(2), 777–778 (2004). ISSN 0021-4922. https://doi.org/10.1143/jjap.43.777. Publisher: IOP Publishing

572. Y. Sang, M. Dubé, M. Grant, Thermal effects on atomic friction. Phys. Rev. Lett. **87** (2001). https://doi.org/10.1103/PhysRevLett.87.174301

573. D. Sarid, D. Iams, V. Weissenberger, L.S. Bell, Compact scanning-force microscope using a laser diode. Opt. Lett. **13**(12), 1057–1059 (1988). ISSN 0146-9592. https://doi.org/10.1364/ol.13.001057

574. L. Scandella, J.-H. Fabian, C.v. Scala, R. Berger, H.P. Lang, Ch. Gerber, J.K. Gimzewski, E. Meyer, Micromechanical thermal gravimetry performed on one single zeolite crystal. Helv. Phys. Acta **71**(Separanda 1), 3 (1998), http://edoc.unibas.ch/dok/A5839462

575. M.R. Scheinfein, J. Unguris, D.T. Pierce, R.J. Celotta, High spatial resolution quantitative micromagnetics (invited). J. Appl. Phys. **67**(9), 5932–5937 (1990). ISSN 0021-8979. https://doi.org/10.1063/1.346018. Publisher: American Institute of Physics

576. A. Schirmeisen, G. Cross, A. Stalder, P. Grütter, U. Dürig, Metallic adhesion and tunnelling at the atomic scale. New J. Phys. **2**, 29 (2000). ISSN 1367-2630. https://doi.org/10.1088/1367-2630/2/1/329. Publisher: IOP Publishing

577. A. Schirmeisen, L. Jansen, H. Hölscher, H. Fuchs, Temperature dependence of point contact friction on silicon. Appl. Phys. Lett. **88**(12), 123108 (2006). ISSN 0003-6951. https://doi.org/10.1063/1.2187575. Publisher: American Institute of Physics

578. I. Schmid, P. Kappenberger, O. Hellwig, M.J. Carey, E.E. Fullerton, H.J. Hug, The role of uncompensated spins in exchange biasing. Europhys. Lett. (EPL) **81**(1), 17001 (2007). ISSN 0295-5075. https://doi.org/10.1209/0295-5075/81/17001. Publisher: IOP Publishing

579. I. Schmid, M.A. Marioni, P. Kappenberger, S. Romer, M. Parlinska-Wojtan, H.J. Hug, O. Hellwig, M.J. Carey, E.E. Fullerton, Exchange bias and domain evolution at 10 nm scales. Phys. Rev. Lett. **105** (2010). https://doi.org/10.1103/PhysRevLett.105.197201

580. M. Schmid, H. Stadler, P. Varga, Direct observation of surface chemical order by scanning tunneling microscopy. Phys. Rev. Lett. **70**, 1441–1444 (1993). https://doi.org/10.1103/PhysRevLett.70.1441

581. M. Schmid, W. Hebenstreit, P. Varga, S. Crampin, Quantum wells and electron interference phenomena in al due to subsurface noble gas bubbles. Phys. Rev. Lett. **76**, 2298–2301 (1996). https://doi.org/10.1103/PhysRevLett.76.2298

582. R. Schmidt, C. Lazo, H. Hölscher, U.H. Pi, V. Caciuc, A. Schwarz, R. Wiesendanger, S. Heinze, Probing the magnetic exchange forces of iron on the atomic scale. Nano Lett. **9**(1), 200–204 (2009). ISSN 1530-6984. https://doi.org/10.1021/nl802770x. Publisher: American Chemical Society

583. R. Schmidt, C. Lazo, U. Kaiser, A. Schwarz, S. Heinze, R. Wiesendanger, Quantitative measurement of the magnetic exchange interaction across a vacuum gap. Phys. Rev. Lett. **106** (2011). https://doi.org/10.1103/PhysRevLett.106.257202

584. M. Schneiderbauer, D. Wastl, F.J. Giessibl, qplus magnetic force microscopy in frequency-modulation mode with millihertz resolution. Beilstein J. Nanotechnol. **3**, 174–178 (2012). ISSN 2190-4286. https://doi.org/10.3762/bjnano.3.18

585. R.S. Schoenfeld, W. Harneit, Real time magnetic field sensing and imaging using a single spin in diamond. Phys. Rev. Lett. **106** (2011). https://doi.org/10.1103/PhysRevLett.106.030802

586. C. Schönenberger, S.F. Alvarado, Understanding magnetic force microscopy. Zeitschrift für Physik B Condensed Matter **80**(3), 373–383 (1990). ISSN 1431-584X. https://doi.org/10.1007/BF01323519

587. C. Schönenberger, S.F. Alvarado, Observation of single charge carriers by force microscopy. Phys. Rev. Lett. **65**, 3162–3164 (1990). https://doi.org/10.1103/PhysRevLett.65.3162

588. C. Schönenberger, S.F. Alvarado, S.E. Lambert, I.L. Sanders, Separation of magnetic and topographic effects in force microscopy. J. Appl. Phys. **67**(12), 7278–7280 (1990). ISSN 0021-8979. https://doi.org/10.1063/1.344511. Publisher: American Institute of Physics

589. A. Schwarz, W. Allers, U.D. Schwarz, R. Wiesendanger, Dynamic-mode scanning force microscopy study of n-inas(110)-(1 × 1) at low temperatures. Phys. Rev. B **61**, 2837–2845 (2000). https://doi.org/10.1103/PhysRevB.61.2837

590. U.D. Schwarz, H. Hölscher, R. Wiesendanger (ed.), *Proceedings of the Third International Workshop on Noncontact Atomic Force Microscopy*, volume S72 of Applied Physics A (Springer, Heidelberg, 2001). https://doi.org/10.1007/s003390100830

591. U.D. Schwarz, O. Zwörner, P. Köster, R. Wiesendanger, Quantitative analysis of the frictional properties of solid materials at low loads. i. carbon compounds. Phys. Rev. B **56**, 6987–6996 (1997). https://doi.org/10.1103/PhysRevB.56.6987, https://link.aps.org/doi/10.1103/PhysRevB.56.6987

592. J. Schwenk, M. Marioni, S. Romer, N.R. Joshi, H.J. Hug, Non-contact bimodal magnetic force microscopy. Appl. Phys. Lett. **104**(11), 112412 (2014). ISSN 0003-6951. https://doi.org/10.1063/1.4869353. Publisher: American Institute of Physics

593. J. Schwenk, X. Zhao, M. Bacani, M.A. Marioni, S. Romer, H.J. Hug, Bimodal magnetic force microscopy with capacitive tip sample distance control. Appl. Phys. Lett. **107**(13), 132407 (2015). ISSN 0003-6951. https://doi.org/10.1063/1.4932174. Publisher: American Institute of Physics

594. J. Scott, S. McVitie, R.P. Ferrier, A. Gallagher, Electrostatic charging artefacts in Lorentz electron tomography of MFM tip stray fields. J. Phys. D: Appl. Phys. **34**(9), 1326–1332 (2001). ISSN 0022-3727. https://doi.org/10.1088/0022-3727/34/9/307. Publisher: IOP Publishing

595. I. Sebastian, H. Neddermeyer, Scanning tunneling microscopy on the atomic and electronic structure of CoO thin films on Ag(100). Surf. Sci. **454–456**, 771–777 (2000). ISSN 0039-6028. https://doi.org/10.1016/S0039-6028(00)00060-1, http://www.sciencedirect.com/science/article/pii/S0039602800000601

596. S.K. Sekatskii, V.S. Letokhov, Nanometer-resolution scanning optical microscope with resonance excitation of the fluorescence of the samples from a single-atom excited center. J. Exp. Theor. Phys. Lett. **63**(5), 319–323 (1996). ISSN 1090-6487. https://doi.org/10.1134/1.567024

597. T.G.T.S.C. Sekhar, *Communication Theory* (McGraw-Hill Education (India) Pvt Limited, 2005). ISBN 978-0-07-059091-5, https://books.google.ch/books?id=C1C6IBiCoXsC

598. Y. Seo, P. Cadden-Zimansky, V. Chandrasekhar, Low-temperature high-resolution magnetic force microscopy using a quartz tuning fork. Appl. Phys. Lett. **87**(10), 103103 (2005). ISSN 0003-6951. https://doi.org/10.1063/1.2037852. Publisher: American Institute of Physics

599. C. Shafai, D.J. Thomson, M. Simard-Normandin, Two dimensional delineation of semiconductor doping by scanning resistance microscopy. J. Vac. Sci. & Technol. B: Microelectron. Nanometer Struct. Proc. Meas. Phen. **12**(1), 378–382 (1994). ISSN 1071-1023. https://doi.org/10.1116/1.587129, https://avs.scitation.org/doi/abs/10.1116/1.587129. Publisher: American Institute of Physics

600. G. Shaw, R.B.G. Kramer, N.M. Dempsey, K. Hasselbach, A scanning Hall probe microscope for high resolution, large area, variable height magnetic field imaging. Rev. Sci. Instr. **87**(11), 113702 (2016). ISSN 0034-6748. https://doi.org/10.1063/1.4967235. Publisher: American Institute of Physics

601. G.S. Shekhawat, V.P. Dravid, Nanoscale imaging of buried structures via scanning near-field ultrasound holography. Science **310**(5745), 89–92 (2005). ISSN 0036-8075. https://doi.org/10.1126/science.1117694, https://science.sciencemag.org/content/310/5745/89. Publisher: American Association for the Advancement of Science _eprint: https://science.sciencemag.org/content/310/5745/89.full.pdf

602. L. Shi, S. Plyasunov, A. Bachtold, P.L. McEuen, A. Majumdar, Scanning thermal microscopy of carbon nanotubes using batch-fabricated probes. Appl. Phys. Lett. **77**(26), 4295–4297 (2000). https://doi.org/10.1063/1.1334658

603. Y. Shibata, S. Nomura, H. Kashiwaya, S. Kashiwaya, R. Ishiguro, H. Takayanagi, Imaging of current density distributions with a Nb weak-link scanning nano-SQUID microscope. Sci. Rep. **5**(1), 15097 (2015). ISSN 2045-2322. https://doi.org/10.1038/srep15097

604. T. Shimizu, J.-T. Kim, H. Tokumoto, Tungsten silicide formation on an STM tip during atom manipulation. Appl. Phys. A **66**(1), S771–S775 (1998). ISSN 1432-0630. https://doi.org/10.1007/s003390051240

605. A.L. Shluger, R.T. Williams, A.L. Rohl, Lateral and friction forces originating during force microscope scanning of ionic surfaces. Surf. Sci. **343**(3), 273–287 (1995). ISSN 0039-6028. https://doi.org/10.1016/0039-6028(95)00841-1, http://www.sciencedirect.com/science/article/pii/0039602895008411

606. H. Siegenthaler, Scanning tunneling microscopy in electrochemistry, in *Scanning Tunneling Microscopy II*, volume 28 of Springer Series in Surface Science ed. by R. Wiesendanger, H.-J. Güntherodt (Springer, Heidelberg, 1992), p. 7. https://doi.org/10.1007/978-3-642-97363-5

607. S. Signoretti, C. Beeli, S.-H. Liou, Electron holography quantitative measurements on magnetic force microscopy probes, in *Proceedings of the International Conference on Magnetism (ICM 2003)*, pp. 272–276:2167–2168 (2004). ISSN 0304-8853. https://doi.org/10.1016/j.jmmm.2003.12.948, http://www.sciencedirect.com/science/article/pii/S030488530301802X

608. G.D. Skidmore, E.D. Dahlberg, Improved spatial resolution in magnetic force microscopy. Appl. Phys. Lett. **71**(22), 3293–3295 (1997). ISSN 0003-6951. https://doi.org/10.1063/1.120316. Publisher: American Institute of Physics

609. V. Skumryev, S. Stoyanov, Y. Zhang, G. Hadjipanayis, D. Givord, J. Nogus, Beating the superparamagnetic limit with exchange bias. Nature **423**(6942), 850–853 (2003). ISSN 1476-4687. https://doi.org/10.1038/nature01687

610. J.C. Slonczewski, Exchange through a tunneling barrier. J. Physique **49**(C8), 1629 (1988), https://hal.archives-ouvertes.fr/jpa-00228986/document

611. D.P.E. Smith, H. Hörber, Ch. Gerber, G. Binnig, Smectic liquid crystal monolayers on graphite observed by scanning tunneling microscopy. Science **245**(4913), 43–45 (1989). ISSN 0036-8075. https://doi.org/10.1126/science.245.4913.43, https://science.sciencemag.org/content/245/4913/43. Publisher: American Association for the Advancement of Science _eprint: https://science.sciencemag.org/content/245/4913/43.full.pdf

612. J. Snauwaert, L. Hellemans, I. Czech, T. Clarysse, W. Vandervorst, M. Pawlik, Towards a physical understanding of spreading resistance probe technique profiling. J. Vac. Sci. & Technol. B: Microelectron. Nanometer Struct. Proc. Meas. Phen. **12**(1), 304–311 (1994). ISSN 1071-1023. https://doi.org/10.1116/1.587158, https://avs.scitation.org/doi/abs/10.1116/1.587158. Publisher: American Institute of Physics

613. I. Yu Sokolov, G.S. Henderson, Atomic resolution imaging using the electric double layer technique: friction vs. height contrast mechanisms. Appl. Surf. Sci. **157**(4), 302–307 (2000). ISSN

0169-4332. https://doi.org/10.1016/S0169-4332(99)00543-7, http://www.sciencedirect.com/science/article/pii/S0169433299005437

614. J.M. Soler, A.M. Baro, N. García, H. Rohrer, Interatomic forces in scanning tunneling microscopy: giant corrugations of the graphite surface. Phys. Rev. Lett. **57**, 444–447 (1986). https://doi.org/10.1103/PhysRevLett.57.444

615. Ch. Sommerhalter, Th.W. Matthes, Th. Glatzel, A. Jäger-Waldau, M.Ch. Lux-Steiner, High-sensitivity quantitative Kelvin probe microscopy by noncontact ultra-high-vacuum atomic force microscopy. Appl. Phys. Lett. **75**(2), 286–288 (1999). ISSN 0003-6951. https://doi.org/10.1063/1.124357. Publisher: American Institute of Physics

616. M.R. Sörensen, K.W. Jacobsen, P. Stoltze, Simulations of atomic-scale sliding friction. Phys. Rev. B **53**, 2101–2113 (1996). https://doi.org/10.1103/PhysRevB.53.2101

617. P.T. Sprunger, L. Petersen, E.W. Plummer, E. Lægsgaard, F. Besenbacher, Giant friedel oscillations on the beryllium(0001) surface. Science **275**(5307), 1764–1767 (1997). ISSN 0036-8075. https://doi.org/10.1126/science.275.5307.1764, https://science.sciencemag.org/content/275/5307/1764. Publisher: American Association for the Advancement of Science _eprint: https://science.sciencemag.org/content/275/5307/1764.full.pdf

618. I. Štich, J. Tóbik, R. Pérez, K. Terakura, S.H. Ke, Tip–surface interactions in noncontact atomic force microscopy on reactive surfaces. Prog. Surf. Sci. **64**(3), 179–191 (2000). ISSN 0079-6816. https://doi.org/10.1016/S0079-6816(00)00015-0, http://www.sciencedirect.com/science/article/pii/S0079681600000150

619. I. Štich, P. Dieška, R. Pérez, Tip–surface interactions in atomic force microscopy: reactive vs. metallic surfaces. Appl. Surf. Sci. (Proceedings of the 4th International Conference on Noncontact Atomic Microscopy) **188**(3), 325–330 (2002). ISSN 0169-4332. https://doi.org/10.1016/S0169-4332(01)00945-X, http://www.sciencedirect.com/science/article/pii/S016943320100945X

620. T. Stifter, O. Marti, B. Bhushan, Theoretical investigation of the distance dependence of capillary and van der waals forces in scanning force microscopy. Phys. Rev. B **62**, 13667–13673 (2000). https://doi.org/10.1103/PhysRevB.62.13667

621. B.C. Stipe, M.A. Rezaei, W. Ho, Single-molecule vibrational spectroscopy and microscopy. Science **280**(5370), 1732–1735 (1998). ISSN 0036-8075. https://doi.org/10.1126/science.280.5370.1732, https://science.sciencemag.org/content/280/5370/1732. Publisher: American Association for the Advancement of Science _eprint: https://science.sciencemag.org/content/280/5370/1732.full.pdf

622. B.C. Stipe, M.A. Rezaei, W. Ho, Localization of inelastic tunneling and the determination of atomic-scale structure with chemical specificity. Phys. Rev. Lett. **82**, 1724–1727 (1999). https://doi.org/10.1103/PhysRevLett.82.1724

623. B.C. Stipe, H.J. Mamin, T.D. Stowe, T.W. Kenny, D. Rugar, Noncontact friction and force fluctuations between closely spaced bodies. Phys. Rev. Lett. **87** (2001). https://doi.org/10.1103/PhysRevLett.87.096801

624. E. Stoll, Resolution of the scanning tunnel microscope. Surf. Sci. **143**(2), L411–L416 (1984). ISSN 0039-6028. https://doi.org/10.1016/0039-6028(84)90540-5, http://www.sciencedirect.com/science/article/pii/0039602884905405

625. T.D. Stowe, T.W. Kenny, D.J. Thomson, D. Rugar, Silicon dopant imaging by dissipation force microscopy. Appl. Phys. Lett. **75**(18), 2785–2787 (1999). ISSN 0003-6951. https://doi.org/10.1063/1.125149. Publisher: American Institute of Physics

626. E.W.J. Straver, J.E. Hoffman, O.M. Auslaender, D. Rugar, K.A. Moler, Controlled manipulation of individual vortices in a superconductor. Appl. Phys. Lett. **93**(17), 172514 (2008). ISSN 0003-6951. https://doi.org/10.1063/1.3000963. Publisher: American Institute of Physics

627. J.A. Stroscio, D.M. Eigler, Atomic and molecular manipulation with the scanning tunneling microscope. Science **254**(5036), 1319–1326 (1991). ISSN 0036-8075. https://doi.org/10.1126/science.254.5036.1319, https://science.sciencemag.org/content/254/5036/1319. Publisher: American Association for the Advancement of Science _eprint: https://science.sciencemag.org/content/254/5036/1319.full.pdf

628. Y. Sugawara, M. Ohta, H. Ueyama, S. Morita, Defect motion on an InP(110) surface observed with noncontact atomic force microscopy. Science **270**(5242), 1646 (1995). https://doi.org/10.1126/science.270.5242.1646

629. Y. Sugimoto, P. Pou, M. Abe, P. Jelinek, R. Pérez, S. Morita, Ó. Custance, Chemical identification of individual surface atoms by atomic force microscopy. Nature **446**(7131), 64–67 (2007). ISSN 1476-4687. https://doi.org/10.1038/nature05530

630. T. Sulchek, R. Hsieh, J.D. Adams, G.G. Yaralioglu, S.C. Minne, C.F. Quate, J.P. Cleveland, A. Atalar, D.M. Adderton, High-speed tapping mode imaging with active Q control for atomic force microscopy. Appl. Phys. Lett. **76**(11), 1473–1475 (2000). ISSN 0003-6951. https://doi.org/10.1063/1.126071. Publisher: American Institute of Physics

631. Y. Suzuki, W. Nabhan, K. Tanaka, Magnetic domains of cobalt ultrathin films observed with a scanning tunneling microscope using optically pumped GaAs tips. Appl. Phys. Lett. **71**(21), 3153–3155 (1997). ISSN 0003-6951. https://doi.org/10.1063/1.120274. Publisher: American Institute of Physics

632. K. Takayanagi, Y. Tanishiro, S. Takahashi, M. Takahashi, Structure analysis of Si(111)-7 × 7 reconstructed surface by transmission electron diffraction. Surf. Sci. **164**(2), 367–392 (1985). ISSN 0039-6028. https://doi.org/10.1016/0039-6028(85)90753-8, http://www.sciencedirect.com/science/article/pii/0039602885907538

633. S.J. Tans, M.H. Devoret, H. Dai, A. Thess, R.E. Smalley, L.J. Geerligs, C. Dekker, Individual single-wall carbon nanotubes as quantum wires. Nature **386**(6624), 474–477 (1997). ISSN 1476-4687. https://doi.org/10.1038/386474a0

634. K. Tao, V.S. Stepanyuk, W. Hergert, I. Rungger, S. Sanvito, P. Bruno, Switching a single spin on metal surfaces by a stm tip: ab initio studies. Phys. Rev. Lett. **103**, 057202 (2009). https://doi.org/10.1103/PhysRevLett.103.057202

635. G. Tarrach, R. Wiesendanger, D. Bürgler, L. Scandella, H.J. Güntherodt, Laser and thermal annealed si(111) and si(001) surfaces studied by scanning tunneling microscopy. J. Vac. Sci. & Technol. B: Microelectron. Nanometer Struct. Proc. Meas. Phen. **9**(2), 677–680 (1991)

636. G. Tarrach, P. Lagos, R. Hermans, F. Schlaphof, Ch. Loppacher, L.M. Eng, Nanometer spot allocation for Raman spectroscopy on ferroelectrics by polarization and piezoresponse force microscopy. Appl. Phys. Lett. **79**(19), 3152–3154 (2001). ISSN 0003-6951. https://doi.org/10.1063/1.1414292. Publisher: American Institute of Physics

637. R. Temirov, S. Soubatch, A. Luican, F.S. Tautz, Free-electron-like dispersion in an organic monolayer film on a metal substrate. Nature **444**(7117), 350–353 (2006). ISSN 1476-4687. https://doi.org/10.1038/nature05270

638. R. Temirov, S. Soubatch, O. Neucheva, A.C. Lassise, F.S. Tautz, A novel method achieving ultra-high geometrical resolution in scanning tunnelling microscopy. New J. Phys. **10**(5) (2008). https://doi.org/10.1088/1367-2630/10/5/053012

639. M. Ternes, C.P. Lutz, C.F. Hirjibehedin, F.J. Giessibl, A.J. Heinrich, The force needed to move an atom on a surface. Science **319**(5866), 1066–1069 (2008). ISSN 0036-8075. https://doi.org/10.1126/science.1150288, https://science.sciencemag.org/content/319/5866/1066. Publisher: American Association for the Advancement of Science _eprint: https://science.sciencemag.org/content/319/5866/1066.full.pdf

640. J. Tersoff, D.R. Hamann, Theory and application for the scanning tunneling microscope. Phys. Rev. Lett. **50**, 1998–2001 (1983). https://doi.org/10.1103/PhysRevLett.50.1998

641. J. Tersoff, D.R. Hamann, Theory of the scanning tunneling microscope. Phys. Rev. B **31**, 805–813 (1985). https://doi.org/10.1103/PhysRevB.31.805

642. J.-P. Tetienne, L. Rondin, P. Spinicelli, M. Chipaux, T. Debuisschert, J.-F. Roch, V. Jacques, Magnetic-field-dependent photodynamics of single NV defects in diamond: an application to qualitative all-optical magnetic imaging. New J. Phys. **14**(10), 103033 (2012). ISSN 1367-2630. https://doi.org/10.1088/1367-2630/14/10/103033. Publisher: IOP Publishing

643. J.-P. Tetienne, T. Hingant, L.J. Martnez, S. Rohart, A. Thiaville, L. Herrera Diez, K. Garcia, J.-P. Adam, J.-V. Kim, J.-F. Roch, I.M. Miron, G. Gaudin, L. Vila, B. Ocker, D. Ravelosona, V. Jacques, The nature of domain walls in ultrathin ferromagnets revealed by scanning nano-

magnetometry. Nat. Commun. **6**(1), 6733 (2015). ISSN 2041-1723. https://doi.org/10.1038/ncomms7733

644. A.A. Thiele, Theory of the static stability of cylindrical domains in uniaxial platelets. J. Appl. Phys. **41**(3), 1139–1145 (1970). ISSN 0021-8979. https://doi.org/10.1063/1.1658846. Publisher: American Institute of Physics

645. R.E. Thomson, J. Moreland, Development of highly conductive cantilevers for atomic force microscopy point contact measurements. J. Vac. Sci. & Technol. B: Microelectron. Nanometer Struct. Proc. Meas. Phen. **13**(3), 1123–1125 (1995). ISSN 1071-1023. https://doi.org/10.1116/1.588221, https://avs.scitation.org/doi/abs/10.1116/1.588221. Publisher: American Institute of Physics

646. T. Thundat, R.J. Warmack, G.Y. Chen, D.P. Allison, Thermal and ambient induced deflections of scanning force microscope cantilevers. Appl. Phys. Lett. **64**(21), 2894–2896 (1994). https://doi.org/10.1063/1.111407

647. U. Tietze, Ch. Schenk, E. Gamm, *Electronic Circuits* (Springer, Heidelberg, 2008). https://doi.org/10.1007/978-3-540-78655-9

648. M. Todorovic, S. Schultz, Magnetic force microscopy using nonoptical piezoelectric quartz tuning fork detection design with applications to magnetic recording studies. J. Appl. Phys. **83**(11), 6229–6231 (1998). ISSN 0021-8979. https://doi.org/10.1063/1.367642. Publisher: American Institute of Physics

649. S.L. Tomlinson, A.N. Farley, S.R. Hoon, M.S. Valera, Interactions between soft magnetic samples and MFM tips. J. Magn. Magn. Mater. **157–158**, 557–558 (1996). ISSN 0304-8853. https://doi.org/10.1016/0304-8853(95)01237-0, http://www.sciencedirect.com/science/article/pii/0304885395012370

650. M. Tortonese, R.C. Barrett, C.F. Quate, Atomic resolution with an atomic force microscope using piezoresistive detection. Appl. Phys. Lett. **62**(8), 834–836 (1993). ISSN 0003-6951. https://doi.org/10.1063/1.108593. Publisher: American Institute of Physics

651. T. Tran, D.R. Oliver, D.J. Thomson, G.E. Bridges, "Zeptofarad" (10^{-21} F) resolution capacitance sensor for scanning capacitance microscopy. Rev. Sci. Instr. **72**(6), 2618–2623 (2001). ISSN 0034-6748. https://doi.org/10.1063/1.1369637. Publisher: American Institute of Physics

652. T. Trenkler, T. Hantschel, R. Stephenson, P. De Wolf, W. Vandervorst, L. Hellemans, A. Malavé, D. Büchel, E. Oesterschulze, W. Kulisch, P. Niedermann, T. Sulzbach, O. Ohlsson, Evaluating probes for "electrical" atomic force microscopy. J. Vac. Sci. & Technol. B: Microelectron. Nanometer Struct. Proc. Meas. Phen. **18**(1), 418–427 (2000). ISSN 1071-1023. https://doi.org/10.1116/1.591205, https://avs.scitation.org/doi/abs/10.1116/1.591205. Publisher: American Institute of Physics

653. T.T. Tsong, *Atom-Probe Field Ion Microscopy* (Cambridge University Press, Cambridge, 1990). https://doi.org/10.1017/CBO9780511599842

654. M. Tsukada, N. Sasaki, M. Gauthier, K. Tagami, S. Watanabe, Theory of non-contact atomic force microscopy, in *Noncontact Atomic Force Microscopy*, ed. by S. Morita, R. Wiesendanger, E. Meyer. NanoScience and Technology (Springer, Berlin, 2002), pp. 257–278. ISBN 978-3-642-56019-4. https://doi.org/10.1007/978-3-642-56019-4_15

655. C.T. Chiang, C. Xu, Z. Han, W. Ho, Real-space imaging of molecular structure and chemical bonding by single-molecule inelastic tunneling probe. Science **344**, 885 (2014), https://science.sciencemag.org/content/344/6186/885

656. T. Uchihashi, T. Ishida, M. Komiyama, M. Ashino, Y. Sugawara, W. Mizutani, K. Yokoyama, S. Morita, H. Tokumoto, M. Ishikawa, High-resolution imaging of organic monolayers using noncontact AFM. Appl. Surf. Sci. **157**(4), 244–250 (2000). ISSN 0169-4332. https://doi.org/10.1016/S0169-4332(99)00534-6, http://www.sciencedirect.com/science/article/pii/S0169433299005346

657. H. Ueyama, M. Ohta, Y. Sugawara, S. Morita, Atomically resolved InP(110) surface observed with noncontact ultrahigh vacuum atomic force microscope. Jpn. J. Appl. Phys. **34**(Part 2, No. 8B), L1086–L1088 (1995). ISSN 0021-4922. https://doi.org/10.1143/jjap.34.l1086, http://dx.doi.org/10.1143/JJAP.34.L1086. Publisher: IOP Publishing

658. H. Unbehauen, *Regelungstechnik I*. Vieweg und Teubner, Wiesbaden (2008). https://doi.org/10.1007/978-3-8348-9491-5

659. I. Utke, P. Hoffmann, R. Berger, L. Scandella, High-resolution magnetic Co supertips grown by a focused electron beam. Appl. Phys. Lett. **80**(25), 4792–4794 (2002). ISSN 0003-6951. https://doi.org/10.1063/1.1489097. Publisher: American Institute of Physics

660. M.C.M.M. van der Wielen, A.J.A. van Roij, H. van Kempen, Direct observation of friedel oscillations around incorporated si$_{ga}$ dopants in gaas by low-temperature scanning tunneling microscopy. Phys. Rev. Lett. **76**, 1075–1078 (1996). https://doi.org/10.1103/PhysRevLett.76.1075

661. N.F. van Hulst, J.A. Veerman, M.F. Garcia Parajo, L.K. Kuipers, Analysis of individual (macro)molecules and proteins using near-field optics. J. Chem. Phys. **112**(18), 7799–7810 (2000). ISSN 0021-9606. https://doi.org/10.1063/1.481385, https://aip.scitation.org/doi/10.1063/1.481385. Publisher: American Institute of Physics

662. P.J.A. van Schendel, H.J. Hug, B. Stiefel, S. Martin, H.-J. Güntherodt, A method for the calibration of magnetic force microscopy tips. J. Appl. Phys. **88**(1), 435–445 (2000). ISSN 0021-8979. https://doi.org/10.1063/1.373678. Publisher: American Institute of Physics

663. D. Vasyukov, L. Ceccarelli, M. Wyss, B. Gross, A. Schwarb, A. Mehlin, N. Rossi, G. Tütüncüoglu, F. Heimbach, R.R. Zamani, A. Kovacs, A. Fontcuberta i Morral, D. Grundler, M. Poggio, Imaging stray magnetic field of individual ferromagnetic nanotubes. Nano Lett. **18**(2), 964–970 (2018). ISSN 1530-6984. https://doi.org/10.1021/acs.nanolett.7b04386, https://doi.org/10.1021/acs.nanolett.7b04386. Publisher: American Chemical Society

664. D. Vasyukov, Y. Anahory, L. Embon, D. Halbertal, J. Cuppens, L. Neeman, A. Finkler, Y. Segev, Y. Myasoedov, M.L. Rappaport, M.E. Huber, E. Zeldov, A scanning superconducting quantum interference device with single electron spin sensitivity. Nat. Nanotechnol. **8**(9), 639–644 (2013). ISSN 1748-3395. https://doi.org/10.1038/nnano.2013.169

665. P. Vettiger, J. Brugger, M. Despont, U. Drechsler, U. Dürig, W. Häberle, M. Lutwyche, H. Rothuizen, R. Stutz, R. Widmer, G. Binnig, Ultrahigh density, high-data-rate NEMS-based AFM data storage system. Microelectr. Eng. **46**(1), 11–17 (1999). ISSN 0167-9317. https://doi.org/10.1016/S0167-9317(99)00006-4, http://www.sciencedirect.com/science/article/pii/S0167931799000064

666. P. Vettiger, M. Despont, U. Drechsler, U. Durig, W. Haberle, M.I. Lutwyche, H.E. Rothuizen, R. Stutz, R. Widmer, G.K. Binnig, The millipede: More than thousand tips for future afm storage. IBM J. Res. Dev. **44**(3), 323–340 (2000). https://doi.org/10.1147/rd.443.0323

667. M.B. Viani, T.E. Schäffer, G.T. Paloczi, L.I. Pietrasanta, B.L. Smith, J.B. Thompson, M. Richter, M. Rief, H.E. Gaub, K.W. Plaxco, A.N. Cleland, H.G. Hansma, P.K. Hansma, Fast imaging and fast force spectroscopy of single biopolymers with a new atomic force microscope designed for small cantilevers. Rev. Sci. Instr. **70**(11), 4300–4303 (1999). ISSN 0034-6748. https://doi.org/10.1063/1.1150069. Publisher: American Institute of Physics

668. J.S. Villarrubia, Morphological estimation of tip geometry for scanned probe microscopy. Surf. Sci. **321**(3), 287–300 (1994). ISSN 0039-6028. https://doi.org/10.1016/0039-6028(94)90194-5, http://www.sciencedirect.com/science/article/pii/0039602894901945

669. J.S. Villarrubia, Scanned probe microscope tip characterization without calibrated tip characterizers. J. Vac. Sci. & Technol. B: Microelectr. Nanometer Struct. Proc. Meas. Phen. **14**(2), 1518–1521 (1996). ISSN 1071-1023. https://doi.org/10.1116/1.589130, https://avs.scitation.org/doi/abs/10.1116/1.589130. Publisher: American Institute of Physics

670. J.S. Villarrubia, Algorithms for scanned probe microscope image simulation, surface reconstruction, and tip estimation. J. Res. Natl. Inst. Stand. Technol. **102**(4), 425–454 (1997). ISSN 1044-677X. https://doi.org/10.6028/jres.102.030, https://pubmed.ncbi.nlm.nih.gov/27805154. Publisher: [Gaithersburg, MD] : U.S. Dept. of Commerce, National Institute of Standards and Technology

671. S. Vock, F. Wolny, T. Mühl, R. Kaltofen, L. Schultz, B. Büchner, C. Hassel, J. Lindner, V. Neu, Monopolelike probes for quantitative magnetic force microscopy: calibration and application. Appl. Phys. Lett. **97**(25), 252505 (2010). ISSN 0003-6951. https://doi.org/10.1063/1.3528340. Publisher: American Institute of Physics

672. S. Vock, Z. Sasvari, C. Bran, F. Rhein, U. Wolff, N.S. Kiselev, A.N. Bogdanov, L. Schultz, O. Hellwig, V. Neu, Quantitative magnetic force microscopy study of the diameter evolution of bubble domains in a $(Co/Pd)_{80}$ multilayer. IEEE Trans. Magn. **47**(10), 2352–2355 (2011). https://doi.org/10.1109/TMAG.2011.2155630

673. S. Vock, C. Hengst, M. Wolf, K. Tschulik, M. Uhlemann, Z. Sasvari, D. Makarov, O.G. Schmidt, L. Schultz, V. Neu, Magnetic vortex observation in FeCo nanowires by quantitative magnetic force microscopy. Appl. Phys. Lett. **105**(17), 172409 (2014). ISSN 0003-6951. https://doi.org/10.1063/1.4900998. Publisher: American Institute of Physics

674. C. Vogler, M. Heigl, A.-O. Mandru, B. Hebler, M. Marioni, H.J. Hug, M. Albrecht, D. Suess, Hysteresis-free magnetization reversal of exchange-coupled bilayers with finite magnetic anisotropy. Phys. Rev. B **102**, 014429 (2020). https://doi.org/10.1103/PhysRevB.102.014429

675. M. Völcker, W. Krieger, H. Walther, Laser-driven scanning tunneling microscope. Phys. Rev. Lett. **66**, 1717–1720 (1991). https://doi.org/10.1103/PhysRevLett.66.1717

676. M. Völker, Laser scanning tunneling microscope, in *Scanning Probe Microscopy: Analytical Methods*, ed. by R. Wiesendanger. Nanoscience and Technology (Springer, Heidelberg, 1998), p. 135. https://doi.org/10.1007/978-3-662-03606-8

677. A.I. Volokitin, B.N.J. Persson, Adsorbate-induced enhancement of electrostatic noncontact friction. Phys. Rev. Lett. **94** (2005). https://doi.org/10.1103/PhysRevLett.94.086104

678. K. von Bergmann, M. Menzel, D. Serrate, Y. Yoshida, S. Schröder, P. Ferriani, A. Kubetzka, R. Wiesendanger, S. Heinze, Tunneling anisotropic magnetoresistance on the atomic scale. Phys. Rev. B **86** (2012). https://doi.org/10.1103/PhysRevB.86.134422

679. Y. Wada, T. Uda, M. Lutwyche, S. Kondo, S. Heike, A proposal of nanoscale devices based on atom/molecule switching. J. Appl. Phys. **74**(12), 7321–7328 (1993). https://doi.org/10.1063/1.354999

680. A. Wadas, P. Grütter, Theoretical approach to magnetic force microscopy. Phys. Rev. B **39**, 12013–12017 (1989). https://doi.org/10.1103/PhysRevB.39.12013

681. A. Wadas, H.J. Güntherodt, Lateral resolution in magnetic force microscopy. application to periodic structures. Phys. Lett. A **146**(5), 277–280 (1990). ISSN 0375-9601, https://doi.org/10.1016/0375-9601(90)90980-3, http://www.sciencedirect.com/science/article/pii/0375960190909803

682. A. Wadas, H.J. Hug, Models for the stray field from magnetic tips used in magnetic force microscopy. J. Appl. Phys. **72**(1), 203–206 (1992). ISSN 0021-8979. https://doi.org/10.1063/1.352159. Publisher: American Institute of Physics

683. K. Wang, A.R. Smith, Three-dimensional spin mapping of antiferromagnetic nanopyramids having spatially alternating surface anisotropy at room temperature. Nano Lett. **12**(11), 5443–5447 (2012). ISSN 1530-6984. https://doi.org/10.1021/nl204192n. Publisher: American Chemical Society

684. L. Wang, Q. Feng, Y. Kim, R. Kim, K.H. Lee, S.D. Pollard, Y.J. Shin, H. Zhou, W. Peng, D. Lee, W. Meng, H. Yang, J.H. Han, M. Kim, Q. Lu, T.W. Noh, Ferroelectrically tunable magnetic skyrmions in ultrathin oxide heterostructures. Nat. Mater. **17**(12), 1087–1094 (2018). ISSN 1476-4660. https://doi.org/10.1038/s41563-018-0204-4

685. J.M.R. Weaver, D.W. Abraham, High resolution atomic force microscopy potentiometry. J. Vac. Sci. & Technol. B: Microelectron. Nanometer Struct. Proc. Meas. Phen. **9**(3), 1559–1561 (1991). https://doi.org/10.1116/1.585423, https://avs.scitation.org/doi/abs/10.1116/1.585423

686. J.M.R. Weaver, L.M. Walpita, H.K. Wickramasinghe, Optical absorption microscopy and spectroscopy with nanometre resolution. Nature **342**(6251), 783–785 (1989). ISSN 1476-4687. https://doi.org/10.1038/342783a0

687. U. Weierstall, J.C.H. Spence, Atomic species identification in STM using an imaging atom-probe technique. Surf. Sci. **398**(1), 267–279 (1998). ISSN 0039-6028. https://doi.org/10.1016/S0039-6028(98)80030-7, http://www.sciencedirect.com/science/article/pii/S0039602898800307

688. C. Weiss, C. Wagner, C. Kleimann, M. Rohlfing, F.S. Tautz, R. Temirov, Imaging pauli repulsion in scanning tunneling microscopy. Phys. Rev. Lett. **105** (2010). https://doi.org/10.1103/PhysRevLett.105.086103

689. K.L. Westra, A.W. Mitchell, D.J. Thomson, Tip artifacts in atomic force microscope imaging of thin film surfaces. J. Appl. Phys. **74**(5), 3608–3610 (1993). https://doi.org/10.1063/1.354498

690. M.-H. Whangbo, G. Bar, R. Brandsch, Description of phase imaging in tapping mode atomic force microscopy by harmonic approximation1Dedicated to Professor H.-J. Cantow on the occasion of his 75th birthday.1. Surf. Sci. **411**(1), L794–L801 (1998). ISSN 0039-6028. https://doi.org/10.1016/S0039-6028(98)00349-5, http://www.sciencedirect.com/science/article/pii/S0039602898003495

691. R. Wiesendanger, H.-J. Güntherodt, *Scanning Tunneling Microscopy II*, volume 28 of Springer Series in Surface Science (Springer, Berlin, 1992). https://doi.org/10.1007/978-3-642-97363-5

692. R. Wiesendanger, H.-J. Güntherodt, *Scanning Tunneling Microscopy III*, volume 29 of Springer Series in Surface Science (Springer, Berlin, 1993). https://doi.org/10.1007/978-3-642-80118-1

693. R. Wiesendanger, H.-J. Güntherodt, G. Güntherodt, R.J. Gambino, R. Ruf, Observation of vacuum tunneling of spin-polarized electrons with the scanning tunneling microscope. Phys. Rev. Lett. **65**, 247–250 (1990). https://doi.org/10.1103/PhysRevLett.65.247

694. R. Wiesendanger, Spin mapping at the nanoscale and atomic scale. Rev. Mod. Phys. **81**, 1495–1550 (2009). https://doi.org/10.1103/RevModPhys.81.1495

695. C.C. Williams, Two-dimensional dopant profiling by scanning capacitance microscopy. Ann. Rev. Mater. Sci. **29**(1), 471–504 (1999). https://doi.org/10.1146/annurev.matsci.29.1.471

696. C.C. Williams, H.K. Wickramasinghe, Scanning thermal profiler. Appl. Phys. Lett. **49**(23), 1587–1589 (1986). https://doi.org/10.1063/1.97288

697. C.C. Williams, W.P. Hough, S.A. Rishton, Scanning capacitance microscopy on a 25 nm scale. Appl. Phys. Lett. **55**(2), 203–205 (1989). ISSN 0003-6951. https://doi.org/10.1063/1.102096. Publisher: American Institute of Physics

698. C.C. Williams, J. Slinkman, W.P. Hough, H.K. Wickramasinghe, Lateral dopant profiling with 200 nm resolution by scanning capacitance microscopy. Appl. Phys. Lett. **55**(16), 1662–1664 (1989). https://doi.org/10.1063/1.102312

699. P. Willke, K. Yang, Y. Bae, A.J. Heinrich, C.P. Lutz, Magnetic resonance imaging of single atoms on a surface. Nat. Phys. **15**(10), 1005–1010 (2019). ISSN 1745-2481. https://doi.org/10.1038/s41567-019-0573-x

700. M. Wilms, M. Kruft, G. Bermes, K. Wandelt, A new and sophisticated electrochemical scanning tunneling microscope design for the investigation of potentiodynamic processes. Rev. Sci. Instr. **70**(9), 3641–3650 (1999). https://doi.org/10.1063/1.1149971

701. J. Wintterlin, J. Wiechers, H. Brune, T. Gritsch, H. Höfer, R.J. Behm, Atomic-resolution imaging of close-packed metal surfaces by scanning tunneling microscopy. Phys. Rev. Lett. **62**, 59–62 (1989). https://doi.org/10.1103/PhysRevLett.62.59

702. P. De Wolf, T. Clarysse, W. Vandervorst, L. Hellemans, W. Ph Niedermann, Hänni, Cross-sectional nano-spreading resistance profiling. J. Vac. Sci. Technol. B **16**(1), 355 (1998). https://doi.org/10.1116/1.589810

703. F. Wolny, T. Mühl, U. Weissker, A. Leonhardt, U. Wolff, D. Givord, B. Büchner, Magnetic force microscopy measurements in external magnetic fields comparison between coated probes and an iron filled carbon nanotube probe. J. Appl. Phys. **108**(1) (2010). https://doi.org/10.1063/1.3459879

704. D. Wortmann, S. Heinze, G. Ph Kurz, S. Blügel, Bihlmayer, Resolving complex atomic-scale spin structures by spin-polarized scanning tunneling microscopy. Phys. Rev. Lett. **86**, 4132–4135 (2001). https://doi.org/10.1103/PhysRevLett.86.4132

705. C.D. Wright, E.W. Hill, Reciprocity in magnetic force microscopy. Appl. Phys. Lett. **67**(3), 433–435 (1995). ISSN 0003-6951. https://doi.org/10.1063/1.114623. Publisher: American Institute of Physics

706. Y. Wu, Y. Shen, Z. Liu, K. Li, J. Qiu, Point-dipole response from a magnetic force microscopy tip with a synthetic antiferromagnetic coating. Appl. Phys. Lett. **82**(11), 1748–1750 (2003). ISSN 0003-6951. https://doi.org/10.1063/1.1560863. Publisher: American Institute of Physics

707. Z. Wu, T. Nakayama, M. Sakurai, M. Aono, Spin-polarized electron tunneling detected using a scanning tunneling microscope, in *TOYOTA Conference on Atomic, Molecular and Electronic Dynamic Processes on Solid Surfaces*, vol. 386, no. 1 (1997), pp. 311–314. ISSN 0039-6028. https://doi.org/10.1016/S0039-6028(97)00319-1, http://www.sciencedirect.com/science/article/pii/S0039602897003191

708. W. Wulfhekel, J. Kirschner, Spin-polarized scanning tunneling microscopy on ferromagnets. Appl. Phys. Lett. **75**(13), 1944–1946 (1999). ISSN 0003-6951. https://doi.org/10.1063/1.124879. Publisher: American Institute of Physics

709. A. Yagil, A. Almoalem, A. Soumyanarayanan, A.K.C. Tan, M. Raju, C. Panagopoulos, O.M. Auslaender, Stray field signatures of Néel textured skyrmions in Ir/Fe/Co/Pt multilayer films. Appl. Phys. Lett. **112**(19), 192403 (2018). ISSN 0003-6951. https://doi.org/10.1063/1.5027602. Publisher: American Institute of Physics

710. T. Yamamoto, Y. Suzuki, M. Miyashita, H. Sugimura, N. Nakagiri, Development of a metal patterned cantilever for scanning capacitance microscopy and its application to the observation of semiconductor devices. J. Vac. Sci. & Technol. B: Microelectron. Nanometer Struct. Proc. Meas. Phen. **15**(4), 1547–1550 (1997). https://doi.org/10.1116/1.589397, https://avs.scitation.org/doi/abs/10.1116/1.589397

711. H. Yang, A. Thiaville, S. Rohart, A. Fert, M. Chshiev, Anatomy of dzyaloshinskii-moriya interaction at Co/Pt interfaces. Phys. Rev. Lett. **115** (2015)

712. K. Yang, W. Paul, S.-H. Phark, P. Willke, Y. Bae, T. Choi, T. Esat, A. Ardavan, A.J. Heinrich, C.P. Lutz, Coherent spin manipulation of individual atoms on a surface. Science **366**(6464), 509–512 (2019). ISSN 0036-8075. https://doi.org/10.1126/science.aay6779, https://science.sciencemag.org/content/366/6464/509. Publisher: American Association for the Advancement of Science _eprint: https://science.sciencemag.org/content/366/6464/509.full.pdf

713. S. Yang, W. Huang, Three-dimensional displacements of a piezoelectric tube scanner. Rev. Sci. Instr. **69**(1), 226–229 (1998). ISSN 0034-6748. https://doi.org/10.1063/1.1148500. Publisher: American Institute of Physics

714. Y. Shen, W. Yihong, Response function study of a new kind of multilayer-coated tip for magnetic force microscopy. IEEE Trans. Magn. **40**(1), 97–100 (2004). https://doi.org/10.1109/TMAG.2003.821128

715. A. Yazdani, B.A. Jones, C.P. Lutz, M.F. Crommie, D.M. Eigler, Probing the local effects of magnetic impurities on superconductivity. Science **275**(5307), 1767–1770 (1997). ISSN 0036-8075. https://doi.org/10.1126/science.275.5307.1767, https://science.sciencemag.org/content/275/5307/1767. Publisher: American Association for the Advancement of Science _eprint: https://science.sciencemag.org/content/275/5307/1767.full.pdf

716. A. Yazdani, C.M. Howald, C.P. Lutz, A. Kapitulnik, D.M. Eigler, Impurity-induced bound excitations on the surface of $i_2sr_2cacu_2O_8$. Phys. Rev. Lett. **83**, 176–179 (1999)

717. S. You, J.-T. Lü, J. Guo, Y. Jiang, Recent advances in inelastic electron tunneling spectroscopy. Adv. Phys.: X **2**(3), 907–936 (2017). https://doi.org/10.1080/23746149.2017.1372215

718. R. Young, J. Ward, F. Scire, Observation of metal-vacuum-metal tunneling, field emission, and the transition region. Phys. Rev. Lett. **27**, 922–924 (1971). https://doi.org/10.1103/PhysRevLett.27.922

719. R. Young, J. Ward, F. Scire, The topografiner: an instrument for measuring surface microtopography. Rev. Sci. Instr. **43**(7), 999–1011 (1972). ISSN 0034-6748. https://doi.org/10.1063/1.1685846. Publisher: American Institute of Physics

720. T.J. Young, M.A. Monclus, T.L. Burnett, W.R. Broughton, S.L. Ogin, P.A. Smith, The use of the PeakForceTMquantitative nanomechanical mapping AFM-based method for high-resolution Young's modulus measurement of polymers. Meas. Sci. Technol. **22**(12), 125703 (2011). ISSN 0957-0233. https://doi.org/10.1088/0957-0233/22/12/125703. Publisher: IOP Publishing

721. A. Zangwill, *Physics at Surfaces*. (Cambridge University Press, Cambridge, 1988). https://doi.org/10.1017/CBO9780511622564

722. V.V. Zavyalov, J.S. McMurray, C.C. Williams, Advances in experimental technique for quantitative two-dimensional dopant profiling by scanning capacitance microscopy. Rev. Sci.

Instr. **70**(1), 158–164 (1999). ISSN 0034-6748. https://doi.org/10.1063/1.1149558. Publisher: American Institute of Physics

723. H. Zhang, L. Wu, F. Huang, Electrochemical microprocess by scanning ion-conductance microscopy. J. Vac. Sci. & Technol. B: Microelectron. Nanometer Struct. Proc. Meas. Phenom. **17**(2), 269–272 (1999). ISSN 1071-1023. https://doi.org/10.1116/1.590549, https://avs.scitation.org/doi/abs/10.1116/1.590549. Publisher: American Institute of Physics

724. R. Zhang, Y. Zhang, Z.C. Dong, S. Jiang, C. Zhang, L.G. Chen, L. Zhang, Y. Liao, J. Aizpurua, Y. Luo, J.L. Yang, J.G. Hou, Chemical mapping of a single molecule by plasmon-enhanced Raman scattering. Nature **498**(7452), 82–86 (2013). ISSN 1476-4687. https://doi.org/10.1038/nature12151

725. S. Zhang, J. Zhang, Q. Zhang, C. Barton, V. Neu, Y. Zhao, Z. Hou, Y. Wen, C. Gong, O. Kazakova, W. Wang, Y. Peng, D.A. Garanin, E.M. Chudnovsky, X. Zhang, Direct writing of room temperature and zero field skyrmion lattices by a scanning local magnetic field. Appl. Phys. Lett. **112**(13), 132405 (2018). ISSN 0003-6951. https://doi.org/10.1063/1.5021172. Publisher: American Institute of Physics

726. Y. Zhang, Y. Luo, Y. Zhang, Y.J. Yu, Y.M. Kuang, L. Zhang, Q.S. Meng, Y. Luo, J.L. Yang, Z.C. Dong, J.G. Hou, Visualizing coherent intermolecular dipole dipole coupling in real space. Nature **531**(7596), 623–627 (2016). ISSN 1476-4687. https://doi.org/10.1038/nature17428

727. X. Zhao, J. Schwenk, A.O. Mandru, M. Penedo, M. Bacani, M.A. Marioni, H.J. Hug, Magnetic force microscopy with frequency-modulated capacitive tip sample distance control. New J. Phys. **20**(1), 013018 (2018). ISSN 1367-2630. https://doi.org/10.1088/1367-2630/aa9ca9. Publisher: IOP Publishing

728. X. Zhao, J. Schwenk, A.O. Mandru, M. Penedo, M. Baćani, M.A. Marioni, H.J. Hug, Magnetic force microscopy with frequency-modulated capacitive tip–sample distance control. New J. Phys. **20**(1), 013018 (2018). ISSN 1367-2630. https://doi.org/10.1088/1367-2630/aa9ca9. Publisher: IOP Publishing

729. X. Zhao, A.-O. Mandru, C. Vogler, M.A. Marioni, D. Suess, H.J. Hug, Magnetization reversal of strongly exchange-coupled double nanolayers for spintronic devices. ACS Appl. Nano Mater. **2**(12), 7478–7487 (2019). https://doi.org/10.1021/acsanm.9b01243. Publisher: American Chemical Society

730. X. Zhao, S.R. Phillpot, W.G. Sawyer, S.B. Sinnott, S.S. Perry, Transition from thermal to athermal friction under cryogenic conditions. Phys. Rev. Lett. **102**, 186102 (2009)

731. Q. Zhong, D. Inniss, K. Kjoller, V.B. Elings, Fractured polymer/silica fiber surface studied by tapping mode atomic force microscopy. Surf. Sci. **290**(1), L688–L692 (1993). ISSN 0039-6028. https://doi.org/10.1016/0039-6028(93)90582-5, http://www.sciencedirect.com/science/article/pii/0039602893905825

732. J.G. Ziegler, N.B. Nichols, Optimum settings for automatic controllers. J. Dyn. Syst. Meas. Control **115**(2B), 220–222 (1993). ISSN 0022-0434. https://doi.org/10.1115/1.2899060

733. O. Zwörner, H.Hölscher, U.D. Schwarz, R. Wiesendanger, The velocity dependence of frictional forces in point-contact friction. Appl. Phys. A **66**(1), S263–S267 (1998). ISSN 1432-0630. https://doi.org/10.1007/s003390051142

Index

© Springer Nature Switzerland AG 2021
E. Meyer et al., *Scanning Probe Microscopy*, Graduate Texts in Physics,
https://doi.org/10.1007/978-3-030-37089-3

Printed in the United States
by Baker & Taylor Publisher Services